Case Studies of Modified Treatment Practices for Disinfection By-product Control

The mission of the Awwa Research Foundation (AwwaRF) is to advance the science of water to improve the quality of life. Funded primarily through annual subscription payments from over 1,000 utilities, consulting firms, and manufacturers in North America and abroad, AwwaRF sponsors research on all aspects of drinking water, including supply and resources, treatment, monitoring and analysis, distribution, management, and health effects.

From its headquarters in Denver, Colorado, the AwwaRF staff directs and supports the efforts of over 500 volunteers, who are the heart of the research program. These volunteers, serving on various boards and committees, use their expertise to select and monitor research studies to benefit the entire drinking water community.

Research findings are disseminated through a number of technology transfer activities, including research reports, conferences, videotape summaries, and periodicals.

Case Studies of Modified Treatment Practices for Disinfection By-product Control

Prepared by:
Stuart W. Krasner and **Srilatha R. Rajachandran**
Metropolitan Water District of Southern California
700 Moreno Avenue
La Verne, CA 91750-3399
John E. Cromwell III
PA Consulting Group
1530 Wilson Boulevard, Suite 400
Arlington, VA 22209-2406
Douglas M. Owen
Malcolm Pirnie, Inc.
104 Corporate Park Drive - 10604
P.O. Box 751
White Plains, NY 10602-0751
and
Zaid K. Chowdhury
Malcolm Pirnie, Inc.
432 North 44th Street
Phoenix, AZ 85008

Sponsored by:
Awwa Research Foundation
6666 West Quincy Avenue
Denver, CO 80235-3098

Published by the

Awwa Research Foundation
Advancing the Science of Water®

IWA Publishing

DISCLAIMER

Published by IWA Publishing, Alliance House, 12 Caxton Street, London SW1H 0QS, UK

Telephone: +44 (0) 20 7654 5500; Fax: +44 (0) 20 7654 5555; Email: publications@iwap.co.uk;
Web: **www.iwapublishing.com**

AwwaRF report number 90946F. Originally published by AwwaRF for its subscribers in 2003.
IWA Publishing version published 2004.

ISBN 1 84339 854 0

CONTENTS

TABLES

FIGURES

FOREWORD

The Awwa Research Foundation is a nonprofit corporation that is dedicated to the implementation of a research effort to help utilities respond to regulatory requirements and traditional high-priority concerns of the industry. The research agenda is developed through a process of consultation with subscribers and drinking water professionals. Under the umbrella of a Strategic Research Plan, the Research Advisory Council prioritizes the suggested projects based upon current and future needs, applicability, and past work; the recommendations are forwarded to the Board of Trustees for final selection. The foundation also sponsors research projects through the unsolicited proposal process; the Collaborative Research, Research Applications, and Tailored Collaboration programs; and various joint research efforts with organizations such as the U.S. Environmental Protection Agency, the U.S. Bureau of Reclamation, and the Association of California Water Agencies.

This publication is a result of one of these sponsored studies, and it is hoped that its findings will be applied in communities throughout the world. The following report serves not only as a means of communicating the results of the water industry's centralized research program but also as a tool to enlist the further support of the nonmember utilities and individuals.

Projects are managed closely from their inception to the final report by the foundation's staff and large cadre of volunteers who willingly contribute their time and expertise. The foundation serves a planning and management function and awards contracts to other institutions such as water utilities, universities, and engineering firms. The funding for this research effort comes primarily from the Subscription Program, through which water utilities subscribe to the research program and make an annual payment proportionate to the volume of water they deliver and consultants and manufacturers subscribe based on their annual billings. The program offers a cost-effective and fair method of funding research in the public interest.

A broad spectrum of water supply issues is addressed by the foundation's research agenda: resources, treatment and operations, distribution and storage, water quality and analysis, toxicology, economics, and management. The ultimate purpose of the coordinated effort is to assist water suppliers to provide the highest possible quality of water economically and reliably. The true benefits are realized when the results are implemented at the utility level. The foundation's trustees are pleased to offer this publication as a contribution toward that end.

Utilities have been in the process of making decisions on whether any modifications to their existing treatment practices are needed to achieve compliance with the new Disinfectants/Disinfection By-Products regulations. A regulatory impact analysis that was prepared for this rule predicted that up to 70 percent of large surface-water systems would need to make some treatment modifications to comply with the requirements of the Rule. This report examines 10 utilities that have implemented treatment modifications to comply with the Rule and documents the modifications made, associated costs, and lessons learned through implementation.

Edmund G. Archuleta, P.E.
Chair, Board of Trustees
Awwa Research Foundation

James F. Manwaring, P.E.
Executive Director
Awwa Research Foundation

ACKNOWLEDGMENTS

The authors of this report are indebted to the following water utilities and individuals for their cooperation and participation in this project:

Newport News Waterworks, Yorktown, Va., Michael L. Hotaling

City of Poway, Poway, Calif., Dan Harrison

Gulf Coast Water Authority, Texas City, Texas, Nancy Mathews

Skagit County Public Utility District, Mount Vernon, Wash., Greg Peterka, Greg Hamilton

Ann Arbor Water Utilities, Ann Arbor, Mich., Janice Skadsen, Larry Sanford, Glenn Maggard

Alameda County Water District, Fremont, Calif., Susan Teefy, Steven Takahashi, Jennifer Chiou

City of Phoenix, Phoenix, Ariz., Carlos Padilla, Laxman Devkota

City of Tampa, Tampa, Fla., James M. Gianatasio

Virginia American Water Company, Hopewell, Va., John J. Hanchak, Brenda Frye

City of Plantation, Plantation, Fla., Mel Entus, Larry Vann

The aforementioned individuals and their staff and colleagues at the participating utilities were an integral part of the research team. Participants from these utilities engaged in a critical review of their experiences. They assisted in documenting their experiences and sharing with the project team. These individuals reviewed the draft case-study write-ups and participated in developing the final versions. The authors of this report gratefully acknowledge the aforementioned utilities and individuals for their invaluable contributions to this project.

In addition, the following individuals at Malcolm Pirnie, Inc., were highly involved in the preparation of selected case studies: Timothy P. Brodeur and Victor Hurlburt (Maitland, Fla.), and Anne Jack (Phoenix, Ariz.).

The authors wish to acknowledge Irwin H. (Mel) Suffet, Professor of Environmental Science and Engineering, University of California at Los Angeles, who was Ms. Rajachandran's faculty advisor.

The authors also thank Awwa Research Foundation for its support—in particular, Jeff Oxenford, Kathleen Faller, and the members of the Project Advisory Committee: Gil Crozes and Stephen Booth (Carollo Engineers), Tom Grubbs (U.S. Environmental Protection Agency), Mark Morgan (Heyward, Inc.), and John Rissel (Brick Utilities).

EXECUTIVE SUMMARY

In 1989, the U.S. Environmental Protection Agency (USEPA) released a "strawman" paper that discussed the possibility of lowering the maximum contaminant level (MCL) for total trihalomethanes (TTHMs). In 1994, the USEPA proposed Stage 1 and Stage 2 of the Disinfectants/Disinfection By-Products (D/DBP) Rule. Stage 1 was promulgated in December 1998 and will be effective by January 2002 for large systems (i.e., serving a population of at least 10,000) that use surface water or groundwater under the direct influence of surface water. In 2000, a Stage 2 Microbial and DBP Federal Advisory Committee developed an Agreement in Principle for a revised Stage 2 D/DBP Rule and for a Long Term 2 Enhanced Surface Water Treatment Rule (LT2ESWTR).

Utilities have been in the process of making decisions on whether any modifications to their existing treatment practices are needed to achieve compliance with the new D/DBP regulations. A regulatory impact analysis that was prepared for the D/DBP Rule predicted that up to 70 percent of large surface-water systems would need to make some treatment modifications to comply with the requirements of the Rule.

RESEARCH OBJECTIVES

The American Water Works Association Research Foundation (AwwaRF) funded this case-study project. As part of this study, 10 utilities that implemented treatment modifications to enable them to comply with the D/DBP regulations were examined. The case-study project is serving as a research continuum to bridge the gap between research and practical application. Project objectives included development of a case-study protocol, documentation of what worked and what did not so others can learn from the problems and the solutions, documentation of the decision-making process, and examination of implementation issues.

APPROACH

The major goal was to study utilities that had implemented treatment modifications that spanned the range of choices that utilities were considering for complying with the D/DBP Rule.

Other utility-selection considerations included (1) geographical representation, (2) different size (flow) categories, and (3) studying utilities with a range of raw-water qualities that presented challenges in complying with the D/DBP Rule. In general, it was important to select utilities for study that had not made their treatment changes too recently, in order that long-term operational experiences could be evaluated.

The case-study protocol utilized an iterative approach, with the participating utilities as partners in documenting the case-study information. Some of the steps in the protocol included:

- Initial contacts were established and a screening interview was conducted with utilities to explain project scope and objectives, utility role and benefits.
- Follow-up mailings to the selected utilities were made, which included information-collection instruments (questionnaires).
- Kick-off or profile interviews were conducted to identify data/information/research that was perceived to be important to document the utility's case study.
- Once the questionnaires had been completed and the data/information received, potential data/information gaps were identified and discussed with utility contacts.
- Follow-up questionnaires were sent to utilities to fill in data and information gaps.
- A final interview was conducted, which encompassed a broader spectrum of issues that may have factored into the decision-making process.
- A draft case-study narrative was developed, which was shared with the utility participant.
- A final narrative was prepared—through a series of interations—which took account of feedback and comments from the utility participants.

A series of questionnaires were developed and used, which asked for information on water quality, operational experiences, implementation issues, costs, and the decision-making process. To facilitate efficient replies to the questionnaires, they were divided into different groups that could be answered by different staff at each utility. Moreover, the utilities were encouraged to send reports, memorandum, etc., in lieu of directly answering each question.

The Internet proved to be a very valuable research source, which complemented the information obtained with the questionnaire. For example, a number of utilities have web sites with relevant information. Other web sites that yielded important information were federal government web sites with U.S. census data (e.g., consumer-demographic information). The Internet also provided access to local newspapers in the cities where the utilities were located, in which treatment changes at the utilities were discussed.

CONCLUSIONS

Multiple Water-Quality Objectives

Meeting multiple water-quality objectives plays an important role in the decision-making process when a utility selects a treatment modification for compliance with the D/DBP Rule. Besides compliance with the D/DBP Rule, utilities must meet other regulatory requirements (e.g., the ESWTR, the Total Coliform Rule, the Lead and Copper Rule) and secondary drinking-water standards (e.g., color, taste and odor, hardness). In addition, there are operational, financial, and engineering issues that affect the selection of treatment technologies. In the case studies in this project, some utilities chose advanced treatment processes (e.g., ozonation, membranes) that would enable them to satisfy other current and anticipated future regulations or other water quality objectives in addition to the D/DBP Rule.

Although many of the utilities in this project use chloramines for secondary disinfection and to minimize DBP formation, a stronger disinfectant (chlorine, chlorine dioxide, ozone) was required for primary disinfection. Moreover, some utilities may need—as part of the LT2ESWTR—to inactivate *Cryptosporidium*, so ozone is being installed.

Although chlorine dioxide and ozone are both alternative disinfectants that can achieve good primary disinfection while typically not producing THMs, some systems will choose ozone—which is more expensive—because of the ability of ozone to destroy certain taste- and odor-causing contaminants in water. Likewise, granular activated carbon (GAC) is being added to filters for taste-and-control, but it is typically not being regenerated with a frequency required for DBP precursor control.

Enhanced coagulation has been used to remove color, which is also a major source of DBP precursors. For highly colored and hard groundwaters, membranes can be a cost-effective means of treating such waters.

Very importantly, many utilities may have site-specific issues that significantly affect the selection of treatment modifications. In a region in which neighboring utilities use chloramines, there are advantages in switching to chloramines, especially if any of the systems have any interconnections. Alternatively, if there are blended distribution systems with chlorinated groundwaters, the use of precursor removal technologies instead of using post-chloramination may be preferable.

Space and retrofit considerations will affect technology choices. Finally, residual disposal issues may affect treatment selection options.

Staged Implementation

Because of the uncertainty of how stringent certain regulations will be and the high costs of advanced treatment technologies, many utilities have implemented treatment modifications in stages. Typically, most utilities have made treatment modifications to meet Stage 1 of the D/DBP Rule that have been cost-effective to meet their site-specific needs and objectives, while continuing to study or implement long-term treatment changes to meet future more stringent regulations.

Typically, enhanced coagulation and/or a switch to chloramines for secondary disinfection are being used as part of a utility's first stage of treatment changes. The next treatment modification that is typically being incorporated is a switch to chlorine dioxide or ozone for primary disinfection. Utilities that are going straight to the more expensive advanced treatment modifications (e.g., ozone, membranes) to meet the Stage 1 D/DBP Rule are typically those that have relatively high levels of DBP precursors (total organic carbon [TOC] and/or bromide) and/or other contaminants (e.g., color, hardness). These utilities recognize that they ultimately will need such technologies to handle future anticipated more stringent regulations.

The major disadvantage to staged implementation is that the treatment process must be re-optimized each time treatment modifications are made. However, this problem can be minimized if the full range of changes in treatment is envisioned in advance and if incremental modifications are made that will be part of and consistent with long-term modifications. Cost is typically an important driver that results in advancing incrementally. Also, avoidance of unknown problems and tradeoffs is another reason for incremental changes.

Financing Options

Multiple water-quality objectives that can be met by implementing the new technology are important in selling the treatment change to the utility management, the City Council, and the consumers. Consumers that have a higher income and/or educational attainment may be more willing to support the implementation of expensive advanced technologies, whereas in other communities there may be more of a tendency to implement modifications in a staged manner so that lower cost modifications can be utilized initially.

In addition to a community's "willingness to pay," these case studies have demonstrated that there may be unique site-specific considerations that affect the financial feasibility of implementing advanced technologies.

Decision-Making Strategies

The decision-making process to implement treatment changes in order to comply with the D/DBP Rule typically began with an examination of anticipated rule requirements, as the regulation has been developed in stages. Information on the advantages and disadvantages of different treatment modifications were then gathered from the literature and from the experiences of other utilities and/or consulting engineering firms.

Most utilities than embarked on a series of bench-, pilot-, and/or full-scale tests to evaluate the cost and feasibility of one or more treatment processes that could be used to comply with the regulations. Based on this information, an implementation strategy was designed to either incrementally meet changes in regulations one at a time or to make a "quantum leap," thus requiring no additional efforts to meet any future anticipated regulations.

RECOMMENDATIONS

Technology Selection

The case studies in this project have demonstrated that there are multiple water-quality objectives and issues that affect technology selection. The decision-making process needs to consider different driving forces, which affect the choice of technology, in order to meet multiple objectives. Use of such a decision-making process presents a starting framework for consideration:

- The first question that must be addressed is whether the DBP levels are less than the Stage 1 (or 2) MCLs or not.
  - Even if the current DBP levels are less than regulatory limits, the consequences of complying with the ESWTR must be considered.
- If a utility needs to make changes to meet the D/DBP Rule, simultaneous compliance with the ESWTR must be examined.
- Moreover, for many utilities, there are other water quality objectives that need to be met, which affect the choice of technology.
  - For example, some utilities will choose ozone, GAC or membranes if there are multiple objectives that can be met with these treatment processes.
  - Alternatively, a utility may be able to better optimize an existing process, especially if this action allows for a cost-effective solution to many issues.

As part of the decision-making process, utility staff and consultants need to consider the control of a wide range of contaminants, as well as trade-off issues associated with each technology. A table that summarizes the ability of different technologies to control contaminants of particular concern to a utility can be an important tool in technology selection. A technology that is the lowest cost but controls the least number of contaminants may not result in the lowest overall cost of treatment modifications when a utility is attempting to address many regulations and water-quality objectives simultaneously.

In addition, a utility must consider constraints and other issues during the decision-making process. In addition to meeting primary and secondary drinking-water standards, there are operational, engineering, and financial issues that must be addressed. In most cases, utilities must consider treatment modifications that pose retrofit issues. Space availability and hydraulics of the plant may limit cost-effective retrofit choices. In other cases, a utility may be either considering an infrastructure replacement or adding additional parallel treatment trains in order to meet new capacity demands. In those cases, the installation of advanced treatment processes may be more cost-effective to consider than in a retrofit situation. Because of the uncertainty of future regulations, a utility must consider future anticipated requirements in addition to promulgated standards during the decision-making process.

Implementation

Once a technology has been selected, there are a number of steps required for the successful implementation of that technology:

- The information used in the decision-making process will need to be summarized and presented to the utility's governing body and to the consumers.
- In order to persuade the governing body to agree to implement the modifications, a series of presentations will need to be made to provide them with the background and the underlying logic for the project.
- A public information program will need to be used to ensure consumer buy-in.
- After the treatment change has been approved, additional training for the operating staff will be needed.
- An implementation plan will need to be developed and supported by a multi-disciplinary, interdepartmental implementation team.
- Contingency planning will need to address how to handle expected and unexpected problems during the implementation of the treatment modifications.
- There may need to be shifts in the overall operational strategy or operating paradigms to make the modifications successful.
- If there are implementation problems or tradeoffs, some will need to be resolved immediately, whereas for others there may be a period of adjustment.
- It will probably take a period of time for the operational staff to be "comfortable" with the new or modified technology and for the treatment plant to operate like an "integral unit," rather than as a treatment plant with a "new process."

FUTURE RESEARCH

Many participating utilities in this case-study project began their decision-making process in the later 1980s or early 1990s. In most cases in this study, decisions were being made and implemented prior to the promulgation of the Stage 1 D/DBP Rule. Moreover, most utilities utilized "placeholder" MCLs in the proposed Stage 2 D/DBP Rule for long-term planning. In the near future, the USEPA will propose a newly negotiated Stage 2 D/DBP Rule.

As other utilities select and implement treatment modifications in order to comply with the promulgated Stage 1 D/DBP Rule and/or the soon-to-be-proposed Stage 2 D/DBP Rule, a research opportunity exists to document the decision-making process in the new millennium. Moreover, with an even greater number of utilities exploring alternative technologies, a great opportunity exists for the sharing and exchange of implementation experiences. For example, with the greater use of the Internet for information exchange, a potential effort could be to establish electronic bulletin boards to facilitate the sharing of the knowledge gained from these experiences.

However, even with the anticipated promulgation of a Stage 2 D/DBP Rule, the uncertainty over the health effects of DBPs and the likelihood of even future DBP regulations remains. Thus, there is still a research need to document how utilities can make decisions in the face of uncertainty.

Utilities are selecting treatment modifications that will enable compliance with the D/DBP Rule, as well as meet multiple water quality objectives. However, in assessing the costs and benefits of the D/DBP Rule, compliance with only the D/DBP Rule and the ESWTR have been considered. A research effort should be considered to develop tools to do a cost/benefit analysis that considers the multiple benefits of installing these treatment modifications. Moreover, such an analysis will also need to more fully consider the direct and indirect costs of tradeoffs associated with the implementation of these treatment changes. Such an analysis that considers the various benefits and costs of these treatment changes will enable utilities, consultants, and the regulators a better means of judging the cost-effectiveness of the implementation of new technology beyond compliance with one (or two) regulations at a time.

CHAPTER 1
INTRODUCTION

REGULATORY BACKGROUND

Disinfectants/Disinfection By-Products (D/DBP) Rule

The Rule Requirements

The goal of the D/DBP Rule is to reduce the level of exposure to disinfectants and DBPs without compromising the microbial safety of drinking water. This Rule replaced the trihalomethane (THM) regulation of 1979, which established a maximum contaminant level (MCL) of 100 μg/L for total THMs (TTHMs) (U.S. Environmental Protection Agency [USEPA] 1979). Stage 1 of the D/DBP Rule included MCLs of 80 μg/L for TTHMs, 60 μg/L for the sum of five haloacetic acids (HAA5), 1.0 mg/L for chlorite, and 10 μg/L for bromate (USEPA 1998a). Compliance with the Stage 1 MCLs for TTHMs and HAAs—as was done for the 1979 THM Rule—is based on a running annual average (RAA). The RAA is an average of values in the distribution system collected over the four quarters of sampling. In addition, the Stage 1 Rule included removal requirements for total organic carbon (TOC). Stage 1 was promulgated in December 1998 and will be effective by January 2002 for large systems (i.e., serving a population of at least 10,000) that use surface water or groundwater under the direct influence of surface water. For all other systems, compliance with Stage 1 of the D/DBP Rule will begin in December 2003.

In 1994, the USEPA proposed Stage 2 of the D/DBP Rule (USEPA 1994a). The 1994 proposed Stage 2 Rule included "placeholder" MCLs of 40 μg/L for TTHMs and 30 μg/L for HAA5. Compliance with the placeholder MCLs was to be on an RAA basis. During the development of the D/DBP Rule, the possibility of lowering the bromate MCL to 5 μg/L (or lower) in Stage 2 was discussed. The cost impacts of the proposed Stage 2 Rule are substantial (Krasner et al. 1996). Consequently, very large utilities (i.e., serving a population of at least 100,000) were required to collect data via an Information Collection Rule (USEPA 1996a), which was used in a second negotiated rule-making process (in 1999-2000) to resolve Stage 2 D/DBP Rule requirements.

In 2000, a Stage 2 Microbial and DBP Federal Advisory Committee developed an Agreement in Principle for a revised Stage 2 D/DBP Rule and for a Long Term 2 Enhanced Surface Water Treatment Rule (LT2ESWTR) (USEPA 2000). As part of the Stage 2 D/DBP Rule, the Advisory Committee agreed that the MCLs for TTHMs and HAA5 will be changed from a system-wide RAA to a locational RAA (LRAA). That is, each location in the distribution system will need to comply with the 80 μg/L TTHM and 60 μg/L HAA5 MCLs on an RAA basis (i.e., averaging throughout the distribution system will no longer be allowed). In terms of the bromate MCL, the Advisory Committee considered two alternatives: retain the Stage 1 bromate MCL of 10 μg/L or revise the bromate MCL to 5 μg/L. The Advisory Committee agreed that the bromate standard would remain at 10 μg/L, noting that this standard should be reviewed as part of the USEPA's 6-year review process. The Stage 2 Rule is to be proposed in 2002 and to be promulgated by 2003.

Regulatory Impact Analysis (RIA) and Best Available Technologies (BATs)

Utilities have been in the process of making decisions on whether any modifications to their existing treatment practices are needed to achieve compliance with the new D/DBP regulations. An RIA that was prepared for the D/DBP Rule predicted that up to 70 percent of systems would need to make some treatment modifications to comply with the requirements of the Rule (Table 1.1) (USEPA 1994a, Krasner et al. 1996). The technologies evaluated in the RIA included precursor removal technologies--i.e., optimized or enhanced coagulation (Reckhow and Singer 1990, Krasner and Amy 1995), granular activated carbon (GAC) (Lykins et al. 1988, McGuire et al. 1991), membranes (Taylor et al. 1987, Fu et al. 1994) or alternative disinfectants--e.g., chloramines (Norman et al. 1980, Symons et al. 1998), ozone (Jacangelo et al. 1989, Reckhow and Singer 1990, Ferguson et al. 1991). Although chlorine dioxide was not specifically examined in this list of compliance choices, some of the systems utilitizing alternative disinfectants will choose chlorine dioxide (Aieta and Berg 1986).

Best available technologies (BATs) are selected as part of the rule-making process. A specific technology (or group of technologies) is chosen that represents a cost-effective and

Table 1.1

Compliance forecast for large surface-water systems*

Treatment technology	Compliance forecast (%)† Stage 1	Stage 2‡
Current	28	21
Chlorine (Cl_2)/chloramines (NH_2Cl)	3	6
Enhanced coagulation	43	15
Enhanced coagulation + Cl_2/NH_2Cl	10	23
Ozone (O_3)/NH_2Cl	5	5
Enhanced coagulation + O_3/NH_2Cl	6	17
Enhanced coagulation + GAC10	3	0
Enhanced coagulation + GAC10 + Cl_2/NH_2Cl	0	1
Enhanced coagulation + GAC10 + O_3/NH_2Cl	0	6
Enhanced coagulation + GAC20	3	4
Enhanced coagulation + GAC20 + O_3/NH_2Cl	0	2
NH_2Cl	0	1
O_3/Cl_2	0	0.1
O_3/diatomaceous-earth filtration	0	0.1
O_3/direct filtration/NH_2Cl	0	0.1

*Included surface-water systems that did not filter.

†Each of the values in this table were rounded to one or two significant figures. When the rounded values are summed, the totals equal 101 percent. When the original values—calculated to an additional significant figure—were used, the totals equal 100 percent.

‡Compliance forecasts based on the placeholder Stage 2 MCLs.

Source: Adapted from USEPA 1994a.

feasible control measure. Other control strategies may be used to comply with a rule, as long as they can meet the same water quality standards as the BAT. Enhanced coagulation or GAC treatment with a 10-min empty bed contact time (EBCT) and a regeneration frequency of 180 days (i.e., GAC10)--with chlorine as both the primary and secondary disinfectant--was chosen as the BAT for Stage 1 of the D/DBP Rule (USEPA 1994a, Krasner et al. 1996). For the 1994 proposed Stage 2 Rule, (1) enhanced coagulation followed by GAC10 or (2) GAC treatment with a 20-min EBCT and a regeneration frequency of 60 days (GAC20)--with chlorine as both the primary and secondary disinfectant--was chosen as the BAT.

In the RIA, it was predicted that most surface-water systems would be required to use enhanced coagulation with and without alternative disinfectants. Because of the wide variety of treatment options, it was predicted that few systems would be required to use GAC in order to comply with Stage 1. Moreover, it was predicted that large surface-water systems would not need to install membranes, whereas it was predicted that membranes would be used by a certain percentage of small systems and groundwater systems in order to comply with the proposed Rule (USEPA 1994a).

The Technology Workgroup in support of the Stage 2 D/DBP Rule development was asked by the Federal Advisory Committee to compare different Stage 2 regulatory options for DBP control. The Technology Workgroup compared the Stage 2 placeholder MCLs of 40 μg/L TTHMs and 30 μg/L HAA5 as RAAs to a Stage 2 option of 80 μg/L TTHMs and 60 μg/L HAA5 as "single highest" allowed values. Both Stage 2 options resulted in large reductions in TTHM and HAA5 average exposure and highest exposure. Both regulatory scenarios under consideration showed similar cumulative probability distributions, and both required similar shifts in treatment technologies to accomplish. The Advisory Committee decided on Stage 2 MCLs of 80 μg/L TTHMs and 60 μg/L HAA5 on an LRAA basis, not a single highest allowed basis. However, as part of the Agreement in Principle, systems that have significant excursions in DBP levels during peak periods will need to refer to USEPA guidance on how to reduce such peaks. In the RIA for the revised Stage 2 D/DBP Rule, it was predicted that more than 50 percent of surface-water systems will switch to chloramines for secondary disinfection in order to meet the LRAAs.

Other Regulations that Effect Treatment-Process Selection for DBP Control

The decision-making process at utilities is also influenced by a number of State and Federal regulations besides the D/DBP Rule. Some of the most important regulations constraining treatment-process selection are discussed in the following paragraphs (Daniel 1998).

Surface Water Treatment Rule (SWTR)

The SWTR established minimum treatment and disinfection requirements for surface waters (USEPA 1989a). The requirements include removal or inactivation of 3 log (99.9 percent) *Giardia* and 4 log (99.99 percent) viruses, a lower MCL for turbidity, and watershed protection for unfiltered systems.

An Interim Enhanced SWTR (IESWTR) included more stringent turbidity-control requirements to improve the physical removal of *Cryptosporidium* (USEPA 1998b). Additional *Cyptosporidium* requirements (e.g., disinfection) were preliminarily proposed in 1994 as an option under the ESWTR (USEPA 1994b).

In the LT2ESWTR, the Advisory Committee agreed to a round of microbial monitoring that will be used to place systems in different "bins" based on average *Cryptosporidium* levels (USEPA 2000). Systems in the different bins will go to a "toolbox" of options for additional removal and/or inactivation requirements. Disinfectants in the toolbox that can be used for *Cryptosporidium* inactivation include chlorine dioxide, ozone, and ultraviolet (UV) disinfection. In the RIA for the LT2ESWTR, it was predicted that surface-water systems that have significant levels of *Cryptosporidium* will utilize ozone or UV disinfection. The disinfection requirements in these Rules present treatment challenges, as utilities are simultaneously required to lower DBP concentrations.

Total Coliform Rule (TCR)

The TCR, which applies to all public water systems, set compliance requirements for the control of coliform bacteria (USEPA 1989b). It had extensive monitoring requirements and posed challenges because of (1) requirements for the maintenance of a disinfectant residual and (2) problems encountered with re-growth in distribution systems. Utilities that have switched to chloramines need to avoid nitrification, which can stimulate re-growth problems (Wolfe et al., 1990). Plants that have implemented ozonation need to control biodegradable organic carbon, which can increase the potential for re-growth (van der Kooij et al. 1982).

Lead and Copper Rule (LCR)

The LCR established regulatory standards for lead and copper, and required utilities to optimize corrosion-control treatment (USEPA 1991). The most common cause of elevated lead at the customer's tap was leaching from household plumbing and fixtures as a result of corrosive water. One of the primary objectives of the regulation was to minimize the corrosivity of the water. All large water systems (serving over 50,000 persons) were required to complete corrosion-control studies by January 1998 and to begin installation of optimal corrosion-control treatment by January 1999 and complete it by January 2000.

MULTIPLE WATER-QUALITY OBJECTIVES

Multiple water-quality objectives usually face utilities that contemplate treatment modifications to their existing facilities (Daniel 1998). Besides compliance with primary drinking water regulations, utilities are also concerned with secondary drinking-water standards for color, taste and odor, total dissolved solids (TDS), etc. Aesthetic issues—such as color and taste and odor of the drinking water—are also associated with public perceptions of safety. Hence, treatment processes at utilities are selected to minimize/eliminate aesthetic issues.

The water-quality goals of a utility also have to take into account emerging chemical and microbial issues that may require future regulations, operational uncertainty (margin of safety, etc.), and speculative uncertainties regarding future regulations (Daniel 1998). In addition, there are a number of operational, financial, and engineering constraints that limit the selection and implementation of treatment technologies. Some of these issues include, but are not limited to, retrofit issues, staffing concerns, plant hydraulics and layout, site-specific conditions, space availability, electrical requirements, access to capital, etc.

CASE-STUDY PROJECTS

In the 1980s and early 1990s, various surveys and studies documented the changes that utilities had made to comply with the 1979 THM regulation (McGuire and Meadow 1988), as well as modifications to comply with future, more stringent disinfection and DBP standards (Krasner et al. 1989, McGuire et al. 1989, Glaze and Weinberg 1993, Randtke et al. 1994). Case studies of 20 utilities that modified their disinfection practices to comply with the THM Rule were evaluated in a project funded by the American Water Works Association Research Foundation (AwwaRF) (George et al. 1990). Utilizing a case-study questionnaire, the latter AwwaRF project examined various alternative disinfectants, their usage, benefits, problem areas, and costs.

Many utilities had to change their treatment to comply with the 1979 THM Rule. In an AwwaRF THM survey (McGuire and Meadow 1988), 727 utilities reported THM data. Of these, 225 utilities reported changes in treatment. A total of 543 treatment changes were reported, reflecting more than one change per respondent. Table 1.2 shows what treatment changes were selected for those systems requiring modification to comply with the THM Rule. The majority of utilities changed their chlorine dosage and/or moved the point of disinfection. A significant number of utilities that responded to the AwwaRF THM survey had changed to chloramines as the primary or secondary disinfectant. Some utilities enhanced their coagulation process or switched to chlorine dioxide for primary disinfection. Only a small number of utilities switched to ozone for primary disinfection or incorporated GAC treatment.

The majority of utilities in the AwwaRF THM survey were able to change their chlorine dosage, change the point of disinfection or switch to chloramines without requiring precursor removal, chlorine dioxide or ozone to comply with the 1979 THM Rule. Alternatively, the D/DBP Rule is requiring precursor removal that may require a different set of treatment modifications (Table 1.1).

The current AwwaRF case-study project examined 10 utilities that implemented treatment modifications to enable them to comply with the D/DBP regulations. The latter case-study project is serving as a research continuum to bridge the gap between research and practical application. Project objectives include development of a case-study protocol, documentation of

Table 1.2

Treatment changes made by utilities to comply with the 1979 THM Rule

Treatment change	Number of treatment changes
Procedure modification	
Improve clarification	41
Change chlorine dosage	122
Change point of disinfection	150
Chloramines	100
Chlorine dioxide	26
Powdered activated carbon	25
Off-line storage	4
Alternate source	2
Aeration	9
Install clarification	4
Ozone	3
GAC	10
Other	47
Total	543

Source: Adapted from McGuire and Meadow 1988.

what worked and what did not so others can learn from the problems and the solutions, documentation of the decision-making process, and examination of implementation issues.

CHAPTER 2
METHODS

UTILITY SELECTION

The selection of appropriate utilities to study was an iterative process, in order to ensure that a representative group of utilities were studied. The major goal was to study utilities that had implemented treatment modifications that spanned the range of choices that utilities were considering for complying with the D/DBP Rule (Table 2.1). Other utility-selection considerations included (1) geographical representation (Figure 2.1), (2) different size (flow) categories (Table 2.2), and (3) studying utilities with a range of raw-water qualities that presented challenges in complying with the D/DBP Rule (Table 2.2). In general, it was important to select utilities for study that had not made their treatment changes too recently, in order that long-term operational experiences could be evaluated.

Table 2.1

Matrix of treatment modifications to study

Treatment modifications	Utilities selected
Alternative disinfectants	
Chloramines	Newport News Waterworks (Va.)
	City of Poway (Calif.)
Chlorine dioxide	Gulf Coast Water Authority (GCWA) (Texas)
	Skagit County (Co.) Public Utility District (PUD) (Wash.)
Ozone	Ann Arbor Water Utilities (Mich.)
	Alameda County Water District (ACWD) (Calif.)
Precursor removal	
Enhanced coagulation	City of Phoenix (Ariz.)
	City of Tampa (Fla.)
GAC*	Virginia American Water Company, Hopewell (Va.)
Membranes	City of Plantation (Fla.)

*This utility operates with biologically active carbon for the removal of DBPs, not precursors.

Figure 2.1 Location of participating utilities in case-study project

Table 2.2

General characteristics of the participating utilities in the case-study project

Participating utility	Population served	Flow* (mgd)	Raw water (average values) TOC (mg/L)	Bromide (mg/L)	Alkalinity (mg/L)	Comments
Newport News	160,000	23	6.5	0.06	30	
Poway	50,000	12	~4	~0.1	~120	
GCWA	92,000	12	4.7	0.12	140	
Skagit Co. PUD	50,000	8	4.3	0.01	14	Originally unfiltered
Ann Arbor	115,000	20	5.8	0.07	230	Softening plant
ACWD	302,000	18	4.6	0.07†	87	
Phoenix	1,100,000	640‡	~3	0.06-0.08§	120	
Tampa	450,000	80	13	0.06	93	Sludge-disposal issue
Hopewell	28,000	20	3.9	NA†	45	
Plantation	78,000	12	~20	0.14	244	Originally softening

*Typical flow at the plant that was modified for DBP control in the case-study project.

†During droughts, bromide has significantly increased.

‡Combined capacity at five treatment plants.

§Range of average values for five treatment plants

At the time that this study was conducted, utilities that had installed GAC had made this treatment modification in order to control synthetic organic contaminants (SOCs) and/or taste and odor. One U.S. utility that had installed GAC for SOC control was able to also remove DBP precursors (i.e., TOC) in their operations. However, they were not available to participate in this project. Therefore, a decision was made to study a utility that had installed GAC for taste-and-odor control, in which the GAC was biologically active and was able to remove DBPs, not precursors.

At the time that this study was conducted, many of the utilities that had installed membranes were Florida groundwaters (e.g., City of Plantation) (Taylor et al. 1987, Bergman 1996). Attempts were made to identify a surface-water system that had installed membranes for

DBP control, but at the time of this study, most surface-water systems had only evaluated membranes at a pilot scale and did not have full-scale installations. However, as noted in Table 1.1, no surface-water utilities had been predicted in the RIA to need membrane treatment in order to comply with the D/DBP Rule.

CASE-STUDY APPROACH

Case-Study Protocol

The case-study protocol utilized an iterative approach, with the participating utilities as partners in documenting the case-study information. Some of the steps in the protocol included:

(1) Establishment of initial contacts and conducting a screening interview with utilities to explain project scope and objectives, utility role and benefits. This step provided an idea as to whether the utility was a good "fit" in terms of project objectives, utility-selection matrix, and mutual benefits.

(2) Follow-up mailings to the selected utilities. The mailings included a project scope of work, a Memorandum of Understanding (MOU) to incorporate a partnership relationship with the utility, and information-collection instruments (questionnaires).

(3) Kick-off or profile interviews. The main objective was to document the plans for data collection and evaluation, and identify data/information/research that was perceived to be important to decision-making.

(4) Data gaps and monitoring needs analysis. Once the questionnaires had been completed and the data received, potential data gaps were identified and discussed with utility contacts. Where necessary, specific monitoring programs were designed to fill these gaps. In addition, follow-up questionnaires were sent to utilities to fill in data and information gaps.

(5) Follow-up interviews. This next level of interviews involved a broader range of utility staff and encompassed a broader spectrum of issues that may have factored into the decision-making. Subjects covered included, but were not limited to, tradeoff issues within the realm of treatment optimization, implementation and operational issues, decision criteria

and constraints. In addition, comparisons were made to the treatment modifications implemented as a result of the 1979 THM Rule.

(6) Development of a draft case-study narrative. A draft description of the case study was assembled, combining the results of the data evaluation and the results of the interview process. This draft was shared with the utility participant.

(7) Iterations leading to a final narrative. The draft narrative was revised to take account of feedback and comments from the utility participant.

Pilot Testing of the Case-Study Protocol

The first three utilities that were studied included a system that had enhanced the coagulation process for the removal of DBP precursors (i.e., Phoenix), a system that had incorporated ozonation (i.e., ACWD), and a system that was using membranes for DBP control (i.e., Plantation). These initial case studies allowed the research team to "pilot test" the case-study protocol. Pilot-testing of the case-study protocol indicated that two approaches may be utilized to develop a case study:

(1) A "bottom-to-top" analysis initially utilized the questionnaires and then concluded with a series of follow-up interviews. In this approach, the research team got the data and other information first and then conducted interviews on the decision-making process. This approach was utilized in the majority of case studies when the research team was unfamiliar with the utility's history of treatment modifications.

(2) Alternatively, a "top-down" approach was used in which a general interview was conducted first, with less reliance on filling in the questionnaires. Then supporting data and information were subsequently obtained to complete the case study. This approach was more suitable in situations when members of the research team had intimate knowledge of the decision-making process at a particular utility or had helped the utility implement their treatment modifications.

QUESTIONNAIRE

Information-Collection Instruments

The first step in the project was the development of a series of questionnaires (Appendix A) covering a wide range of treatment scenarios, which asked for information on water quality, operational experiences, implementation issues, costs, and the decision-making process. To facilitate efficient replies to the questionnaires, they were divided into different groups that could be answered by different staff at each utility. Moreover, it was emphasized to the utility staff that the questionnaires served as a guide to the types of information needed by the research team. The utilities were encouraged to send reports, memoranda, etc., in lieu of directly answering each question.

Water Quality Information

In order to evaluate compliance choices with the D/DBP Rule, certain water quality data for the raw and finished water—before and after treatment-modifications—were needed. This information was obtained from the staff of the water quality laboratory at the participating utility. Some of the water quality parameters targeted in this section included, in addition to DBP data, indicators of DBP precursor levels such as TOC and bromide; temperature; pH; alkalinity; hardness; turbidity; and pathogens. It was important to document other water quality parameters that affected the choice of treatment modifications.

Treatment-Modification Information

This section examined the treatment-modification selection process with information provided by staff in the process-engineering group at the utility. Information was obtained on bench-, pilot-, and full-scale evaluations of new or modified treatment processes. Information was collected on the range of treatment choices evaluated by the utility, not just on the final process(es) selected for implementation.

Operations Information

Information on the water-treatment process was obtained from the operations staff. This included information on the disinfection process, coagulation or softening, the use of advanced precursor-control technologies, and other specific technologies to control DBP levels.

In terms of disinfection (for chlorine, chloramines, chlorine dioxide or ozone), information on application points, disinfection doses, residuals, contact times, etc., were obtained. It was important to document overall disinfection capability before and after the implementation of treatment modifications. If ozone, chlorine dioxide or hypochlorite solutions were used, information on control strategies for minimizing the levels of oxyhalides was obtained. In addition, for ozonation, information on biofiltration was collected.

The coagulation or softening process was evaluated for chemicals used, dosages, coagulation or softening pH, addition of pH-adjusting chemicals, etc. If GAC was used, information was collected on the EBCT and regeneration frequency, as well as criteria for re-generating the GAC. If membranes were used, information was collected on membrane type, composition, operating pressure, biofouling experiences and control measures, etc.

Implementation Information

This section dealt with questions designed to elicit an accurate documentation of problems experienced in implementing treatment modifications, as well as the potential "tradeoffs" involved. In the case where complex treatment/operation processes were involved, it was important to document what steps utilities have taken to ensure compliance (e.g., operator training, written protocols, etc.).

Site-specific cost information was also covered in this section. Information was obtained on capital costs and annual operating costs.

Information on the Decision-Making Process

An important aspect of the case-study project was to document the decision-making process that these utilities underwent to decide upon which modifications to make in order to comply with the D/DBP Rule. Questions were asked on the important driving factors that shaped the decision; areas of uncertainty; major trip wires in terms of operation and reliability and how these affected the decision-making; anticipated and unanticipated factors in terms of technology; consumer support; compliance issues; selection issues and future strategies.

Internet Resources

The Internet proved to be a very valuable research source, which complemented the information obtained with the questionnaire. A number of utilities have web sites with information on treatment processes used, water quality obtained, and background information on water resources utilized. A number of them address consumers' concerns about water quality on their web sites.

In some cases, information about the utility could be found through the web site of the city they are located in. For example, the web site of the city could have links to the water department. The information obtained included consumer confidence reports, a water quality master plan detailing how the department plans to keep abreast of water quality regulations, water supply and demand data, annual reports containing information on issues such as administration, personnel, water quality and finance.

Other web sites that yielded important information were federal government web sites with U.S. census data. These data, obtained for regions supplied by the utility under study, examined consumer demographic factors such as age, household income, education, cost of housing, etc. The consumer is a key player in decisions taken by the utility, as most treatment modifications result in an increase in water rates that have to be paid by the consumer. Demographic information such as income, cost of housing, and education may indicate how supportive a utility customer might be of treatment modifications and consequent rate changes.

The Internet also provided access to local newspapers in the cities where the utilities were located. Local newspaper articles were found in which treatment changes at utilities were discussed for the edification of the public.

INTERVIEW PROCESS

Initial Contact, Solicitation of Interest, and Screening Interview

The project team used "talking points" (see below) to explain the project scope and objectives, as well as the utility role and benefits. This step provided an idea as to whether the utility was a good fit in terms of project objectives, utility-selection matrix, and mutual benefits.

Talking Points

The project team summarized the objectives and benefits to be derived from the outputs intended from the case-study project in general and from individual case studies. In making contacts with candidate utilities, it was important to offer a quick summary. The key points included:

(1) Benefit to participants: an external peer review that might offer a few new insights.

(2) Benefit to utilities in general: a collection of case studies revealing how the selected utilities have approached a significant technical and managerial decision that many utilities are facing.

(3) Benefit to the national regulatory program:

  (a) improved understanding of compliance behavior in response to Stage 1 standards will beget a better understanding of the impacts of Stage 2 options;

  (b) and a better understanding of the utility response to compliance requirements in general—including the development and refinement of a research protocol for conducting case studies of compliance decision-making.

Establish a Partnership Relationship

The second item of business in approaching candidate utilities was allaying any fears they may have about airing their laundry publicly or about revealing their decision process. The key points (which were reflected in the MOU) included:

(1) Participating utilities become a part of the research team, including an in-kind contribution. If the project team were to simply roll into town as the "researchers" and study the utilities as their "subjects," there would be little benefit to the utility. Much of the internal benefit to the utility resulted simply from their studying their own decision process, something they probably might not otherwise have taken the time to do.

(2) Participants had to be willing to engage in this critical review of their experience, but it was not intended to be "critical." It was about utilities documenting their experiences and sharing and learning from each other.

(3) The full technical and managerial dimensions of this decision-making process will never be properly appreciated in the standard setting process if they are not documented.

(4) Utility participants were intended to be "participants," not "subjects." Participants reviewed the draft case-study write-ups and participated in developing the final version.

Kick-Off or Profile Interview

The objective was to inventory in detail what information, data, and research was developed or used and to obtain copies of reports, summaries, and files that may be available. This was tailored to where the utility was in the decision-making process:

(1) "Before," just getting started: prospective case-study protocol; less data available. Inquiries focused on documenting the plans for data collection and evaluation to see how information gaps were perceived and what was perceived to be important to the decision making.

(2) "During," before the decision had been made: half-and-half protocol; inquiries may determine that some data, information, and research were available, but may not be final basis for the decision.

(3) "After," post mortem: inquiries focused on identifying all the data, information, and research that were used to reach a decision.

Although most of the utilities selected for study had completed their decision-making process in terms of compliance with the Stage 1 D/DBP Rule, many of the participating utilities were at different places in the decision-making process in terms of compliance with a potential Stage 2 D/DBP Rule.

Second-Tier Interviews

After a review of the available data and information, second-tier interviews were conducted to investigate the relationships between available data, information, and research and the specific technical questions the utility was trying to answer in making technical decisions believed necessary to the evaluation of technical feasibility and cost.

Third-Tier Interviews

The next level of interviews involved a broader range of utility staff and encompassed a broader spectrum of issues that may have been factored into the decision-making process. Key players involved in the decision-making process were identified. A telephone-administered executive interview was conducted with the key-decision makers at participating utilities at a point after the rest of the project data had been collected and digested. The issues discussed included:

(1) Trade-offs within the realm of treatment optimization: This sometimes included issues extending all the way from the source to the tap. Questions were asked on how the technology switch potentially affected other things at that utility and how were these conceivable side effects considered and weighed.

(2) Implementation and operational issues: This encompassed necessary shifts in instrumentation and operator training, as well as the operational demands of fluctuations in influent conditions that challenged the plant in different ways.

(3) Imposed decision criteria and constraints: This was an open set of questions that bridged into the managerial side. It sometimes included water-resource issues, state and other federal regulations, dealing with governing bodies and consumers, economic and cost considerations, etc.

Questions were asked on the important driving factors that shaped the decision; areas of uncertainty; major trip wires in terms of operation and reliability and how these affected the decision-making process; anticipated and unanticipated factors in terms of technology; consumer support; compliance issues; selection issues and future strategies.

STATISTICAL ANALYSES

A number of statistical methods were used to examine the data in the case-study project. Some of the data were examined using box-and-whisker plots (Figure 2.2). The bottom and top of the box correspond to the 25th and 75th percentiles, respectively, which is referred to as the interquartile range. The whiskers (vertical lines below and above the box) range from the 10th to the 90th percentiles, with outliers shown as individual data points above or below the whiskers. (Note, for Figures 3.10 – 3.11 in the Poway case study and for Figures 3.15 – 3.25 in the GCWA case study, the individual dots above and below the whiskers represent the 95th and 5th percentiles, respectively.) A solid and a dashed line—which pass horizontally through the box—correspond to the median (50th percentile) and mean (average) values, respectively. If there is a normal distribution, the median and mean values will be quite similar. In box-and-whisker analyses, the data can be segmented into different groups—in which a separate box-and-whisker plot is generated for each group—in order to visually examine the similarities and/or differences in each group's distribution.

For a number of the parameters under investigation, the minimum, mean, and maximum concentrations of the parameters were examined. When a DBP was not detected (ND), the NDs were set to zero in computing DBP class sums, as is currently done when computing TTHMs for compliance monitoring. However, when individual DBPs were examined (e.g., bromate), NDs were sometimes set at one-half of the detection limit (DL) as a compromise, as the DBP may be present at some value between zero and the DL.

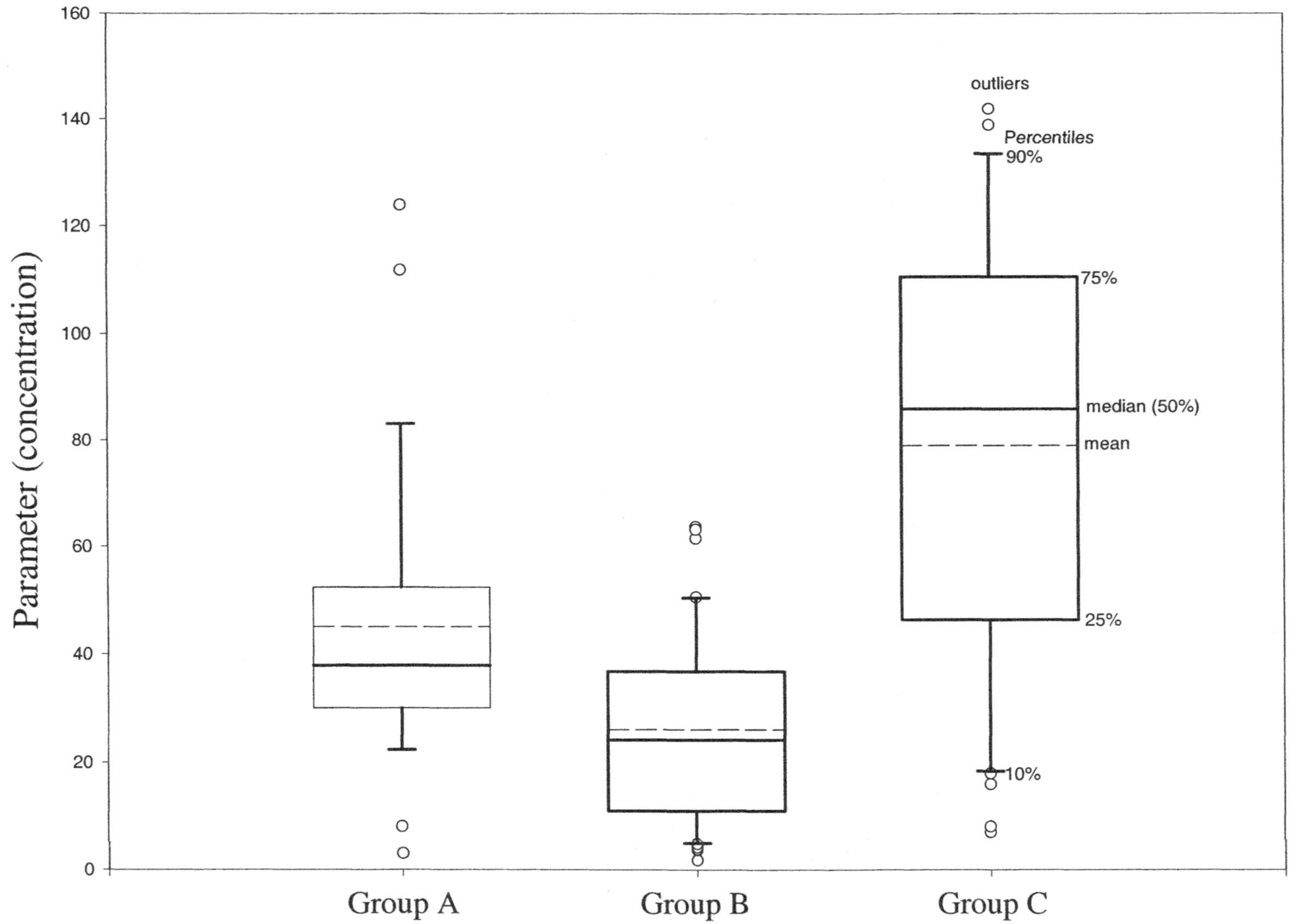

Figure 2.2 Guide to box-and-whisker plots

CHAPTER 3
CASE STUDIES FOR ALTERNATIVE DISINFECTANTS

INTRODUCTION

Alternative disinfectants to chlorine for the control of DBPs (in the Stage 1 D/DBP Rule) include chloramines, chlorine dioxide, and ozone. An alternative disinfectant for the control of microbial pathogens (e.g., *Cryptosporidium* in the LT2ESWTR) is UV. Six case studies for alternative disinfectants were conducted, with two case studies per disinfectant choice, except for UV (Table 2.1).

CHLORAMINES: NEWPORT NEWS

Summary

Newport News Waterworks (NNWW) operates two conventional water treatment plants, which formerly used chlorine for primary and secondary disinfection. The source-water TOC is relatively high, so NNWW converted to chloramines for secondary disinfection in order to comply with Stage 1 of the D/DBP Rule. Because chloramines were not expected to enable NNWW to meet future anticipated stricter disinfection and DBP regulations, ozone was being installed. Sequential implementation of chloramines and ozone was being used to cost effectively meet disinfection and DBP regulations.

Selection of the ammonia feed system was based on a cost analysis and on retrofit/siting issues. The finished water pH was raised to a level that was optimal for monochloramine formation. A storage tank in the distribution system with a very long detention time was taken out of service to avoid nitrification problems.

NNWW conducted an extensive campaign to inform the public of the chloramine conversion. Special attention was made to kidney dialysis centers and to fish owners. During the first month of the chloramine conversion, NNWW did trend analysis of customer call-in activity and used automation tools, which facilitated taking corrective action.

THM and ultimately HAA concentrations were reduced as expected. In addition to switching to post-chloramination, contact time with free chlorine in the plant was reduced to minimize HAA formation. The higher finished water pH with the pebble lime feed system resulted in an increase in the finished water turbidity. Therefore, NNWW was converting to a hydrated lime feed system to handle this problem.

At NNWW, the use of post-chloramination was allowing for a cost-effective means of meeting the Stage 1 MCLs for THMs and HAAs. This incremental change in treatment is consistent with the future use of ozone to meet new and/or future anticipated disinfection and DBP regulations.

Background

Case History

NNWW operates the Harwood's Mill Water Treatment Plant (HMWTP) and the Lee Hall Filtration Plant, which supply drinking water to approximately 350,000 people on the lower Virginia Peninsula. The HMWTP—on which this case study was based—was placed in service in 1989, has a rated capacity of 31 mgd (typical flow = 23 mgd), and serves approximately 160,000 people.

Traditionally, the principal treatment goals at NNWW have been turbidity removal through chemical coagulation/settling/filtration and disinfection with chlorine; and removal of color, iron and manganese for aesthetic considerations. HMWTP treats water from the Harwood's Mill Reservoir, which is fed from the Chickahominy River (typically 55 percent of the source water blend), Diascund Reservoir (typically 15 percent), and local runoff (typically 30 percent). The source-water TOC is relatively high (~6-7 mg/L on average). NNWW converted to chloramines for secondary disinfection in January 1998 in order to comply with Stage 1 of the D/DBP Rule.

Process Selection

Bench and pilot testing indicated that a 30 to 50 percent reduction in THM levels could be achieved by switching secondary disinfection from free chlorine to chloramines, which would enable compliance with the Stage 1 MCLs of 80 μg/L TTHMs and 60 μg/L HAA5. The previously proposed placeholder Stage 2 MCLs of 40 μg/L TTHMs and 30 μg/L HAA5 could be met by the installation of ozonation/biofiltration followed by chloramines.

Conceptual studies, preliminary design, and pilot testing was conducted by a team consisting of NNWW staff and consultants from Malcolm Pirnie, Inc., and Montgomery Watson, Inc. The pilot plant was staffed with personnel from NNWW. A 1-year evaluation was conducted in 1995-96.

Coagulation studies: The evaluation included DBP precursor (TOC) removal experiments by coagulation. The coagulants evaluated were alum and polyaluminum chloride. The coagulation pH evaluated was 5.5 to 6.5. Sulfuric acid was evaluated for pH reduction and lime and caustic soda were evaluated as pH-increasing chemicals.

Evaluation of alternative disinfectants: Chloramines were evaluated for Stage 1 DBP control and ozone was evaluated for Stage 2 DBP control. Two parallel ozone treatment trains were studied, one to achieve 0.5 log of *Giardia* inactivation and the other to achieve 2 logs of *Cryptosporidium* inactivation. The intent was to bracket the lower and upper ends of expected ozone usage. Because *CT* values for *Cryptosporidium* inactivation were not established at the time of the pilot testing, the test procedure was to maintain a constant *CT* of approximately 4 mg/L-min through the high ozone dose contactor. This contactor provided more than one order of magnitude of inactivation greater than the SWTR requirements for 0.5-log Giardia inactivation.

Advanced precursor-removal studies: GAC was tested (at the bench scale) as part of NNWW's compliance with the ICR. Nanofiltration was also pilot tested. Although effective for TOC removal, neither process was deemed feasible for NNWW because of unfavorable economics.

Parameters that Influenced Process Design

The parameters and treatment objectives that influenced process design were particle/cyst removal and microbial inactivation, maintenance of a disinfectant residual within the distribution system to provide secondary disinfection, DBP control, iron and manganese removal, taste-and-odor control, and color removal. Of these, those associated with regulations most likely to drive the potential implementation of alternative treatment processes were the following:

- Removal/inactivation of microbial contaminants; i.e., meeting existing primary disinfection requirements of the SWTR and possible future requirements contained in an ESWTR; while
- Minimizing the formation of DBPs.

To achieve a balance between microbial and DBP control, NNWW needed to do the following:

- Increase the removal of DBP precursors (TOC); and/or
- Use an alternative disinfectant, which does not form DBPs in concentrations greater than the anticipated regulatory limits.

The Decision-Making Process

The decision-making process relied on extensive pilot testing and NNWW staff's knowledge of the regulations and their best guesses for the future direction (Figure 3.1). Both the water production manager and the Director were extensively involved in the negotiated rulemaking process convened to develop the new DBP standards, and this knowledge gave them a fairly high degree of certainty of the direction of future regulations.

The chloramine conversion project was a small part (in terms of dollars, not significance) of a larger decision process—which included a new replacement plant for Lee Hall and the installation of ozone at both NNWW plants—in order to meet likely Stage 2 compliance requirements, going beyond Stage 1. Given the level of staff knowledge of future regulatory directions and the need for a major plant replacement, it did not make sense to convene a

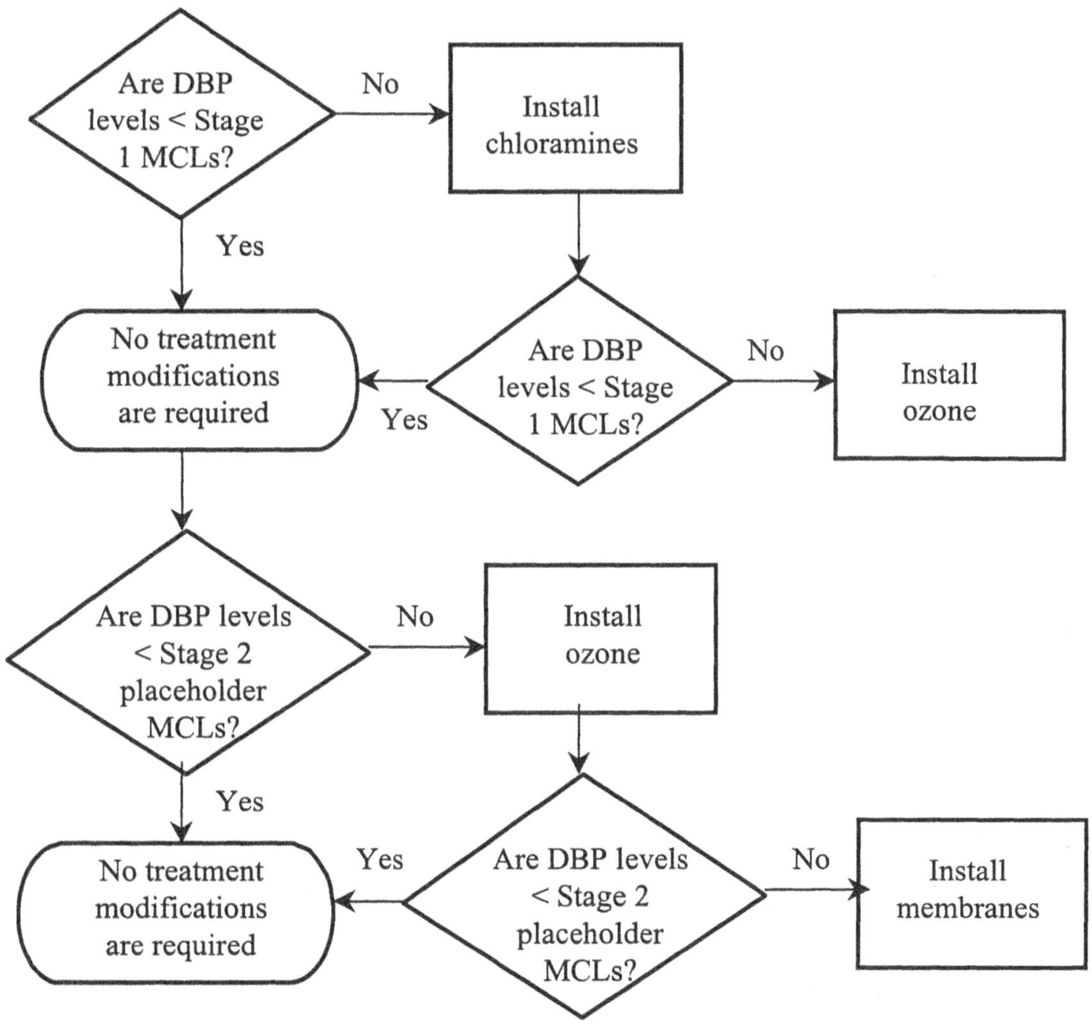

Figure 3.1 The decision-making process at NNWW: sequential implementation of chloramines and ozone (based on ozone being less costly technology than membranes).

strategic effort to meet Stage 1 DBP goals. Rather, Stage 1 compliance was considered from the start as part of the larger need for a higher level of compliance, which was nominally targeted around the previously proposed placeholder values of 40 μg/L for TTHMs and 30 μg/L for HAA5.

A strategic brainstorming process involving numerous experts in the field was convened in 1995 to assess options and lay the groundwork for a pilot testing phase. Although GAC was tested at bench scale and membranes were piloted, ozonation with post-chloramination was the preferred option from the start. The major reasons for this included capital savings and operating cost savings compared to membranes and GAC.

The decision-making process recognized that ozone would not be installed in time for the Stage 1 D/DBP Rule. Thus, the chloramine conversion project was fast-tracked to meet Stage 1.

Driving Forces

There were several driving forces in the decision-making process:

- The utility's knowledge of the regulatory process gave them a strong reason to plan for Stage 2 compliance rather than just Stage 1 compliance.
- The need to replace a plant made it even more imperative to adopt the longest view of the future.
- Ozone/chloramines offered capital savings because there was room for it within existing structures.
- By contrast, the future cost of membrane systems was more uncertain.
- Due to low bromide levels, it was possible to avoid side effects of significant bromate formation from ozonation.
- Ozone offered the option of providing 2 logs of *Cryptosporidium* inactivation, which was considered a bonus because the original objective was DBP control.
- Once the ozone decision was taken, the post-chloramination part of the plan was simply moved forward to obtain Stage 1 compliance.

Constraints

Many typical constraints were not in evidence in this case study:

- Lack of knowledge of the ultimate direction of regulations was not a constraint because the staff had excellent knowledge that levels beyond Stage 1 were inevitable.
- The threat of forming bromate was not a constraint due to low bromide levels.
- Process control required to effectively implement an ozone/chloramines strategy under varying raw water challenge conditions were not extraordinary and posed no significant constraints.

- Managing the transition to chloramines and managing a chloraminated distribution system were not constraints due to the fact the water production manager already had operational experience with chloramines at another utility.

Technology Checklist—Anticipated Factors

- THMs in the distribution system were reduced by 33 percent (from 90 to 60 μg/L) as expected from pilot testing.
- The transition to chloramines went very smoothly at both the operational level and the public relations level.
- Alkalinity and pH control was addressed with a new hydrated lime system, and it has performed smoothly as expected.
- There have been no problems with nitrification, as expected, although one ground storage tank was removed from service to avoid such problems developing.

Technology Checklist—Less Well Anticipated Factors

The initial drop in HAA concentrations was not as dramatic as was hoped. Prior to the conversion to chloramination, there was no benefit in reducing the chlorine contact time in the plant, because the DBP reaction continued in the distribution system. Before chloramination, typically all process units were kept in service all the time. After the conversion to chloramination, filters were removed from service to reduce contact time in the plant with free chlorine to further minimize HAA formation. This required additional work with operators to become accustomed to the practice of rotating filters on and off for this purpose

The higher finished water pH with the pebble lime feed system resulted in an increase in the finished water turbidity. NNWW was converting the plant to a hydrated lime feed system, which should significantly reduce their finished water turbidity.

Compliance Issues

NNWW has been very fortunate that their overall ozone/chloramines compliance strategy will allow them to deliver 2 logs of *Cryptosporidium* inactivation without adverse effects from bromate formation and will meet the previously proposed Stage 2 placeholder targets. The use of chloramines should also help meet new requirements for controlling peak by-product occurrence levels in the distribution system.

Key Players

The key players involved in the decision-making process were NNWW staff and consultants from Malcolm Pirnie, Inc., and Montgomery Watson, Inc. As noted above, key decision makers at NNWW included the water production manager and the Director, who were extensively involved in the regulatory process.

A series of presentations were made to the City Council to provide them background and the underlying logic for the project. Many approved of the project because of the water quality benefits. An extensive public relations campaign was undertaken to educate and inform NNWW's customers as to the water quality benefits from chloramination. The impact on rates was quite small and was a non issue. The larger issues were with fish owners and kidney dialysis patients, whom NNWW staff extensively counseled.

NNWW serves water to the cities of Newport News, Hampton, and Poquoson, and portions of James City and York counties. Demographic information from 1990 was obtained for the City of Newport News. The median household income was $27,469. The median cost of housing was $84,400. The median age of the population was 29 years.

Selection Issues and Future Strategies

Ammonia feed systems: An extensive amount of conceptual design information was prepared for evaluating the installation of ammonia feed systems. Some of the key issues are discussed below:

- The normal chlorine to ammonia-nitrogen ratio (on a weight basis) to be used was 4.75:1 based on the stoichiometric optimum for monochloramine formation of 5:1. A range of alternative values was recommended for developing chemical feeder sizing. A range of ratios from 3:1 to 6:1 was considered for flexibility, which (1) avoided too much free ammonia (and potential nitrification problems) on one end and (2) avoided approaching the region of breakpoint chlorination on the other end. A feed ratio of 4.75:1 results in a slight excess of ammonia. This ratio allowed NNWW to reliably feed a ratio of 5:1 or less, after accounting for process variation. They wanted to be sure that they did not go past breakpoint.
- Adequate storage of treatment chemicals is important for maintaining plant treatment reliability. Based on discussions with chemical suppliers for the three types of ammonia chemicals being considered at NNWW, shipping locations were within a few hours to a one-day trip from the HMWTP.
- Potential system siting options for the ammonia were evaluated:
    - The first option was use of the existing sulfuric acid feed system for ammonia because the acid feed system was rarely used. The advantages were that (1) the containment area, tanks, metering pumps, etc., were already installed; (2) it was the lowest cost; and (3) no new building was required. The disadvantages were that (1) it eliminates pH adjustment capability if needed in the future and (2) the existing acid facilities may not have the proper design criteria for ammonia.
    - A second option involved installation of the ammonia feed system in the existing chemical feed building in the location of the glassy phosphate feed system that was unused. The advantages were that (1) there was easy access for chemical unloading using the existing covered area; (2) there was the possible reuse of the existing metering pumps and solution lines from the glassy phosphate system; and (3) no new building was required. The disadvantage was that it required construction of containment walls in the existing building and slab.
    - The final option was a completely new ammonia storage and feed facility located behind the existing filter building. The advantages were that (1) the system was located closer to the chemical feed points and (2) NNWW could use the most cost-effective chemical without existing facility constraints. The disadvantages

were that (1) this was the most expensive option due to the need for a new building and (2) it would occupy space that may have been reserved for other future processes.

- Option 2 was initially selected for further evaluation. The principal reason for this choice was the ability to locate ammonia storage and feed facilities with all of the other chemicals currently used and the potential need for sulfuric acid in the future to meet the enhanced coagulation requirements of the D/DBP rule.
  - However, in the subsequent cost analysis (see Economic Analysis section below), option 3 was used (1) for the anhydrous ammonia alternative to keep storage and feed systems of this gaseous chemical outside of any existing facilities for safety purposes and (2) for the ammonium sulfate alternative because the large storage tanks required for this system would not fit into the existing glassy phosphate feed area. (Although neither of these options was selected, the decision-making process on how they were evaluated is presented for completeness.)
  - An ammonia storage and feed system using 20 percent aqueous ammonia was selected for the HMWTP. Reasons for selecting aqueous ammonia (in addition to the lowest cost option) included (1) system simplicity and reliability, (2) overall better safety compared with anhydrous ammonia, (3) easy dosing control, (4) short start-up time, and (5) ease of retrofit in the existing facilities.

Distribution-system pH: From a distribution system water quality perspective, the pH needed to be maintained in a range from 7.5 to 8 for optimum monochloramine formation and for optimum performance of the existing corrosion inhibitor.

Future strategies: As discussed above, ozone will be installed to meet future anticipated disinfection and DBP Rule requirements.

Economic Analysis

Ammonia Cost Analysis

The cost of each chemical was obtained from suppliers of the products at the time the conceptual design information was prepared. The costs included shipping charges and were based on the pounds of active ingredient, ammonia-nitrogen: anhydrous ammonia at \$0.144, aqueous ammonia (20 percent solution) at \$0.185, and ammonium sulfate (35 percent solution) at \$1.06. These costs were based on the quotes received at the time of the analysis. These chemical costs were used to determine the best overall feed system on a present worth of costs basis.

System components for all three types of ammonia were evaluated. Common components, as well as unique ones, for each type of ammonia included the following: Metering pump costs were obtained for hydraulic diaphragm pumps. Costs for fiberglass reinforced plastic tanks were also obtained and were approximately \$1.00 per gallon of capacity. Costs for steel pressure tanks for anhydrous ammonia storage were obtained, and the cost for those tanks was approximately \$4.00 per gallon. The price for a complete ammoniator system for anhydrous ammonia (including 100 percent redundancy) with evaporator, ammoniator, controls, and other appurtenances was estimated at \$50,000 for the NNWW's plants. Costs for concrete slabs and walls were assumed to be \$370 and \$630 per cubic yard, respectively. Mechanical equipment installation was based on 45 percent of the equipment cost for the ammoniator system and 22 percent of metering pump and tank costs for the liquid systems. Electrical and instrumentation installation was based on 15 percent of the system cost subtotal. Miscellaneous items such as instrumentation and controls, solution feed lines, diffusers, etc., were not included because they should be comparable between alternatives.

The alternative system components and their costs are summarized in Table 3.1. The annual chemical costs in this table are based on the average plant flow and average dose of ammonia-nitrogen. These annual costs were brought back to a present worth of costs using a six percent interest rate and a 20-year period so that the total present worth of each alternative could be compared. This analysis indicated that the 20 percent aqueous ammonia storage and feed system was the best option at HMWTP.

Capital and Operations Cost Information

The approximate cost of the 20 percent aqueous ammonia feed system was approximately \$1 million for two treatment plants, so approximately \$500,000 per plant. This included two bulk tanks for 30 days of storage, one day tank, three chemical metering pumps, two vertical mixers (for dispersion), and lots of double-walled piping. (This cost is greater than the cost in Table 3.1, because the amounts in the table were relative costs that did not include items common to all alternatives. Miscellaneous items such as instrumentation and controls, solution feed lines, diffusers, etc., were not included in developing the costs in Table 3.1 because they should be comparable between alternatives.) NNWW funds projects using municipal bonds for twenty years at approximately six percent, which works out to an annual payment of approximately \$90,000 for the two plants.

There were increased chemical costs due to ammonia and increased lime usage. Both of these costs were \$40,000 per year for a total of \$80,000 for both plants or approximately \$40,000 per plant per year. However, there was no significant change in the use of chlorine when chloraminating. The increase in electrical costs had been nominal. The amount of labor expended on the ammonia feed system had been small, but much labor had been expended on lime (clogging problems). In fact, a new hydrated lime feed system was being installed at the time of this case study.

Treatment Facilities

Water Supply

The raw water supply for the HMWTP is the Harwood's Mill Reservoir. This reservoir received water from two major sources plus local runoff. The Chickahominy River supplied between 50 and 60 percent (average = 55 percent) of the raw water; the Diascund Reservoir supplied between 10 and 20 percent (average = 15 percent); and local runoff accounted for 20 to 40 percent (average = 30 percent) of the raw water supplied.

Table 3.1

Ammonia system cost comparison for HMWTP*

Item	Cost†
Anhydrous ammonia:	
Pressurized storage tanks (two 7,000 lb)	$14,000
Ammoniator equipment	$50,000
New outdoor tank pads	$10,400
New building for equipment and controls	$30,000
Subtotal	$104,400
Mechanical equipment installation	$22,500
Electrical/instrumentation installation	$14,100
System cost	$141,000
(Annual chemical cost [$/yr])	($7,014)
Present worth of annual chemical cost (6 percent for 20 years)	$80,446
Total present worth of costs‡	$221,446
Ammonium sulfate (35 percent solution):	
Fiberglass reinforced plastic storage tanks (two 10,000 gal)	$20,000
Metering pumps in existing filter building	$10,000
New outdoor tank storage area	$24,000
Subtotal	$54,000
Mechanical equipment installation	$6,600
Electrical/instrumentation installation	$8,100
System cost	$68,700
(Annual chemical cost [$/yr])	($51,628)
Present worth of annual chemical cost (6 percent for 20 years)	$592,168
Total present worth of costs	$660,868

(continued)

Table 3.1 (Continued)

Item	Cost[†]
Aqua ammonia (20 percent solution):	
Fiberglass reinforced plastic storage tanks (two 5,000 gal)	\$10,000
Conversion of existing metering pumps	\$12,000
Interior containment area	\$19,900
Subtotal	\$41,900
Mechanical equipment installation	\$4,840
Electrical/instrumentation installation	\$6,285
System Cost	\$53,025
(Annual chemical cost [\$/yr])	(\$9,011)
Present worth of annual chemical cost (6 percent for 20 years)	\$103,350
Total present worth of costs	\$156,375

[*]The annual chemical costs in this table were based on the average plant flow (i.e., 20 mgd) and average dose of ammonia-nitrogen (i.e., 0.8 mg/L).

[†]These were relative costs that did not include items common to all alternatives. These costs were not actual construction costs.

[‡]The total present worth of costs equaled the sum of the system costs and the present worth of annual chemical costs.

HMWTP: Before Treatment Modifications

The HMWTP had a rated capacity of 31 mgd, with a typical flow of 23 mgd. The plant included preoxidation basins, in-line flash mixers, upflow solids contact flocculator/clarifier (Super-pulsator), granular media filters (sand, anthracite, and garnet), and finished water storage (Figure 3.2). Chemical feed facilities included potassium permanganate for oxidation of iron and manganese, alum for coagulation, polymer for coagulant aid and filter aid, PAC for taste-and-odor control, chlorine for disinfection, lime for pH adjustment, zinc orthophophate for corrosion control, and hydrofluosilisic acid for fluoridation.

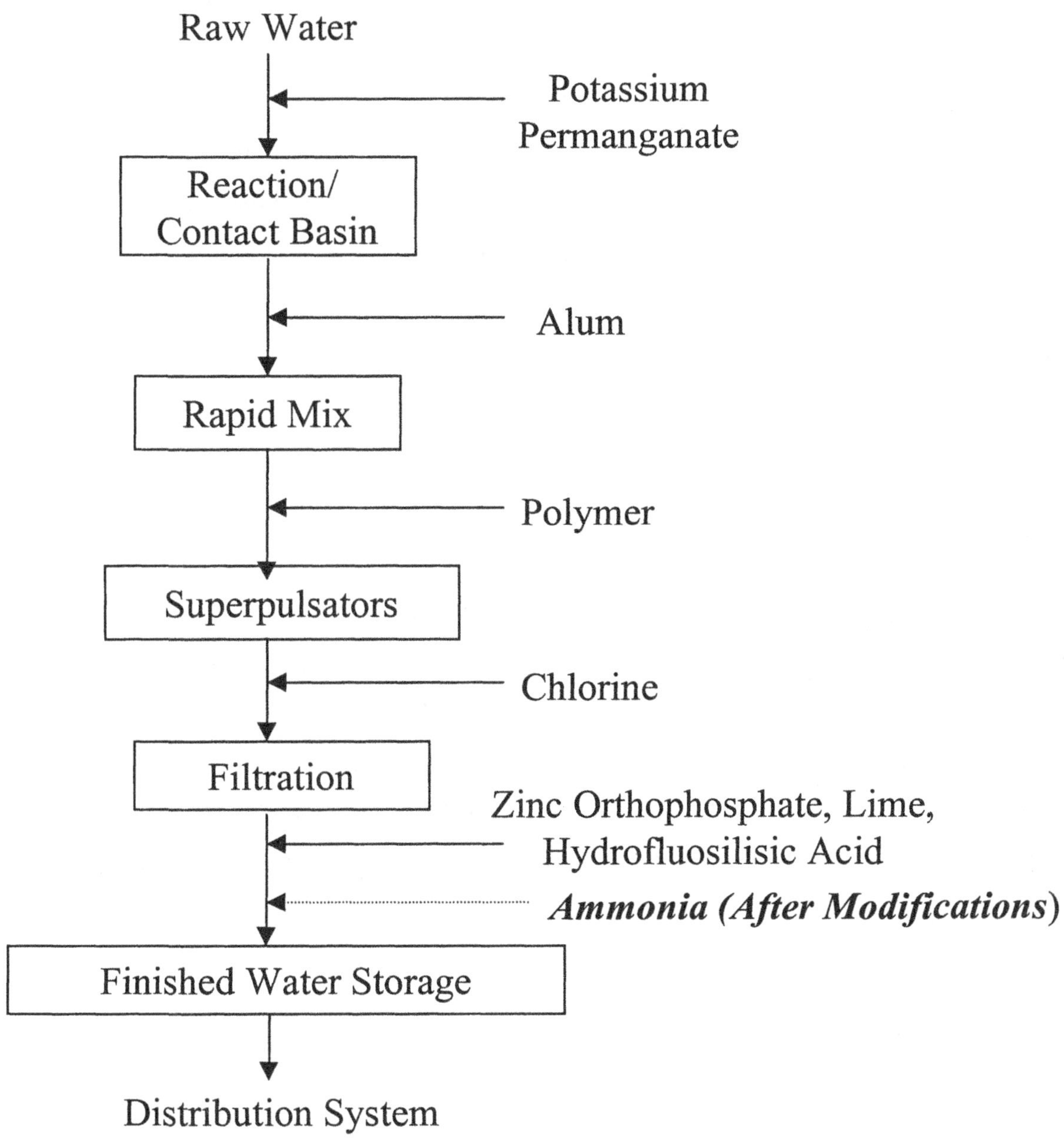

Figure 3.2 Treament process at HMWTP, before and after modifications

Addition of potassium permanganate and PAC typically occurred during the spring and fall seasons when rainfall and subsequent surface runoff to the reservoirs was greater. The water was coagulated with alum (35-60 mg/L, average = 45 mg/L) and a cationic polymer (Allied Colloids Percol LT 22s, 0.08-0.2 mg/L, average = 0.1 mg/L) as a coagulant aid. The pH of coagulation ranged between 5.0 and 7.0 (average = 6.2).

Free chlorine was used for primary and secondary disinfection. The HMWTP averaged 10 logs of *Giardia* inactivation in 1997, with a range of 2.2 to 22 logs (Figure 3.3). The highest levels of inactivation were achieved in the summer months.

Lime was added to the filtered water. The target pH for the finished water was 7.0. Corrosion control included the addition of 2.0 mg/L of zinc orthophosphate, which provided 0.2 mg/L of zinc and 0.6 mg/L of total phosphate. Hydrofluosilisic acid was also added.

HMWTP: After Treatment Modifications

The coagulation practices were unchanged. Ammonia feed facilities were installed for the conversion of secondary disinfection from free chlorine to chloramines (Figure 3.2).

The ammonia was fed immediately after the filters, prior to the clearwell. For a filter flow of 23 mgd and with all eight filters in service, the filter T_{10} was 19 min. This was the amount of contact time with chlorine that was achieved at a typical plant flow. Additional *CT* was achieved in the clearwell with chloramines. HMWTP averaged a total (between the filters and the clearwell) of 3.2 logs of *Giardia* inactivation in the portion of 1998 shown in Figure 3.4, post-chloramine implementation, with a minimum of 2.0 and a maximum of 4.8 logs. The finished water pH was raised from 7.0 (before implementing chloramination) to 7.5 (post implementation). The corrosion inhibitor dose was not changed from 2.0 mg/L as product.

Operations

NNWW is a member of the Partnership for Safe Water. At the HMWTP, the filtered-water turbidity goal is 0.10 NTU.

Implementational and Operational Issues

Public Information Campaign

An important aspect of a conversion to chloramines is informing the public, especially fish owners and kidney dialysis centers. Before information was released to the public, NNWW

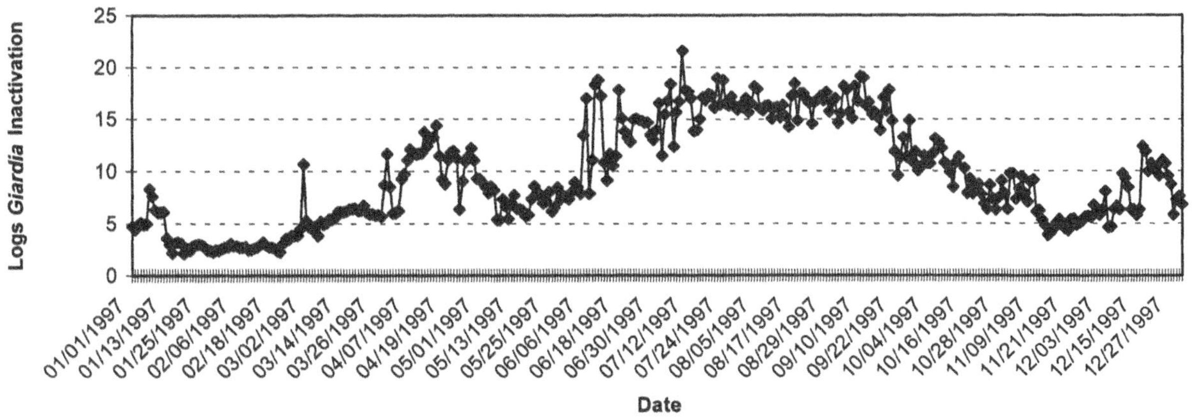

Figure 3.3 *Giardia* inactivation credits at HMWTP in 1997: free chlorine for primary and secondary disinfection

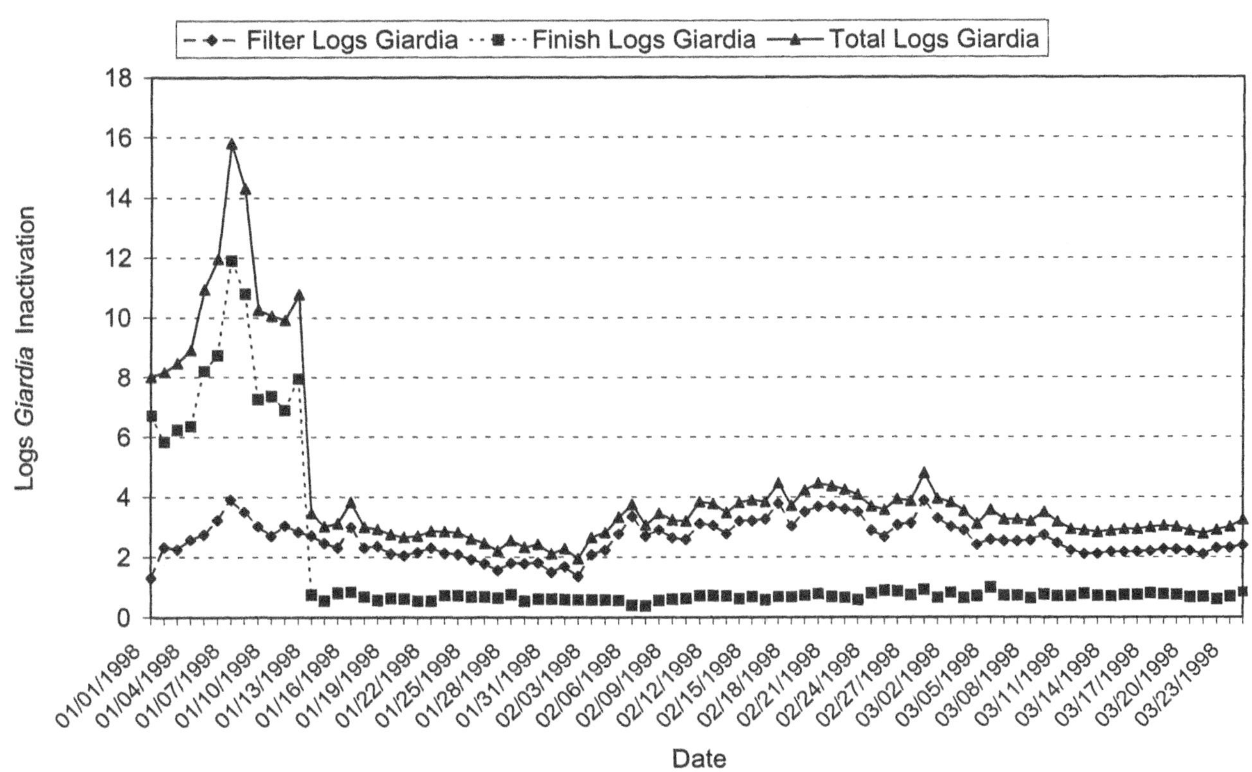

Figure 3.4 *Giardia* inactivation credits at HMWTP in 1998: switch to chloramines for secondary disinfection on January 12th (chlorine was secondary disinfectant in first part of month)

plant staff met with customer service staff, meter readers, and water service representatives to review the new treatment process and intended schedule, answer potential questions that customers may have, and become familiar with the facts.

Four chloramine educational workshops were held in 1997 at the HMWTP for invited targeted groups: local military, local medical, and local government. Each workshop included an overview of the new process and intended schedule, a review of the public information plan, and a question-and-answer session. The program was presented by the Water Quality Laboratory Supervisor, the Project Manager, and a Public Information Specialist.

Medical facilities that were invited to the workshops and those who did not attend were sent letters, which included the following: "Like chlorine, chloramines must be removed from water used in kidney dialysis machines. It is important that dialysis patients and their dialysis centers and operators be aware of the change."

NNWW mailed inserts to all of its billing customers in July and August of 1997. The insert addressed the following questions: What are chloramines?, How will it affect me?, and Why the change? In addition to addressing kidney dialysis patients, the insert noted that "Chlorine, chloramines, and ammonia have to be removed from water before use with tropical fish. Most pet stores sell water conditioners for chloraminated water. "

For those customers who do not receive water bills, NNWW contracted with a subsidiary of the local newspaper to deliver in September 1997 the same billing inserts to the doorsteps of customers in apartment and mobile home parks in the NNWW service area. Where doorstep delivery was not allowed, an adequate amount of inserts were provided to the apartment office.

In addition, NNWW made chloramine conversion information at several community events throughout 1997. Furthermore, NNWW provided and delivered color posters announcing the chloramine conversion to all area pet stores. The poster included information on precautions that tropical fish owners should take.

Moreover, stories on the chloramine conversion appeared in 1997 and in January 1998 in the local newspapers. A news release was faxed to local television and radio stations for coverage prior to the startup. The message included the following: "The most important points are that fish owners need a new 'chloramine remover' which can be purchased in local pet stores, dialysis centers and area hospitals are fully aware of the change, and that most customers WILL NOT notice a difference at all."

During 1997, letters were also sent to targeted audiences informing them of the upcoming conversion. Letters were sent to various businesses, including pet stores, restaurants, grocery stores, schools, seafood processors, and large users. In December 1997 and January 1998, letters were sent to local self-dialysis patients, large users, seafood processors, and pet stores announcing the actual conversion date. In December 1997, workshop attendees (and invited facilities that did not attend) were faxed a letter announcing the actual conversion date.

In December 1997, NNWW employees were briefed on the date of startup and were provided information on what to anticipate during the startup. In the two weeks prior to the conversion, NNWW bought display advertising in five local newspapers (four advertisements were taken in the main newspaper). Information was provided to all area government access channels for their bulletin boards.

Finally, during the first month of the chloramine conversion, NNWW did trend analysis for customer call-in activity. Automation tools were developed and made available for desktop computers. By blending their Customer Work System and Mapping System, NNWW was able to visually see problems occurring throughout the service area, on an hourly basis, and take corrective action. During the week of the actual chloramine conversion, NNWW customer service branch extended its telephone hours to answer any customer questions.

Nitrification Issues

The possibility of nitrification forced NNWW to examine the amount of detention time in the distribution system. One ground storage tank with a very long detention time was removed from service. It now stores water that would be used only for emergencies, such as fire flows, and it employs a re-chlorinating system, which was upgraded at the same time as the ammonia feed project. NNWW worked very closely with its wholesale customers during the conversion. One system, which has its own plant and distribution system, but relies on NNWW as it backup, converted to chloramines at the same time as NNWW. Other wholesale customers performed an extensive flushing procedure at the time of conversion, as did NNWW.

Operator Training and Start-Up

The operators received manufacturer's training near the conclusion of the ammonia feed project. The staff also received training from management on NNWW's water quality goals: i.e., logs of Giardia inactivation, DBP levels, chlorine residual goals in the distribution system, chlorine to ammonia ratio, proper pH, etc. Additional training was not required, as the initial training was adequate.

A period of one month was required for the operational staff to be "comfortable" with the chloramine conversion. A period of three months was required for the treatment plant to operate like an "integral unit," rather than as a treatment plant with a new process. The chloramine conversion did not significantly change operational needs at the plant and no additional staff were added. As a result of the chloramine conversion, more operator laboratory tests were required for free and total chlorine and free and total ammonia.

Operational Issues

The most significant change came in the operation of the plants. Prior to the conversion to chloramination, there was no benefit to reducing the chlorine contact time in the plant, because the DBP reaction continued in the distribution system. Before chloramination, typically all process units were kept in service all the time. After the chloramine conversion, filters were removed from service to reduce contact time with free chlorine to primarily minimize HAA formation. This operation has had an effect on operations and maintenance, as out-of-service units need different levels of care. The increased loading did not cause any other problems.

The higher finished water pH with the pebble lime feed system resulted in an increase in the finished water turbidity, which was considered mainly an aesthetic problem. It has also increased the amount of maintenance on the lime system. Therefore, NNWW was converting from a pebble lime feed system to a hydrated lime feed system, which should significantly reduce their finished water turbidity and reduce the maintenance effort involved with keeping the system functioning.

NNWW's goals are to meet the 80 μg/L TTHM and 60 μg/L HAA5 MCLs with a 20 percent safety factor (i.e., TTHMs $\leq$64 μg/L and HAA5 $\leq$48 μg/L). Changes are made in the treatment process if the 64 μg/L TTHM or 48 μg/L HAA5 goals are not achieved (i.e., reduce the chlorine contact time if feasible, optimize coagulation, etc.).

System Performance

Water Quality

Most of the water-quality and DBP data for the case study were from the ICR (Tables 3.2 and 3.3). Data from July 1997 through January 1998 were used to represent before the chloramines conversion, whereas data from February though August 1998 were used to represent after the chloramines conversion.

TOC: During the time period studied, the influent TOC ranged from minimums of 6-7 mg/L to maximums of 9 mg/L (the averages were 7-8 mg/L). Historically, the TOC typically ranged from 4 to 8 mg/L (average = 6 mg/L) and has gotten as high as 12 mg/L. Generally, high TOC concentrations coincided with heavy rainfall, which typically occurred in the winter and spring. During the time period studied, the filter-effluent TOC ranged from minimums of 2-3 mg/L to maximums of 4-5 mg/L (the averages were 3-4 mg/L).

During the time period studied, the TOC removal ranged from minimums of 43-48 percent to maximums of 60-68 percent (the averages were 52-58 percent). NNWW treats a low-alkalinity water (<60 mg/L), so TOC removals of 45 percent would be typically required (USEPA 1998).

UV: During the time period studied, the influent UV ranged from minimums of 0.12-0.18 cm^{-1} to maximums of 0.22-0.30 cm^{-1} (the averages were 0.16-0.24 cm^{-1}).

SUVA was calculated using the formula UV x 100/DOC; it was assumed that the influent TOC was equivalent to the DOC. (Historical DOC data ranged from 4 to 7 mg/L, average = 6 mg/L, which were close in value to [and somewhat lower than] the historical TOC data.) The SUVA values in the "before" data set (i.e., 1.5 to 2.8 L/mg-m, average = 2.2 L/mg-m) were lower than the values in the "after" data set (i.e., 3.2 to 3.9 L/mg-m, average = 3.6 L/mg-m). Historical SUVA values have ranged from 1.7 to 4.4 L/mg-m (average = 3.1 L/mg-m).

Table 3.2

HMWTP water quality: comparison of before and after chloramine conversion

	Influent		Finished Water	
Water quality	Before*	After†	Before*	After†
TOC, mg/L				
Minimum	6.7	5.6	3.2	2.3
Average	7.7	6.7	3.6	2.7
Maximum	8.8	9.4	4.6	3.6
TOC removal, %				
Minimum			43	48
Average			52	58
Maximum			60	68
UV, cm^{-1}				
Minimum	0.122	0.178		
Average	0.165	0.236		
Maximum	0.218	0.301		
SUVA, L/mg-m				
Minimum	1.5	3.2		
Average	2.2	3.6		
Maximum	2.8	3.9		
Bromide, mg/L				
Minimum	0.04	<0.02		
Average	0.06	0.03		
Maximum	0.13	0.04		
THMFP,‡ μg/L				
Minimum	301		NS§	NS
Average	361		NS	NS
Maximum	449		NS	NS

(continued)

Table 3.2 (Continued)

	Influent		Finished Water	
Water quality	Before*	After†	Before*	After†
Temperature, °C				
Minimum	8.0	8.0		
Average	19	20		
Maximum	29	29		
pH				
Minimum	7.0	6.8	6.8	7.3
Average	7.3	7.2	7.0	7.8
Maximum	7.6	7.4	7.4	8.3
Alkalinity, mg/L as $CaCO_3$				
Minimum	26	24	18	28
Average	30	40	25	38
Maximum	38	51	28	46
Ammonia, mg/L as NH_3-N				
Minimum	<0.02			
Average	0.03			
Maximum	0.05			
Turbidity, NTU				
Minimum	2.5	1.9	0.22	0.35
Average	3.7	5.3	0.28	0.48
Maximum	5.4	17	0.33	0.69
Total Coliforms, # /100 mL				
Minimum	<1			
Average	17			
Maximum	50			

(continued)

Table 3.2 (Continued)

Water quality	Influent Before*/After†	Finished Water Before*	Finished Water After†
Giardia, # cysts/100 L			
Minimum	<30	NS	NS
Maximum	102	NS	NS
Cryptosporidium, # oocysts/100 L			
Minimum	<5	NS	NS
Maximum	32	NS	NS

*Data for TOC, UV, SUVA, bromide, ammonia, temperature, pH, alkalinity, turbidity, and coliforms from ICR (July 1997-January 1998). Data for THMFP from 1997. Data for *Giardia* and *Cryptosporidium* from ICR (1998).

†Data for TOC, UV, SUVA, bromide, ammonia, temperature, pH, alkalinity, and turbidity from ICR (February 1998-August 1998). Data for *Giardia* and *Cryptosporidium* from ICR (1998).

‡THMFP conditions: chlorine dose (mg/L) = 3*TOC (mg/L), pH = 7.0, temperature = 22°C, incubation time = 7 days. See Table 3.3 for instantaneous THM data.

§NS = Not sampled.

Most of these SUVA values are indicative of moderate levels of humic substances in the raw water (Edzwald and Van Benschoten 1990). Because humic substances are well removed during the coagulation process (Owen et al. 1993b), the settled waters—which were the first point of chlorination at HMWTP—should have been relatively low in humic substances.

Bromide: The bromide concentrations in the before data set (i.e., 0.04 to 0.13 mg/L, average = 0.06 mg/L) were higher than the values in the after data set (i.e., <0.02 to 0.04 mg/L, average = 0.03 mg/L). Nonetheless, HMWTP treats a water that is typically low in bromide.

Temperature: During the time period studied, the water temperature ranged from a minimum of 8.0 to a maximum of 29°C (the averages were 19-20°C). The temperatures were comparable in both study periods, as each included warm- and cold-temperature seasons.

Table 3.3

HMWTP DBP data: comparison of before and after chloramine conversion

DBPs	Finished water[*] Before[†]	After[‡]
TTHMs, μg/L		
Minimum[§]	73	45
Average[§]	90	62
Maximum[§]	145	108
HAA5, μg/L		
Minimum[§]	28	33
Average[§]	61	51
Maximum[§]	83	59
HAA6, μg/L		
Minimum[§]	29	36
Average[§]	67	54
Maximum[§]	91	63
TOX, μg Cl^-/L		
Minimum[§]	263	185
Average[§]	340	213
Maximum[§]	447	260

[*]DBP data from four locations in the distribution system, one of which was at maximum detention time.

[†]Data from ICR (August 1997-October 1997).

[‡]Data from ICR (Febuary 1998-August 1998 for THMs and TOX, February 1998-May 1998 for HAAs).

[§]Minimum and maximum values in the distribution system during the period covered. Average values are not running annual average values; rather they are average values for all points collected during the period covered.

pH: The influent pH ranged from minimums of 6.8-7.0 to maximums of 7.4-7.6 (the averages were 7.2-7.3). Prior to the chloramine conversion, the finished-water pH goal was 7.0 (the range achieved in this study period was 6.8-7.4, average = 7.0). After the chloramine conversion, the finished-water pH goal was 7.5 (the range achieved in this study period was 7.3-8.3, average = 7.8).

Turbidity: During the time period studied, the influent turbidity ranged from a minimum of 2 to maximums of 5-17 NTU (the averages were 4-5 NTU). The finished-water turbidity ranged from minimums of 0.2-0.4 to maximums of 0.3-0.7 NTU (the averages were 0.3-0.5 NTU). Subsequent to the ICR (in 1999), finished-water turbidity ranged from 0.02 to 0.2 NTU.

Microorganisms and disinfection: During the ICR, *Cryptosporidium* was detected at very low levels in the source water. Coliform levels peaked during a heavy runoff event, whereas there was no apparent pattern in the *Giardia* and *Cryptosporidium* data.

When the HMWTP used free chlorine for primary and secondary disinfection, a very high level of *Giardia* inactivation (up to 22 logs) was achieved at certain times of the year (e.g., in the summer months) (Figure 3.3). This level of inactivation greatly exceeded what was required in the SWTR. By switching to post-chloramination, the level of inactivation was unchanged during the primary disinfection step (through the filtration process), and the total level of inactivation at the plant was more consistent (and reasonable) through the year (Figure 3.4). The *Giardia* inactivation achieved was reasonable for a free chlorine system, not to mention that more kill is better, other things equal.

THMs: During the time period studied, THMs in the chlorinated distribution system (including locations at maximum detention times) ranged from 73 to 145 μg/L (average = 90 μg/L). During the time period studied, THMs in the chloraminated distribution system ranged from 45 to 108 μg/L (average = 62 μg/L). Subsequent to the ICR (in 1999), the range for the NNWW distribution system (for both treatment plants) was 28 to 89 μg/L. The use of chloramines as the secondary disinfectant has enabled NNWW to meet the 80 μg/L MCL on a running annual average basis.

HAAs: During the time period studied, HAA5 (and HAA6) in the chlorinated distribution system ranged from 28 (29) to 83 (91) μg/L (average = 61 [67] μg/L). During the time period studied, HAA5 (and HAA6) in the chloraminated distribution system ranged from 33 (36) to

59 (63) μg/L (average = 51 [54] μg/L). The HAA levels measured initially after the chloramine conversion were relatively high. Subsequently, the running annual average was lowered to 31 μg/L. The use of chloramines in the distribution system and the control of the free chlorine contact time in the plant were required to enable NNWW to meet the 60 μg/L HAA5 MCL on a running annual average basis.

TOX: During the time period studied, TOX in the chlorinated distribution system ranged from 263 to 447 μg/L (average = 340 μg/L). During the time period studied, TOX in the chloraminated distribution system ranged from 185 to 260 μg/L (average = 213 μg/L). The use of chloramines reduced the formation of both known DBPs (e.g., the THMs and HAAs) as well as unknown DBPs (as measured by TOX).

Summary of System Performance

At the HMWTP, the use of enhanced coagulation, intermediate chlorination, and post-chloramination are allowing NNWW a cost-effective means of meeting the Stage 1 MCLs for THMs and HAAs. Ozonation was being installed to meet new and/or future anticipated disinfection and DBP regulations.

CHLORAMINES: CITY OF POWAY

Summary

The City of Poway, located in San Diego County (Calif.), treats and distributes potable water to a retail customer base of approximately 50,000. The water treatment plant (WTP) receives water from two sources: (1) Lake Poway, which collects local runoff, and (2) untreated imported water from the San Diego County Water Authority (CWA). The latter water is a blend of Colorado River Water (CRW) imported from the California/Arizona border and State Project Water (SPW) imported from Northern California. Imported water can also be stored in Lake Poway. At any given time, the WTP may treat 100 percent CWA water or 100 percent Lake Poway water (a blend of local and imported waters).

The regulatory driver in the decision to convert to chloramines as a secondary disinfectant was to reduce the formation of chlorinated DBPs. The City was already in a position in which they were required by the California Department of Health Services (DHS) to upgrade a portion of their filtration facilities. Therefore, this capital improvement could be included into a design and construction project that was currently underway.

On the surface, this case study appears to be as straightforward an example of a conversion to chloramines as might exist anywhere, but a few features distinguish it. The Operations staff was intent on studying the implementation issues thoroughly and that effort paid off in trouble-free implementation. But the by-product control strategy had to be revisited as a result of the manner in which the finished-water clearwell and proposed regulatory changes factored into by-product formation.

Background

Case History

The City of Poway, located in San Diego County, treats and distributes potable water to a retail customer base of approximately 50,000. The Lester J. Berglund WTP is the City's one treatment facility, with a rated capacity of 24 mgd and an average daily production of approximately 12 mgd. Figure 3.5 illustrates a process flow schematic for the plant.

The WTP receives water from two sources: (1) Lake Poway, which collects local runoff, and (2) imported untreated water from the San Diego CWA. This water is a blend of CRW imported from the California/Arizona border and SPW imported from Northern California. Imported water also can be stored in Lake Poway. At any given time, the WTP may treat 100 percent CWA water or 100 percent Lake Poway water (a blend of local and imported waters).

Since the plant was initially constructed in 1972, expanded in the late 1970s, and upgraded in 1996, it has provided excellent particle removal and has produced water that met all drinking water requirements enforced by the California DHS. The City was able to meet the existing 100 μg/L requirement for TTHMs (until the Stage 1 D/DBP Rule is effective in 2001) while simultaneously meeting the disinfection requirements of the SWTR using chlorine for both

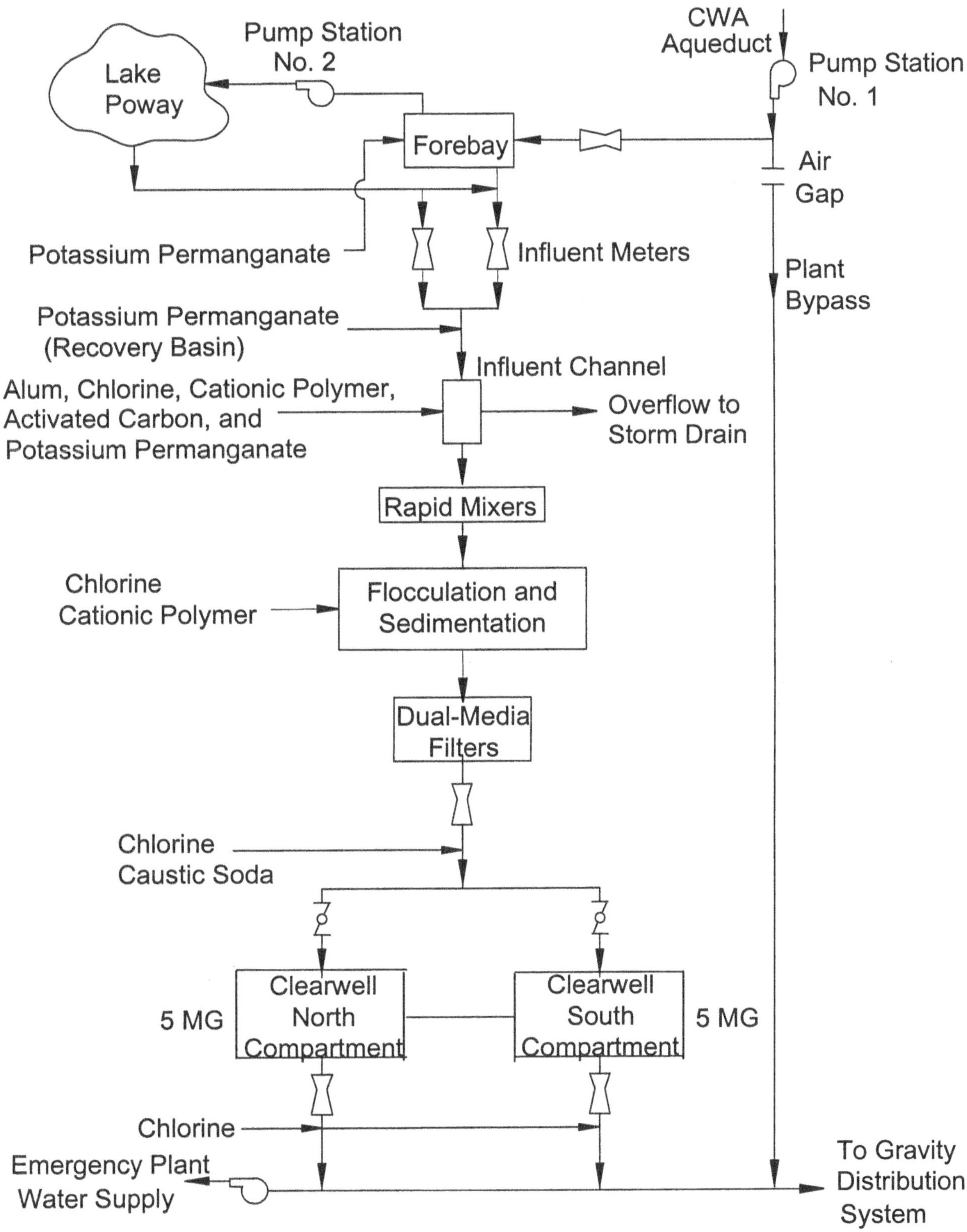

Figure 3.5 Lester J. Berglund WTP, City of Poway, flow schematic

primary and secondary disinfection. With the advent of Stage 1 of the recently finalized D/DBP Rule, and the IESWTR, the City had to reconsider their disinfection options. In addition, the WTP's primary imported raw water source was changing. Historically, the WTP had received an imported blend of approximately 90 percent CRW and 10 percent SPW, as delivered by CWA. This imported blend had recently exceeded 30 percent SPW and was expected to ultimately change to a blend containing at least 50 percent SPW (and possibly up to 60 percent SPW). The SPW source is higher in DBP precursor material (both natural organic matter [NOM] and bromide), which translates into higher DBP production (Krasner et al. 1994).

Based upon treatability evaluations, Poway decided to proceed with implementing chloramines as a secondary disinfectant to reduce chlorinated DBP formation. It was decided that free chlorine would continue to be used as the primary disinfectant to inactivate pathogens, and that the free chlorine residual would be quenched with ammonia at either (1) the outlet from the on-site clearwell storage or (2) the combined filter effluent. Currently, ammonia is applied at the outlet from the on-site clearwell storage. After additional DBP formation data have been collected, Poway will elect whether to move the point of ammonia application to the combined filter effluent.

Process Selection

Between the summer of 1994 and the spring of 1995, the City of Poway sponsored bench-scale treatability studies to evaluate the effect of various chlorine disinfection practices on DBP formation. The treatability studies focused upon:

- Water quality requirements and goals. System-wide water quality requirements were identified based upon existing and anticipated regulations.
- Existing plant performance. Available water quality and operations data were evaluated to identify the present levels of DBPs and variations in raw and treated water quality.
- Alternative treatment strategies for reducing the concentration of DBPs while still achieving adequate disinfection. Enhanced coagulation was also evaluated.

Figure 3.6 illustrates the running annual average of quarterly samples collected for TTHM compliance from 1989 to the time at which Poway switched to chloramines for secondary disinfection. The quarterly average during this period was below the current TTHM MCL of 100 μg/L for all but four individual quarters. During those four quarters, the source water for the treatment plant contained a higher percentage of Lake Poway Water than was typical in the source water blend. This period corresponded to heavy rainfall in the San Diego area, which increased the presence of DBP precursor material resulting from runoff.

The running annual average of the quarterly TTHM data, however, was increasing over time. In general, the running annual average increased from approximately 60 μg/L in the early 1990s to approximately 90 μg/L in 1997 and 1998. This was due to the gradually increasing proportion of SPW in the CWA aqueducts (in 1989 through 1992, there was a drought in San Diego and virtually all water was imported; the City of Poway received CRW almost exclusively at that time). SPW contains more DBP precursor material (i.e., NOM and bromide) than does CRW, so it is expected that increases in the fraction of SPW in the CWA system could cause increased formation of DBPs following chlorination.

Many large systems typically recommend a 20 percent margin of safety in meeting the MCL to allow for operational variations. Thus, Poway's TTHM treatment goal was 64 μg/L (80 percent of the Stage 1 MCL of 80 μg/L for TTHMs). Because of experience in Southern California with almost identical source waters indicating a TTHM-to-HAA5 formation ratio of at least 2:1 (Cheng et al. 1995), Poway believed that TTHMs would control their DBP compliance. (That is, the bromide in the source water favors the formation of other HAAs that are not included in HAA5 [Cowman and Singer 1996].) Although this is appropriate under Stage 1 of the D/DBP Rule, Poway may be required to re-evaluate if additional brominated HAAs are regulated in the future.

When Poway evaluated the anticipated TTHM MCL of 80 μg/L and the treatment goal of 64 μg/L in late 1993, the samples collected had running annual averages exceeding both the anticipated MCL and the treatment goal for all four quarters in 1993. In addition, the treatment goal had been exceeded for the last quarter in 1992. Clearly, alternative treatment strategies needed to be considered to meet the anticipated Stage 1 MCL for TTHMs. In addition, the enhanced coagulation requirements in the anticipated Stage 1 D/DBP Rule had to be addressed. Consequently, Poway commissioned a treatment study.

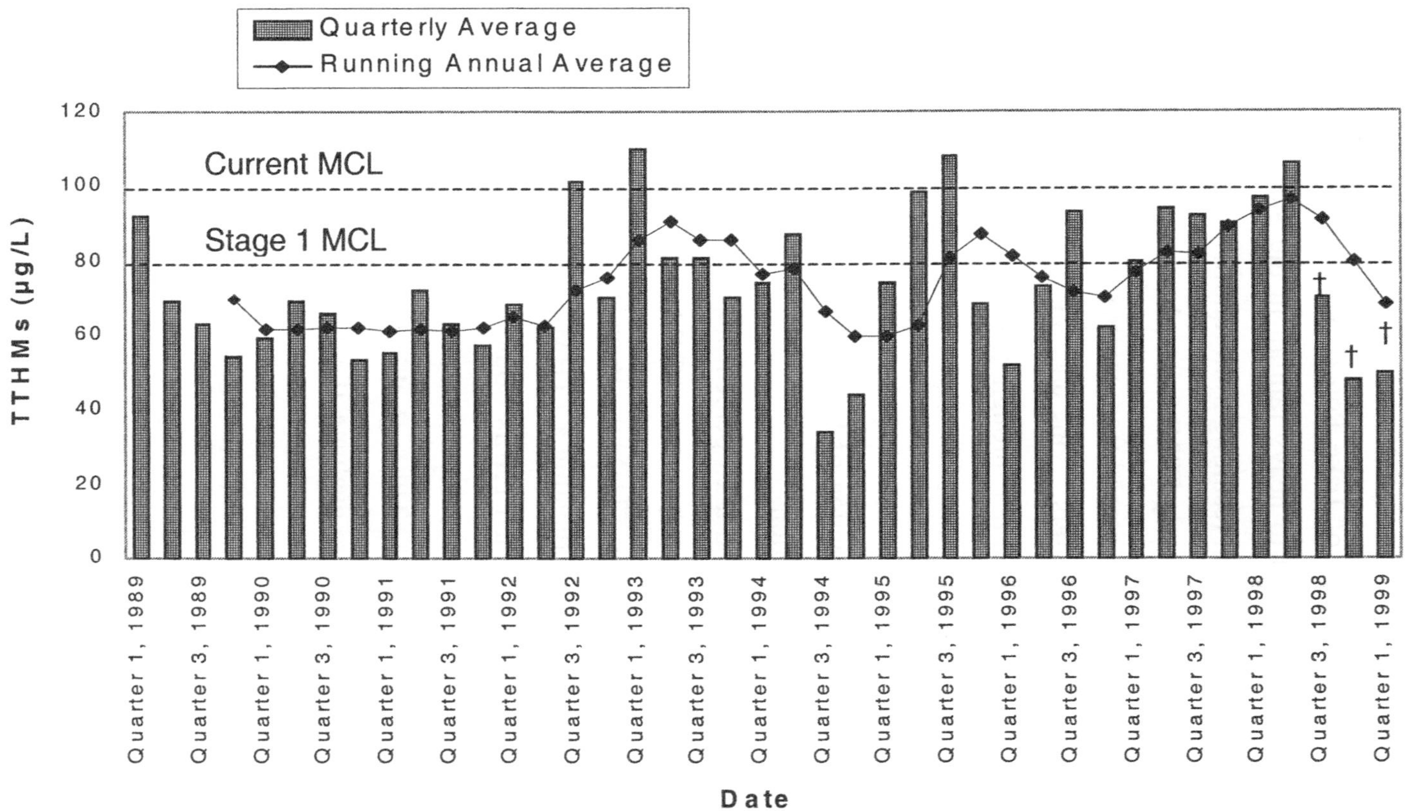

*Running annual averages based upon quarterly data.
†Chloramine conversion occurred between the third and fourth monitoring quarter in 1998.

Figure 3.6 Historical TTHM compliance data at the City of Poway*

Enhanced Coagulation

TOC removal requirements for step 1 of the enhanced coagulation treatment technique increase with increasing source-water TOC concentration and decrease with increasing source-water alkalinity (Krasner and Amy 1995). These percent removal requirements recognized the trend that TOC removal is more difficult as alkalinity increases and/or TOC decreases. The relative coagulability (by comparing the shape of the coagulation curve and the extent of TOC removal) for the source waters treated at the Berglund WTP is illustrated in Figure 3.7. As the alkalinity of the source waters decreases, TOC removal by coagulation improves. Also, as the TOC concentration of the source water increases, TOC removal by coagulation improves.

The source water for the Berglund WTP that will require the highest TOC removal percentage under step 1 is the blend representative of future aqueduct water when CWA operates with a 60 percent SPW/40 percent CRW blend. This blend has the highest TOC and lowest alkalinity, resulting in the potential to remove the highest percentage of TOC. Further, the higher TOC and bromide concentrations imply the greatest potential for forming DBPs; thus the 60 percent SPW/40 percent CRW blend requires the highest coagulant dose to meet enhanced coagulation requirements and the DBP MCLs. For these reasons, Poway based their performance and design criteria on the 60 percent SPW/40 percent CRW blend.

Residual TOC versus coagulant dose is illustrated in Figure 3.8 for the 60 percent SPW/40 percent CRW blend during a one-day evaluation in August 1994 focusing on the treatability of this source water. For this source-water blend, the alkalinity was 108 mg/L (as $CaCO_3$) and the TOC was 3.4 mg/L. The TOC of the blend on this day was lower than measured for this blend when compared to other sources during previous evaluations (the TOC for SPW varies based upon rainfall and transport characteristics, including storage). With these water qualities, step 1 of the 1994 proposed enhanced coagulation treatment technique (being used as guidance at the time of the evaluation) required a 30-percent reduction in TOC during coagulation and sedimentation, prior to achieving *CT* credit for disinfection (USEPA 1994a). (In the 1998 promulgated rule, the step 1 TOC removal requirement for a water with the latter TOC and alkalinity was changed to 25 percent [USEPA 1998a]. In addition, *CT* credit for prechlorination was now being allowed [USEPA 1998a].) Thus, a settled-water TOC of

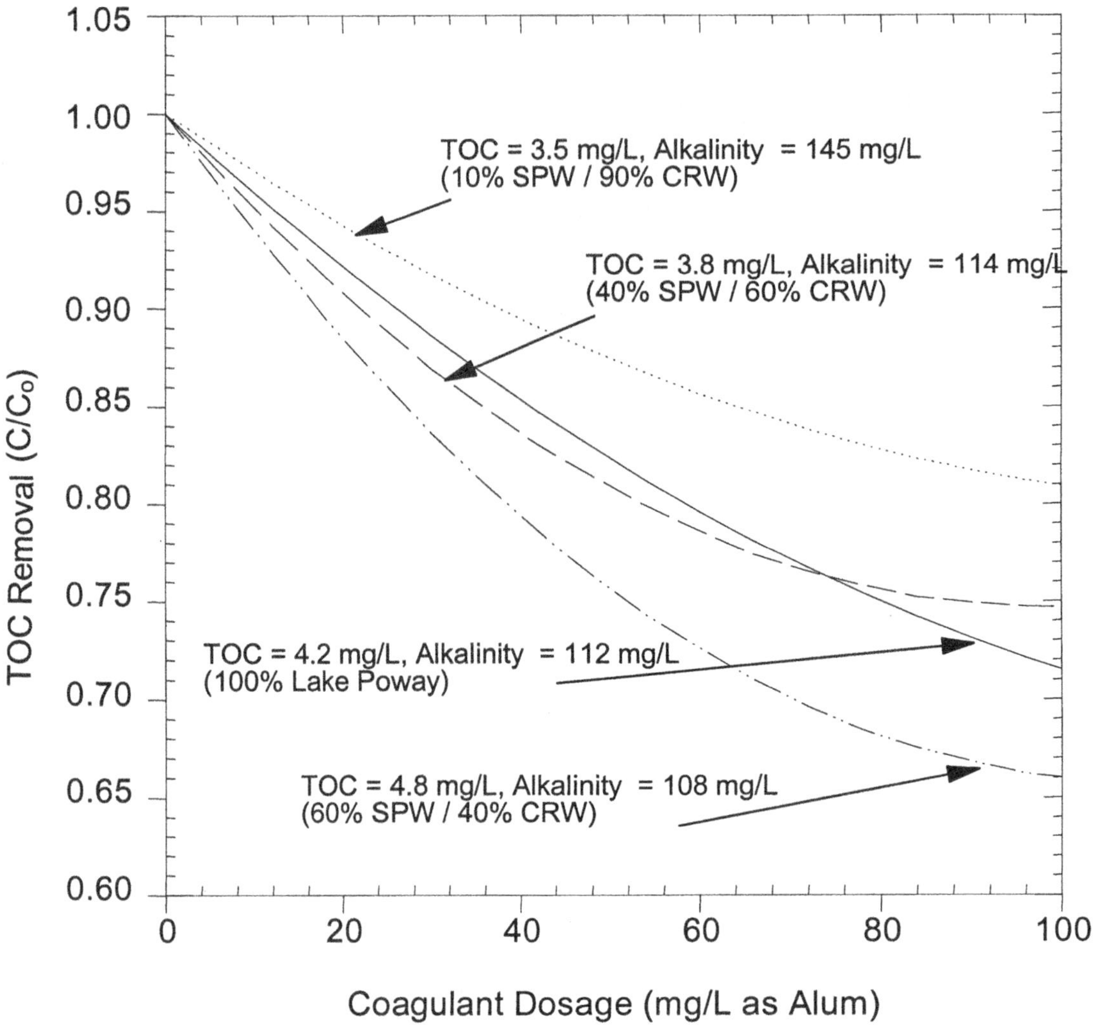

Figure 3.7 Effect of source-water alkalinity and TOC on TOC removal

2.4 mg/L would have been required for this water. Figure 3.8 illustrates that a treated-water TOC of 2.4 mg/L could not be achieved with an alum dose less than or equal to 60 mg/L.

The enhanced coagulation requirements were not intended to require high coagulant dosages with marginal TOC removal benefit. Step 2 of enhanced coagulation defined a point of diminishing returns for coagulant addition for TOC removal (Krasner and Amy 1995). The point of diminishing returns for alum coagulation was defined as the dosage at which 0.3 mg/L of TOC is removed per 10 mg/L of alum added.

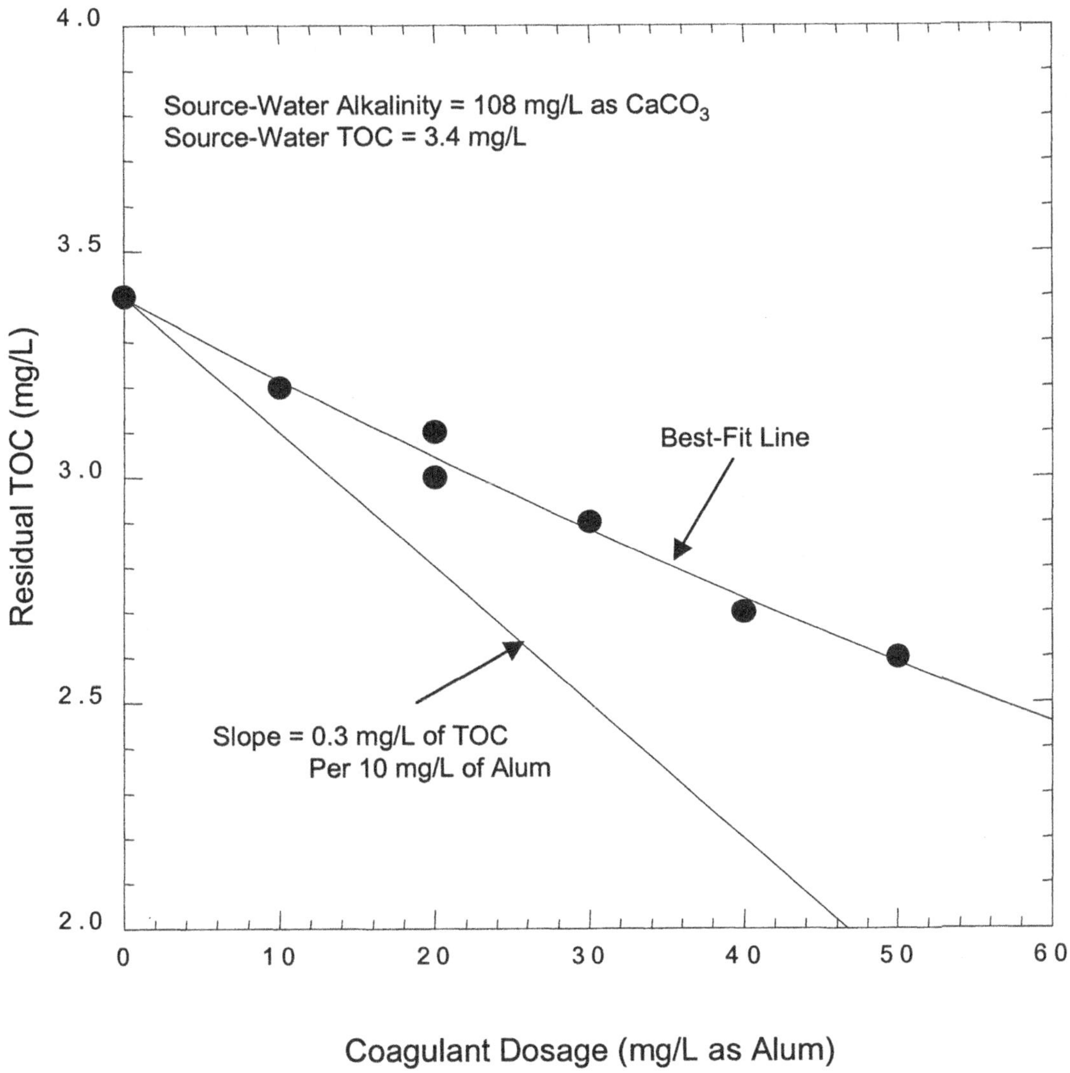

Figure 3.8 Evaluation of TOC removal by enhanced coagulation in a 60 percent SPW/ 40 percent CRW source-water blend (the line with a slope of 0.3 mg/L of TOC per 10 mg/L of alum is the point of diminishing returns used in the Step 2 evaluation)

The point of diminishing returns is also illustrated in Figure 3.8. The percent of TOC removal achieved at this point is defined, per approval by the State, as the alternative percent TOC removal that achieves compliance. The slope for the point of diminishing returns slope in Figure 3.8 is never tangent to the TOC/coagulant dosage curve; therefore, this water is not amenable to TOC removal by enhanced coagulation. Consequently, the worst-case water is not subject to enhanced coagulation. A review of Figure 3.7 indicates that none of the source-water blends would be subject to the treatment technique, as the "steepest" slope was associated with the 60/40 blend requirements and that curve was never tangent to the point of diminishing returns slope criterion. (This was confirmed in subsequent jar tests performed over the following year on various source-water blends throughout San Diego County.)

Subsequent to this evaluation, the rule was modified to include a criterion for specific ultraviolet absorbance (SUVA) (USEPA 1998a). SUVA represents the measured UV value at 254 nanometers divided by the dissolved organic carbon (DOC) concentration (SUVA has units of L/mg-m). If this value is less than 2.0 in the raw water prior to treatment, then the water is not amenable to enhanced coagulation (Krasner et al. 1997). Further, if at any time the treated water (without oxidant addition) falls below a value of 2.0, then the operating conditions (e.g., coagulant dosage) used to reach that point may be used as the alternative compliance point. As shown in Table 3.5 later in this case study, raw waters exhibited a SUVA (based on TOC) less than 2.0. Because the ratio of DOC/TOC is relatively high in these waters (>90 percent) (Owen et al. 1993b), the data indicate that this water is not amenable to the enhanced coagulation requirements.

Coagulation Requirements to Meet DBP Goals

As discussed previously, the running annual average for TTHMs in the Berglund WTP distribution system, prior to switching to chloramines in the last quarter of 1998, was greater than the Stage 1 MCL of 80 μg/L (which will be effective in 2001). DBPs were studied as a part of the treatability testing so that alternative treatment techniques could be evaluated for their ability to reduce TTHMs and meet the Stage 1 D/DBP MCLs, as well as meet the requirements of enhanced coagulation. The TTHM and the HAA5 measurements collected during the treatability testing indicated that, for the source waters treated at the Berglund WTP, TTHMs were the limiting factor for determining treatment strategies, i.e., TTHMs were closer to surpassing the treatment goals than was HAA5. Therefore, the discussion in the remainder of this section focuses on TTHM reduction.

The TTHM concentrations expected for various treated-water qualities were based upon a previous water quality modeling study (Malcolm Pirnie, Inc. 1993; Owen et al. 1993a). In that study, equations were developed for predicting the formation of individual and total THMs. These equations were developed based on over 900 records of eight California source waters ranging from 100 percent SPW to 100 percent CRW. These equations were based on TOC, UV, bromide, pH, and temperature of the water being chlorinated. The chlorine dose, contact time, and ammonia concentration were also input to the equations.

The settled water and chlorination conditions from the bench-scale enhanced coagulation testing were used in these equations to estimate TTHM production for three disinfection scenarios:

- Free chlorine (24-hour simulated distribution system [SDS] chlorine contact time).
- Chlorine/chloramine disinfection, 12-mgd average plant flow (T_{10} of 8 h in the clearwell).
- Chlorine/chloramine disinfection, 24-mgd maximum plant flow (T_{10} of 4 h in the clearwell).

The models were also used to illustrate the effects of temperature on TTHM formation. TTHM formation for three of the four source waters tested is summarized in Table 3.4 for temperatures of 15°C (winter), 25°C (average), and 30°C (summer).

The following observations can be made based on the model predictions summarized in Table 3.4.

- TTHM formation increased as the chlorine contact time was increased.
- TTHM formation decreased as the coagulant dosage was increased and, consequently, the pH was lowered. Data were not collected to evaluate the impact of lowering pH on HAA formation.
- TTHM formation increased as the percentage of SPW in the source-water blend was increased.
- TTHM reduction by coagulant addition improved as the amount of SPW in the source water increased, i.e., the percent reduction of DBP precursor material by coagulation improved with increasing proportions of SPW in the blend. This is because CRW has a higher alkalinity, which acts as a buffer that limits pH depression by coagulant addition (or acid

Table 3.4

TTHM formation for source waters tested: effect of coagulant dose on TTHM reduction

			Temperature = 15°C (59°F) TTHMs (µg/L)			Temperature = 25°C (77°F) TTHMs (µg/L)			Temperature = 35°C (95°F) TTHMs (µg/L)		
Water	Alum dose (mg/L)	Coagulation pH	4 h	8 h	24 h	4 h	8 h	24 h	4 h	8 h	24 h
10% SPW/90% CRW	25	7.6	34	40	53	43	52	68	51	61	80
	35	7.5	33	39	52	42	50	66	49	59	78
	50	7.3	32	38	50	40	48	64	48	57	75
40% SPW/60% CRW	20	7.6	37	44	58	48	57	76	55	66	88
	30	7.4	35	41	55	45	54	71	52	62	82
	40	7.2	33	39	52	43	51	68	50	59	78
60% SPW/40% CRW	20	7.7	46	55	72	58	70	92	69	82	109
	30	7.5	42	50	66	53	64	84	63	75	99
	40	7.3	39	47	62	50	60	79	59	70	93

addition). In addition, the NOM present in CRW is different and more difficult to remove than is the NOM in SPW (Krasner and Amy 1995).

- Consequently, it was more difficult to removal DBP precursor material from sources with higher percentages of CRW.
- TTHM formation increased as temperature increased.

For an 8-h chlorine contact time (representative of ammonia addition after the clearwell at a 12-mgd plant flow), the predicted TTHM levels remain at or below the 64 μg/L treatment goal for all source-water blends, coagulation conditions, and temperatures tested, with the following exceptions:

- The 60 percent SPW/40 percent CRW blend coagulated with 20 mg/L alum at a pH of 7.7 and a temperature of 25°C.
- The 40 percent SPW/60 percent CRW blend coagulated with 20 mg/L alum at a pH of 7.6 and a temperature of 35°C.
- The 60 percent SPW/40 percent CRW blend coagulated with 20, 30, and 40 mg/L alum with pH levels of 7.7, 7.5, and 7.3, respectively, and a temperature of 35°C.

The 60 percent SPW/40 percent CRW coagulated with 20 mg/L alum at a pH of 7.7 and a temperature of 35°C was predicted to exceed the Stage 1 MCL of 80 μg/L TTHMs. So, the limiting-case source water for coagulation was predicted to be 60 percent SPW/40 percent CRW during the summer time (35°C).

Enhanced Coagulation Requirements Versus MCL Compliance

Water treatment has traditionally focused upon particulate removal. Although the organic concentration of the source water controls the coagulant dosage to a certain extent, lower coagulant dosages can be used for lower turbidity waters as compared to dosages that may be required based upon improving NOM (DBP precursor) removal. All sources and potential sources used by the Berglund WTP are relatively low turbidity waters.

In addition to the enhanced coagulation TOC-removal requirements, the MCLs for TTHMs and HAA5 must be met. An algorithm, which can be used to determine the recommended coagulant dosage, is illustrated in Figure 3.9. The analysis begins by determining the coagulant dosages to meet (1) the step 1 percent TOC removal requirements for enhanced coagulation and (2) the step 2 point of diminishing returns. The smaller of the step 1 and step 2 dosages can then be used to move to the next decision, i.e., whether the step 1 or step 2 dosage will satisfy the Stage 1 MCLs for DBPs. This algorithm is used to compare the dosages for TOC removal and DBP compliance.

The treatability testing indicated that the limiting-case source water (60 percent SPW/ 40 percent CRW blend) did not have to meet the requirements of enhanced coagulation because it was not amenable to TOC removal by coagulation. Therefore, the coagulant dosage selected to meet the requirements of the D/DBP Rule was based on meeting the MCLs for DBPs.

For the limiting-case of the 60 percent SPW/40 percent CRW blend and a 12-mgd plant flow during the summer time, a 20 mg/L alum dose was recommended to meet the MCL for TTHMs. This 20 mg/L alum dose must be in combination with a reduced free chlorine contact time to meet the MCL for TTHMs; that is, the disinfection strategy must be changed from a chlorine/chlorine (primary/secondary) disinfection strategy to a chlorine/chloramine disinfection strategy. With the switch to secondary disinfection with chloramines, the 20 mg/L alum dose is a maximum dose that should only be required with the limiting-case blend during summer time temperatures. For blends with 40 percent SPW or less, an average alum dose of 10 mg/L (in combination with chlorine/chloramine disinfection) was predicted to be adequate to meet the MCLs for the DBPs.

Parameters that Influenced Process Design

The most significant parameters that influenced the process design were the concentration and variation of precursor compounds (e.g., TOC, UV, and bromide) in the source waters. As stated previously, there are various sources available to the City that reflect different concentrations of DBP precursors. The CRW source has the lowest concentration of DBP precursors for the sources available to the City of Poway, and for that reason a free chlorine disinfection strategy was feasible. As plans for imported water blends indicated a substantial

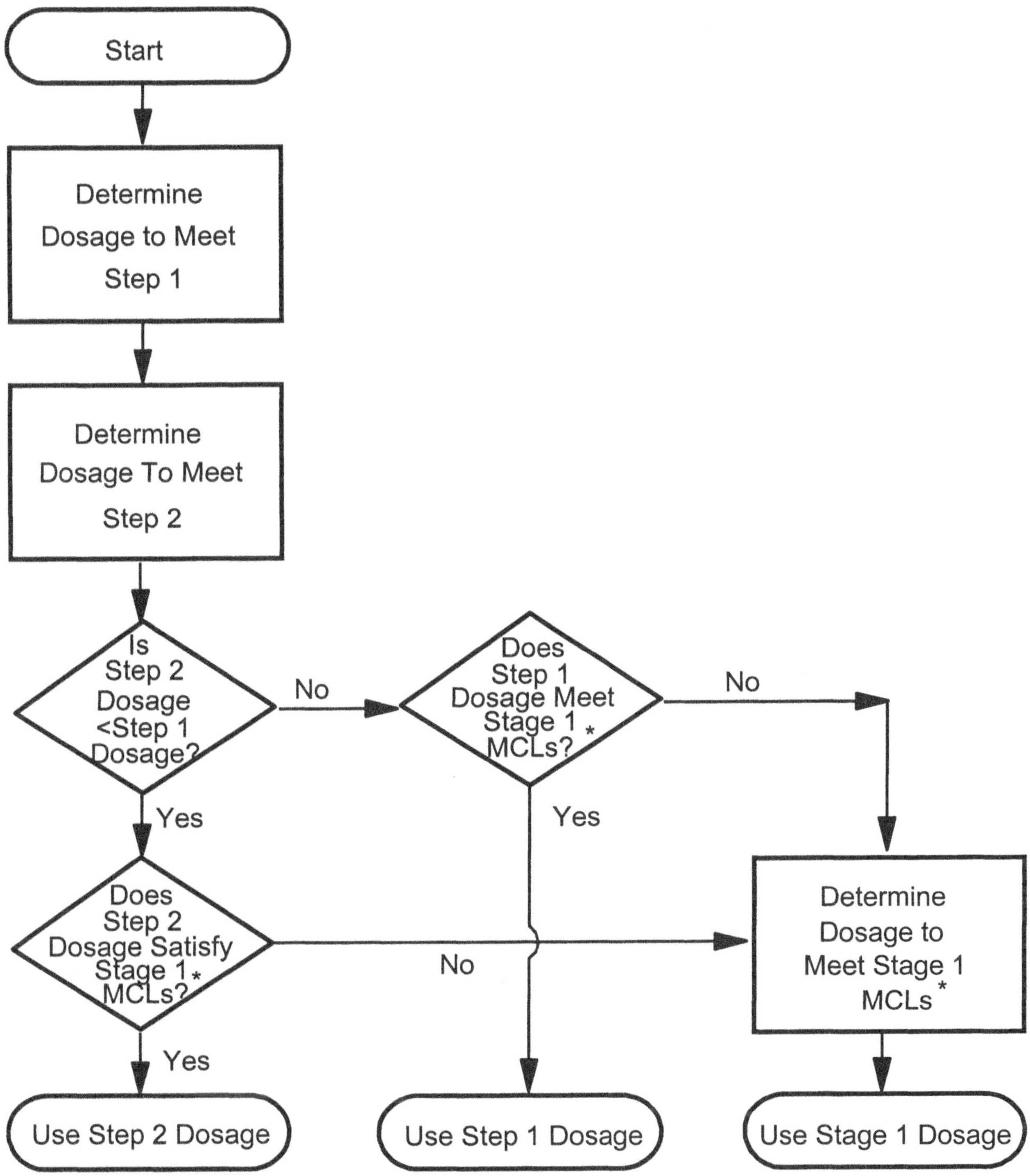

*Assuming ammonia addition following clearwell.

Figure 3.9 Decision algorithm for coagulant dosage

increase in the percentage of SPW, the ability to use free chlorine for secondary disinfection became more compromised. The escalating cost of imported water also built pressure to maximize the use of the local source, although a significant percentage of the water stored in Lake Poway is of imported origin.

Once chloramines were selected as a method to reduce DBP concentrations, the proper chlorine-to-ammonia mass ratio needed to be selected. A mass ratio of 5:1 chlorine to ammonia-nitrogen (1:1 molar ratio) is such that all ammonia will be converted to monochloramine, an effective disinfectant for a distribution-system residual. At ratios higher than 5:1, dichloramine and trichloramine are formed, which result in a loss of residual and taste-and-odor problems (Krasner and Barrett 1985). Many utilities, however, use lower ratios (3:1 or 4:1) to provide some excess ammonia to minimize the risk of dichloramine or trichloramine formation if chlorinated groundwater is used for peaking or if there are minor variations in chlorine or ammonia feed rate.

Excess ammonia in warm climates, however, encourages the growth of nitrifying bacteria, particularly in ground storage tanks (Wolfe et al. 1990). These bacteria, in turn, deplete the residual, thus promoting the growth of other bacteria in the distribution system. Once the residual starts to decline as a result of nitrification, it is very difficult to reverse and large areas of the system can suffer significant bacterial growth and non-detectable residuals.

Consequently, the City targeted a 5:1 chlorine-to-ammonia-nitrogen mass ratio to form chloramines. This is consistent with the practice of other systems throughout Southern California.

The Decision-Making Process

The City of Poway had an agreement with DHS to demolish their existing Hardinge (shallow-bed, automatic backwash) filters and replace them with a conventional dual-media design. The Hardinge filters treated only a portion of the flow. Because the City was going to be constructing these new filters, it was decided to make other modifications as described in this case study to prepare for compliance with Stage 1 of the D/DBP Rule and the IESWTR.

The Operations Department initiated the effort. The Water Treatment Supervisor and Water Utilities Manager solicited the treatability efforts and coordinated the funding through the

Engineering Department. When the treatability studies were completed, the recommendations were approved by the Operations Manager and forwarded to the Engineering Division. The Engineering Division then included the design work required to achieve compliance into the design contract for the facilities upgrade.

The City Council provided approval of the budget for the facility modifications.

Driving Forces

The philosophy of the Operations Department towards DBP control was that it made good sense to apply a very thorough planning and research effort in evaluating the choices for by-product control. More than any other factor, it was the practical wisdom of the Operations staff that caused Poway to act ahead of the final Stage 1 regulatory mandate. It was their experience that allowing plenty of time to research the problem fully is the best way to avoid being forced into an on-the-spot decision driven by imminent compliance pressures. They felt this was especially true in the case of the Stage 1 Rule that was to require simultaneous optimization of both coagulation/filtration and disinfection processes. Poway was further challenged by the unique mix of source waters to be treated, implying an even more compelling need for careful research to define the best strategy.

At the time of the evaluation, the City was already modifying a portion of their filtration facilities to meet requirements mandated by DHS. Consequently, the chemical storage and feed modifications required to address DBP concerns could be incorporated into an ongoing design and construction package at the WTP. This provided another very practical reason to take up the by-product issue ahead of the regulatory hammer.

The ultimate regulatory driver for the modifications was the July 1994 proposal of the D/DBP Rule (USEPA 1994a) and ESWTR (USEPA 1994b). The ability to meet these requirements was compromised by the City's use of free chlorine for both primary and secondary disinfection, as well as a changing imported source water favoring higher concentrations of organic DBP precursors and bromide. These factors led to a desire to investigate enhanced coagulation and the use of chloramines as a secondary (residual) disinfectant. This disinfection strategy was compatible with most of the systems in San Diego County, which receive various blends of SPW and CRW in their source water. The opportunity to learn from others' experience was another compelling factor that moved the analysis in the direction of a specific treatment strategy. The leadership of the Operations staff also brought with it a healthy sensitivity to operating costs per thousand gallons. This may have also promoted the favoring of chloramines as a cost-effective solution.

Constraints

Within the current operations of the treatment plant, sludge disposal was achieved via discharge to the sanitary sewer system. The key to sludge disposal under a Stage 1 compliance strategy was tied to the outcome of research to define the approach to meeting the enhanced coagulation treatment technique requirement. As described elsewhere in this paper, it turned out that a coagulant dose of 10 mg/L was sufficient to meet enhanced coagulation criteria for the expected range of source-water blends. If a 20 mg/L dose had been required, it is possible that drying beds would have been needed for which there was currently inadequate space at the plant site.

Other water purveyors in the area were using chloramines, so there was a good history of operational issues associated with chloramine use. Therefore, compatibility of residuals for emergency connections was not a problem. In addition, a range of things that can go wrong in a change-over to chloramines was observable in the experience of near-by systems that had made the transition treating a similar supply. The potential risk of operational constraints during implementation were therefore minimized before-the-fact. Operations staff was able to approach implementation with much greater confidence than would have been the case without having taken advantage of the benefit of these lessons learned. Uncertainties about implementation were therefore much less of a constraint to decision making than otherwise.

Public notification was required to kidney dialysis patients, tropical fish stores and hobbyists, and pool owners.

Technology Checklist - Anticipated Factors

The planning, design, operational, and notification requirements for chloramines were fairly well understood by the staff. Staff from the Metropolitan Water District of Southern

California (MWDSC), a large wholesaler of raw and treated water in Southern California, provided a presentation to the City concerning conversion issues. The design firm of Malcolm Pirnie provided operator training on chloramine chemistry, water quality issues (e.g., DBPs, regrowth), equipment, and feed rates.

The major implementation issues were expected to be in managing the operational realities of chloramines in the distribution system. Thanks to the extensive advance work studying the experience of others in the region and elsewhere, there were virtually no operational problems during implementation. The staff was prepared with a fully developed plan for managing reservoir hydraulics, an understanding of system turnover, a flushing plan, and a nitrification contingency plan. There were no problems with heterotrophic plate count (HPC) bacteria or coliforms.

To provide the operations staff with flexibility during the conversion process, the ammonia-feed system incorporated a number of design features to facilitate flexibility in operation. The system was designed so that the operator could select the targeted chlorine-to-ammonia ratio through the SCADA system. Chlorine-to-ammonia-nitrogen ratios of 4:1, 4.5:1, 4.75:1 and 5:1 were programmed into the system and could be changed through a software "selector switch." The City found that using the lower operating ratio (4:1) was effective during the conversion from free chlorine to chloramines and that a higher ratio was more appropriate for steady-state operation. Ultimately, the operations staff found that a chlorine-to-ammonia-nitrogen ratio of 4.75:1 provided a slight excess of ammonia and responded appropriately to the transient changes in chlorine residual and clearwell effluent flow that were used to calculate the ammonia dosage.

In a trick of fate, public notification of the changeover was released months prior to the actual changeover. The initial public reaction involved only a handful of complaints that were determined to be groundless because the change had not in fact occurred. When the changeover did in fact take place, there were no compliants at all.

Technology Checklist - Less Well Anticipated Factors

The major unanticipated factor was related to the downstream relocation of the point of chlorination to just before filtration, and the subsequent placement of the ammonia facilities at the exit of the clearwell. This configuration was a result of the Stage 1 Rule proposed in 1994 that prohibited *CT* credit with prechlorination before TOC removal was achieved (USEPA 1994a). The Poway plant did not have sufficient *CT* unless the ammonia was added after the clearwell, thereby allowing *CT* credit with chlorine in the clearwell. But, because Poway happens to have a rather large clearwell, the net effect of the change has been that the summer TTHM levels are running higher than anticipated. Recognizing this problem, there are currently efforts underway to move both the point of chlorination and the point of ammonia addition back upstream to reduce by-product formation. Poway can take *CT* credit from raw water chlorination through filtration, which is now allowed under the final Stage 1 Rule (USEPA 1998a). Poway was looking for options that did not require baffling their clearwell to improve T_{10}. Because the clearwell has a floating cover, baffling of their 10 million gallon clearwell would require significant reconstruction.

From a regulatory perspective, contact time is determined by T_{10}, the point at which 90 percent of the water is in contact with the disinfectant residual for time T. Under ideal (plug flow) conditions, T_{10} approximates the average detention time, known as T_{50}. Under non-ideal circumstances, baffling factors are used to account for short-circuiting in treatment facilities. Large unbaffled finished water clearwells with low effective (0.1 to 0.2) baffling factors will have a higher (5 to 10 times) average detention time (T_{50}). When unbaffled clearwells are used to attain *CT*, the higher average detention time will result in higher formation of DBPs. Although there are more plant data and more sophisticated models of plant treatment processes available to the policy process all the time, there is still only sketchy characterization of the nature of effective post-treatment storage and baffling factors. This example shows how easily everything about an RIA can be thrown off if that one variable is not understood, and how microbial disinfection and DBP formation issues are interrelated.

There were two other factors that were not well anticipated regarding the ammonia addition. These problems included the calcination of the ammonia solution lines and the inability of the ammonia metering pumps to deliver the proper chemical dose.

First, because the ammonia flow rate is low (less than 500 mL/min) and discharges into a buried water main from the ammonia-feed building, the system was designed to use carrier water to assist with the dispersion of the ammonia into the treated water. The carrier water for the aqua ammonia feed solution has a relatively high hardness and TDS. The design called for the chemical feed system to deliver the ammonia as a 30-percent solution using carrier water (1 gpm maximum flow) that was softened.

The City did not install softening equipment on the carrier water at the time of construction. It was postulated that when a sufficient volume of carrier water was used, the pH would be below the threshold to cause precipitate softening. However, when the system was commissioned, precipitative softening occurred in the ammonia solution feed lines at the 1 gpm carrier water flow.

Ultimately, it was determined that the carrier water needed to be softened. A softener was installed for the carrier stream prior to adding the ammonia solution. This eliminated the precipitation in the solution feed lines.

Second, the City changed from the 30-percent ammonia solution to a 19.5-percent feed solution to address fire and safety concerns, after the design of the ammonia-feed system was completed. The 19.5-percent solution has a lower vapor pressure and does not require the use of a pressurized storage tank and chemical metering system for storage and delivery.

When the plant operators commissioned the meter pump, they found the chemical metering pumps were unable to deliver the correct amount of ammonia solution to the treated water. The source of this problem was attributed to the type of metering pump used and the change of the ammonia-feed system from pressurized flow to gravity flow. The positive displacement metering pump, when operated under gravity flow conditions, would draw the ammonia solution into the pump under reduced pressure, resulting in off-gasing of the ammonia and poor dosage control. Peristaltic pumps were installed to eliminate this problem.

Compliance Issues

The most important compliance issues for the City of Poway were:

- Maintaining compliance with the SWTR and IESWTR.
- Maintaining compliance with the current MCL for TTHMs of 100 μg/L, as well as the newly finalized MCLs of 80 μg/L for TTHMs and 60 μg/L for HAA5. The City of Poway adopted goals for TTHMs and HAA5 of 64 and 48 μg/L, respectively (80 percent of the MCLs).

With the potential increase in coagulant dosage through enhanced coagulation, the City was also concerned about maintaining settled- and finished-water quality. The combined filter effluent from the WTP is consistently <0.1 NTU, and individual filter turbidities are also typically below 0.1 NTU.

Key Players

The key players involved in the decision-making process were the Water Treatment Plant Supervisor, the Water Utilities Manager, and the Engineering Department Manager. The City of Poway and their consultant worked with DHS to obtain approval for the conversion from free chlorine to chloramine disinfection. Public notification was provided three months before the conversion to ensure that the public was aware of the impact.

Selection Issues and Future Strategies

The City of Poway has been pleased with the effect of the chloramine conversion on TTHM and HAA5 concentrations. The TTHM quarterly averages for the fourth quarter of 1998 and first quarter of 1999 were 48 and 50 μg/L, respectively, and individual measurements ranged between 46 and 55 μg/L. These should be compared to values of 90 and 97 μg/L for the identical period one year earlier when free chlorine was used for secondary disinfection. For the first quarter of 1999, HAA concentrations averaged 27 μg/L and individual values ranged between 26 and 28 μg/L.

The City also monitors for free and total chlorine residual, free ammonia, nitrate, nitrite, total and fecal coliforms, and HPC bacteria on a weekly basis at several locations in the distribution system. Chloramine levels averaged 2 mg/L in the distribution system and ranged between 1.3 and 2.3 mg/L. Coliform bacteria have been absent, and HPC bacteria have been less than 50 cfu/mL. Nitrite concentrations have all been less than 0.005 mg/L (5 μg/L). Although these measurements indicated a successful conversion, the true test will come when results are made available for these parameters collected in the late summer and fall of 1999, when water temperatures increase to 30°C or higher (such data were not available at the time this report was prepared). This is the key period for nitrification.

The current levels of TTHMs indicate that Poway would have had difficulty meeting the previously proposed Stage 2 placeholder MCL of 40 μg/L if that value had been agreed upon. HAA5 measurements indicate a potentially narrow compliance with the previously proposed Stage 2 placeholder MCL of 30 μg/L if that MCL had been agreed upon. At the time of the decision-making process at Poway, these placeholder values were still part of a range of possible new standards. However, the Stage 1 MCLs were maintained with a change in the compliance requirement to an LRAA, and Poway should be able to achieve compliance with the Stage 2 D/DBP Rule as outlined in the Agreement in Principle (USEPA 2000). Clearly, the conversion to chloramines has achieved the following objectives:

1. Reduced the TTHM and HAA5 concentrations substantially below the Stage 1 MCLs, with a considerable margin of safety.
2. Reduced the variation in DBP concentration in the distribution system. This is important because the compliance monitoring method should be modified in Stage 2 in a spatial fashion.
3. If additional brominated HAAs are regulated in the future, HAA compliance may become a problem.

In the future, beyond the aforementioned relocation of the ammonia-feed point, the next step to significantly lower DBPs or improve disinfection capability would likely include ozone disinfection. The issue of bromate formation would have to be closely monitored, however, because the higher blend of SPW in the imported source results in higher bromide concentrations in the source water. In addition, the local surface-water contribution in San Diego has a relatively high bromide concentration (0.2 to 0.4 mg/L), possibly as a result of the geological formations that were at one time submerged below what is now the Pacific Ocean.

Treatment Facilities

Water Supply

The source-water qualities for the enhanced coagulation studies are presented in Table 3.5. These data are based upon on-site measurements in June 1994 and correlations between UV and TOC, which included both San Diego and MWDSC studies. The 10/90 blend of SPW/CRW had a lower NOM content--as illustrated by the lower TOC concentration (3.5 mg/L) and UV measurement (0.041 cm^{-1})--followed in ascending order by the 40/60-SPW/CRW blend, Lake Poway, and then the 60/40-SPW/CRW blend. The 60/40-SPW/CRW blend had the lowest alkalinity of the water sources, followed in ascending order by Lake Poway, 40/60-SPW/CRW blend, and then the 10/90-SPW/CRW blend.

Table 3.5

Raw water qualities in Poway case study

Water Source	TOC (mg/L)	UV (cm^{-1})	SUVA (L/mg-m)	Alkalinity (mg/L as $CaCO_3$)	Bromide (μg/L)	pH
10% SPW/90% CRW	3.5	0.041	1.2	145	114	8.0
40% SPW/60% CRW	3.8	0.048	1.3	114	147	8.0
100% Lake Poway	4.2	0.058	1.4	112	142	8.5
60% SPW/40%CRW	4.8	0.073	1.5	108	166	8.1

Lester J. Berglund WTP

Figure 3.5 illustrates a process flow schematic for the plant before the implementation of the treatment modifications. The current placement of the ammonia facilities is at the exit of the clearwell.

Implementation and Operational Issues

Two operational issues discussed previously under "less well-anticipated factors" were (1) precipitative softening occurring in the ammonia solution feed lines and (2) the reduced accuracy of positive displacement ammonia-feed pumps as a result of the low vapor pressure. Another issue that became apparent upon implementation was the lack of a chloramine residual in a portion of the distribution system.

Immediately after the chloramine conversion was made, storage reservoir levels were lowered so that they could be filled with chloraminated water entering the system. Operators measured the chloramine residual at these storage tanks and elsewhere throughout the system to evaluate how rapidly was the success of the chloramine conversion. It was found that a certain section of the distribution system was measuring a free chlorine, rather than a chloramine, residual. Upon investigation, it was found that a "T" pipe fitting shown with a blind flange on the as-built drawings was actually connected and was feeding water to portions of the distribution system. The initial ammonia-feed point was downstream of this "T." Consequently, it was necessary to move the ammonia-injection point upstream of the "T" so that all distributed water had a chloramine residual.

System Performance

Water Quality

The City of Poway does not perform routine sampling of the source water. Much of this sampling is conducted by MWDSC on the individual SPW and CRW supplies, and the blended sources are monitored for standard parameters (alkalinity, hardness, pH, etc.) by the San Diego CWA. This water quality is relatively well understood throughout Southern California and in the literature (Krasner et al. 1994, Cheng et al. 1995).

On two different days, however, grab samples were collected from various points in the treatment train during the bench-scale treatability studies. On each day, the influent blend to the Berglund WTP consisted of 41 percent SPW and 59 percent CRW. Grab samples were collected for UV, TOC, pH, and chlorine residual analyses from the following locations:

- Plant-influent channel
- Rapid-mix effluent
- Flocculation/sedimentation-basin effluent
- Combined-filter effluent
- Clearwell effluent (chlorinated)

Results from these process-profile analyses are summarized in Table 3.6.

Turbidity. As expected, for both performance evaluations, turbidity increased noticeably through rapid mix, decreased after settling, then decreased again following filtration. Filtered-water turbidity was excellent, measuring approximately 0.1 NTU or less.

Figure 3.10 illustrates raw-, settled-, and filtered-water turbidity for a one-year period from August 1998 through July 1999. These values confirm the process-profile results and demonstrate the relatively high quality filtered-water turbidity produced at Poway. It is notable that the 75th percentile settled-water turbidity is below 1.0 NTU, resulting in a relatively low solids loading on the filters.

UV. UV is a surrogate measure used to quantify the amount of DBP precursor material present in a given water. In both process profiles, the UV measurements indicate that the majority of the DBP precursor removal during the treatment process took place during coagulation with alum (rapid mix). The UV measurements were nearly constant between the rapid-mix effluent and the flocculation/sedimentation-basin effluent. The second greatest reduction in UV measurements occurred during filtration, likely a result of removing some colloidal particles. No change in UV occurred in the clearwell.

TOC. Generally, UV and TOC measurements followed the same relative trends, e.g., as UV decreased, so did TOC. TOC, however, is a sensitive measurement and is subject to more variation than UV. In general, it appears that there was an approximate 15 to 20 percent removal through the treatment process. This level of removal agrees well with previous evaluations performed for the City of San Diego.

Table 3.7 summarizes the monthly average values for TOC measured in the source-water blend provided by MWDSC to the San Diego CWA. The TOC from August 1998 to July 1999 had been in a relatively tight band, generally between 2.5 and 3.1 mg/L. The source water used by Poway can be slightly higher in TOC, depending upon the contribution from Lake Poway.

Table 3.6

City of Poway plant-process profile evaluations

Sample location (test identification)	Turbidity (NTU)	UV (cm^{-1})	TOC* (mg/L)	SUVA (L/mg-m)	pH	TTHMs (µg/L) Instantaneous	24 h	HAA5 (µg/L) Instantaneous	24 h
Process-performance evaluation number 1 (June 7, 1994)									
Plant influent	2.30	0.049	3.9	1.3	8.0	NM†	NM	NM	NM
Rapid-mix effluent	3.30	0.035	3.3	1.1	7.8	NM	NM	NM	NM
Sedimentation-basin effluent	0.83	0.035	3.3	1.1	7.7	NM	NM	NM	NM
Combined-filter effluent	0.03	0.033	3.2	1.0	7.8	22.6	NM	NM	NM
Clearwell effluent (chlorinated)	1.10‡	NM	--	--	7.8	23.4	44.3	18.8	36.2
Process-performance evaluation number 2 (June 16, 1994)									
Plant influent	1.60	0.061	4.3	1.4	8.3	3.4	NM	NM	NM
Rapid-mix effluent	5.20	0.048	3.8	1.3	7.7	NM	NM	NM	NM
Sedimentation-basin effluent	1.00	0.049	3.9	1.3	7.8	NM	NM	NM	NM
Combined-filter effluent	0.12	0.040	3.5	1.1	7.9	23.8	NM	NM	NM
Clearwell effluent	0.12	0.040	3.5	1.1	7.9	51.4	63.7	20.9	37.3

*Generated from UV relationship with TOC.

†NM = Not measured.

‡Suspected sampling error. High turbidity measurement may be due to air entrainment or an improperly flushed sampling line.

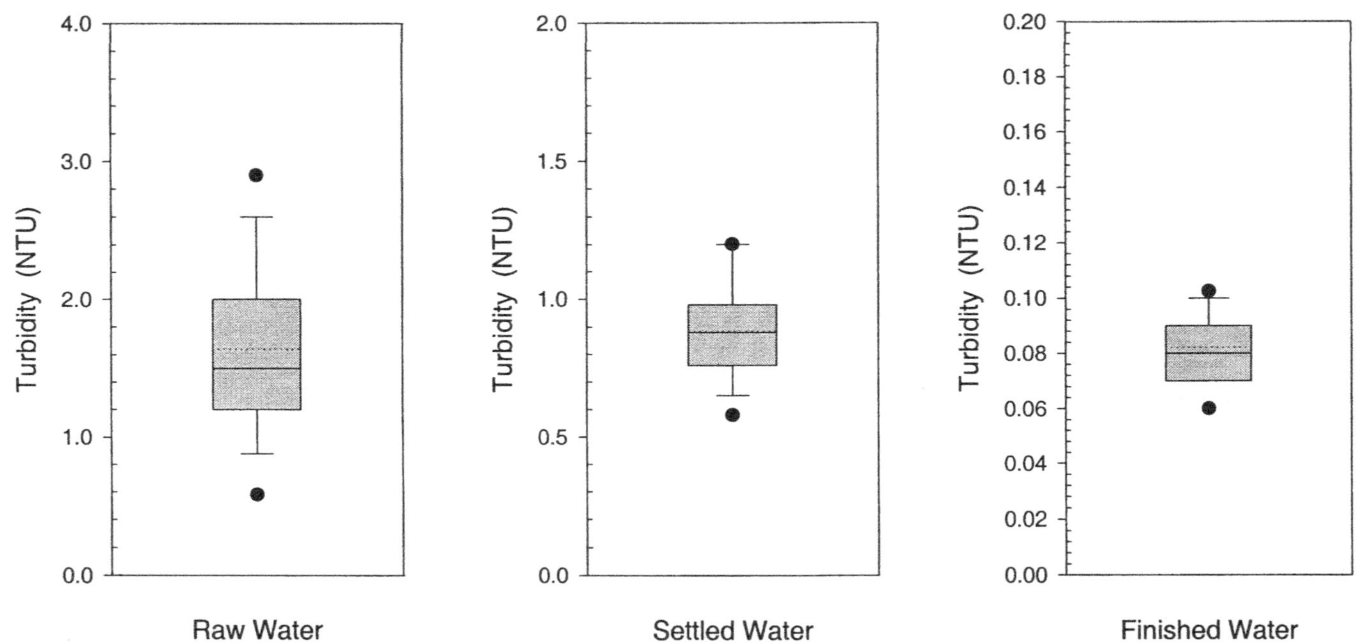

*Period of record = August 1998 to July 1999.

Figure 3.10 Raw-, settled-, and finished-water turbidity at Lester J. Berglund WTP*

Table 3.7

Monthly average source-water quality characteristics

Date	Total hardness (mg/L as $CaCO_3$)	Total alkalinity (mg/L as $CaCO_3$)	Bromide (mg/L)	TOC (mg/L)	SPW blend (%)
August 1998	240	117	0.08	2.64	20
September 1998	256	120	0.07	2.56	15
October 1998	217	108	0.09	2.60	34
November 1998	264	122	0.08	2.55	10
December 1998	276	123	0.07	2.56	5
January 1999	282	128	0.07	2.94	0
February 1999	287	130	0.07	2.86	0
March 1999	270	125	0.06	2.86	9
April 1999	238	116	0.08	2.91	25
May 1999	242	119	0.07	3.01	22
June 1999	243	117	0.07	3.03	22
July 1999	240	116	0.08	2.97	23

Alkalinity, pH, and hardness. For the process-performance evaluations, pH decreased from 8.0 or 8.3 to a pH of 7.8 or 7.7 during rapid mix, upon addition of coagulant. Because the source waters were relatively well buffered and the coagulant dosage was relatively low, the pH did not decrease significantly upon the addition of coagulant and, consequently, did not benefit DOC removal. Between rapid mix and the combined-filter effluent, the pH increases slightly to 7.8 or 7.9, just prior to pH adjustment with caustic.

Figure 3.11 illustrates raw- and finished-water pH for three months illustrative of varying source-water blends (October 1998 = 34 percent SPW; January 1999 = 0 percent SPW; July 1999 = 23 percent SPW). Despite the variation in source-water blends, the raw- and finished-water pH remained in a well-defined range.

The values in Table 3.7 also indicate relatively minor variations in total hardness and alkalinity in the source-water blends. In general, the source water is relatively well-buffered with moderate hardness.

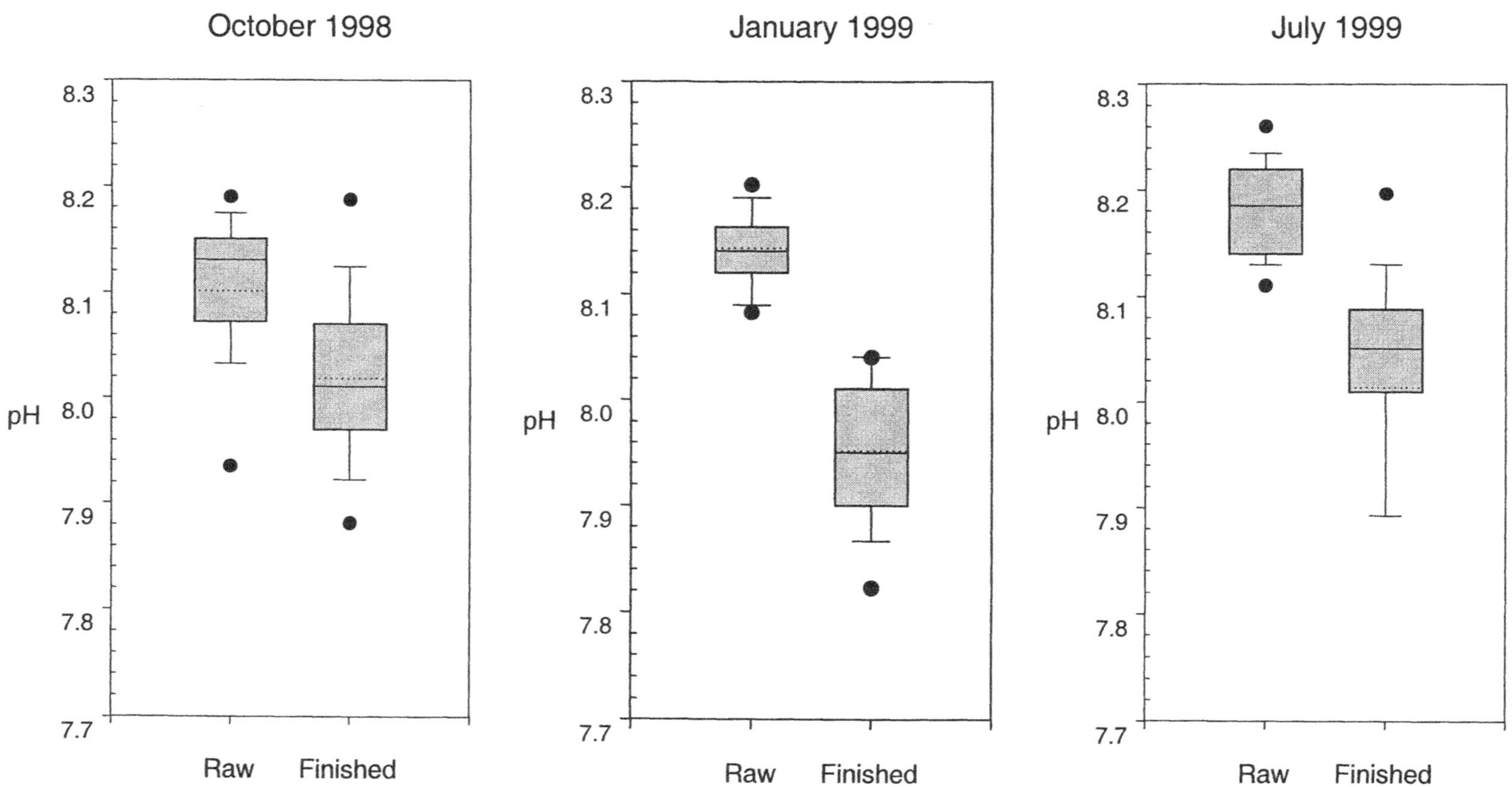

[*] Months represent combination of available water sources. See Table 3.4 for percent SPW contribution.

Figure 3.11 Raw- and finished-water pH for selected months at Lester J. Berglund WTP[*]

DBPs. Combined filter-effluent samples were collected for each performance evaluation and were quenched to measure instantaneous TTHMs. The instantaneous TTHM concentrations in the combined-filter effluent were 23 and 24 μg/L. Chlorinated clearwell-effluent samples were quenched either instantaneously or after a 24-h holding time. The instantaneous TTHM concentrations were 23 and 51 μg/L, reflecting some differences in clearwell storage times at the time of sampling in the latter sample. After a 24-h holding time, these concentrations increased to 44 and 64 μg/L, respectively.

Clearwell-effluent samples were also analyzed for HAA5. The instantaneous HAA5 concentrations were 19 and 21 μg/L for the two performance evaluations. After a 24-h holding time, the HAA5 concentrations increased to 36 and 37 μg/L. These results, indicating that HAA5 is approximately 50 to 100 percent of TTHM concentrations, agree well with other studies performed for Southern California utilities (Malcolm Pirnie, Inc. 1993; Cheng et al. 1995). Because HAA5 does not include mixed chlorobromo compounds, the total HAA concentrations (nine HAA compounds) are likely higher (Cowman and Singer 1996), but are not regulated at this time.

Historical TTHM running averages were shown in Figure 3.6 and discussed previously. The effect of implementing chloramination in reducing TTHM and HAA5 concentrations was also discussed under *Selection Issues and Future Strategies.*

Summary of System Performance

The City of Poway has been able to meet their objectives for DBP compliance by implementing chloramines for secondary disinfection. After overcoming some minor equipment and hardware issues, the chloramine conversion has gone smoothly. The process was straightforward to control and did not require unreasonable supervision.

In terms of water quality, TTHM and HAA5 concentrations have been reduced to well below Stage 1 limits. To date (at the time this report was prepared), no nitrification problems had been measured and bacterial control in the distribution system was complete. The critical period is in the late summer and early fall, when water temperatures can exceed 30°C in storage reservoirs. This condition will favor nitrification as well as increase DBP production. During this period, the plant will increase coagulant dosages to reduce DBP formation (up to 20 mg/L alum from typical dosages of 10 mg/L), and the chloramine residual leaving the plant may be increased to maintain a residual of at least 2 mg/L throughout the distribution system. (Currently, some portions of the system have a minimum chloramine residual of 1.3 mg/L.)

Because of the increase in TTHM and HAA5 concentrations during the summer months, Poway is embarking upon moving the point of ammonia addition to after the filters to reduce free chlorine contact time through the clearwell. Poway will achieve *CT* compliance with free chlorine through the conventional treatment processes, as now permitted by the final Stage 1 D/DBP Rule (USEPA 1998a). This will further reduce the variability and concentration of TTHMs and HAA5.

The TTHM and HAA5 values measured to date indicate that additional treatment modifications may be required to meet Stage 2 of the D/DBP Rule. These modifications will include evaluations of ozone, membranes, and UV disinfection. Bromate will be a constraint when using ozone in the relatively high bromide source waters (up to 0.2 to 0.4 mg/L in Lake Poway). There is some control over using imported or natural sources, so the City may exercise these options to optimize treatment.

CHLORINE DIOXIDE: GULF COAST WATER AUTHORITY

Summary

The Gulf Coast Water Authority (GCWA) wholesales water to seven municipalities in Texas between Houston and Galveston. GCWA also pumps raw water to industry and conveys water treated by the City of Houston through a dedicated pipeline from Houston to Galveston. GCWA's treatment plant uses the Brazos River as a source, with moderate-to-high TOC, hardness, alkalinity, and bromide.

The driving forces in the decision to convert to chlorine dioxide as a primary disinfectant included: (1) improved disinfection, (2) reduction in chlorinated DBPs, and (3) reduction in tastes and odors. Further, these benefits could be achieved at a relatively low cost using chlorine dioxide, compared to other more capital-intensive alternatives such as GAC and membranes. The use of ozone was precluded because of high concentrations of bromide in the source water, resulting in undesirable brominated DBP formation.

Background

Case History

GCWA wholesales treated water from their Thomas S. Mackey WTP to seven municipalities in Texas between Houston and Galveston. The population served through these seven municipalities is approximately 92,000. The communities served by GCWA are mostly urban, with industry as the primary employer. The chemical industry is located in this area; Union Carbide has a large facility in Texas City. The population-weighted median income for wage earners was approximately $27,500.

The systems served by GCWA are consecutive systems and conduct their own monitoring in the distribution system for regulatory compliance. The Texas Natural Resources Conservation Commission (TNRCC) provides compliance monitoring for TTHMs for all applicable systems in Texas (systems with populations greater than 10,000). Four of the seven municipal systems served by GCWA have populations less than 10,000. Therefore, TTHM data are not collected for these systems by TNRCC. These systems, however, perform their own monitoring for other distribution system water quality parameters (e.g., chlorine residual, coliforms).

The current rated capacity of the plant is 25 mgd. The average and maximum daily demands from the plant are approximately 12 and 20 mgd, respectively. GCWA also pumps raw water to industry and conveys treated water through a dedicated pipeline from Houston to Galveston.

The plant was put into service in 1981 and has met all state and federal drinking water requirements since operation began. The treated-water goals for industry are similar to those for potable customers.

Figure 3.12 illustrates a process schematic of the treatment plant. Treatment processes include raw water pumping, chemical addition including lime softening, upflow solids contact/clarification, recarbonation, filtration, disinfection with free chlorine, and finished water pumping.

In 1983, however, it became apparent that some finished water quality goals were not being completely met by the treatment scheme at that time. Chloramine disinfection replaced

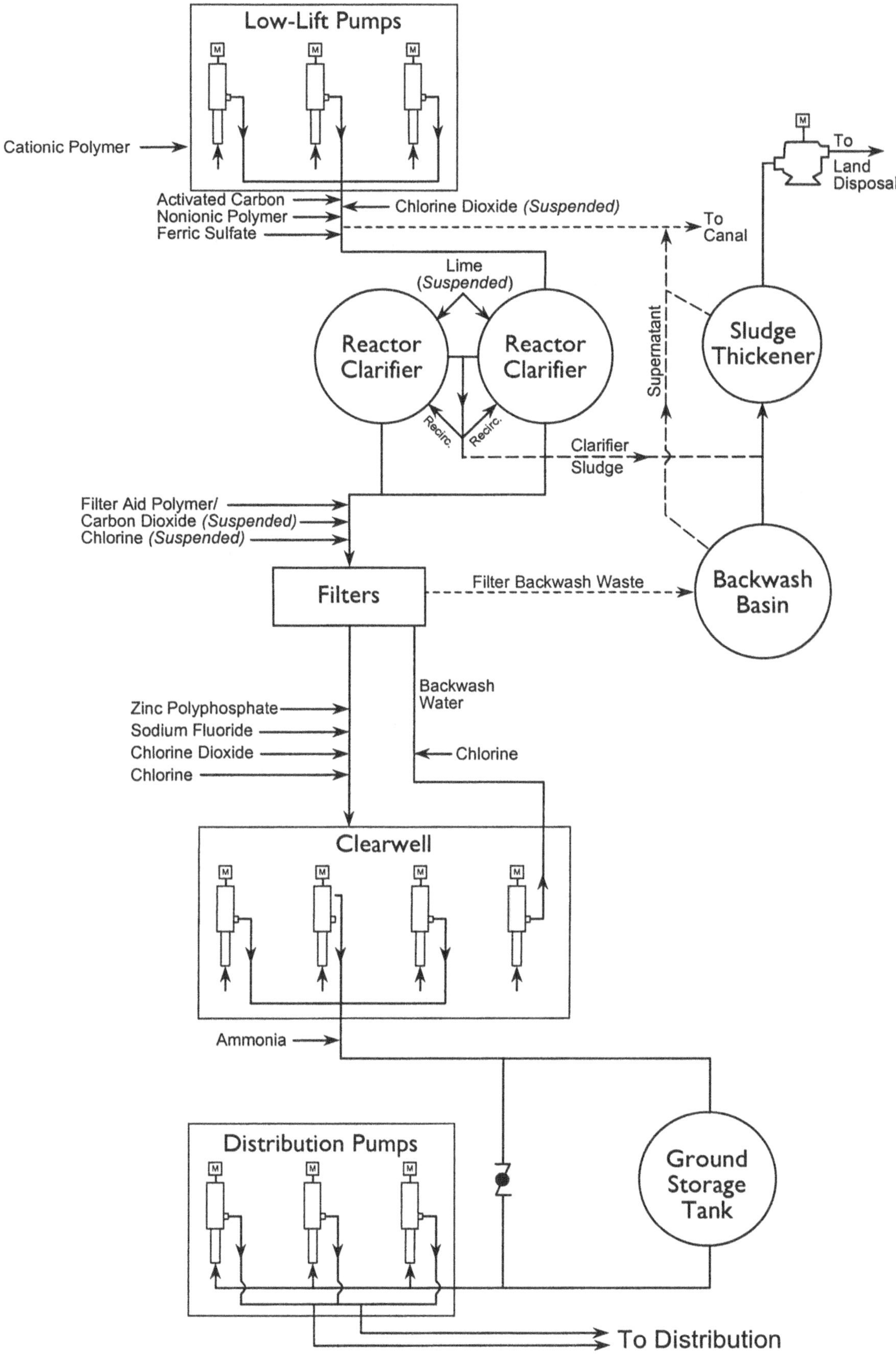

Figure 3.12 GCWA treatment plant schematic

free chlorine disinfection in 1983, before the SWTR, to meet the newly effective requirements for TTHMs. After implementing chloramines, other water quality issues arose. Taste-and-odor complaints began in May 1983, and the oxidant/disinfection scheme seemed ineffective in dealing with these issues. Further, GCWA was concerned about the ability of chloramines alone to provide the level of disinfection they desired for their source water, even though they were in compliance with applicable regulations at that time. Therefore, GCWA undertook an evaluation of a host of disinfectant alternatives, including combinations of chlorine dioxide, free chlorine, and chloramines.

After evaluation, it was decided that chlorine dioxide provided the best overall approach to satisfy the finished water quality objectives. Chlorine dioxide was first implemented in 1984. A final disinfection strategy, using chlorine dioxide in the raw and filtered water with chloramines to maintain a residual in the distribution system, was implemented at the end of 1985. Following promulgation of the SWTR, GCWA discontinued the use of chlorine dioxide in the raw water, maintained chlorine dioxide as the primary disinfectant, following filtration, with chloramines as the secondary disinfectant. Chlorine dioxide was used intermittently as a pre-oxidant.

Process Selection

As stated previously, GCWA conducted a thorough evaluation, including full-scale process testing, to arrive at a final decision to implement chlorine dioxide. Several phases were conducted in a sequential fashion, because the WTP could not be operated in a parallel mode under different oxidation and disinfection strategies. Because chlorine dioxide was implemented before the SWTR, the following phases in Table 3.8 are described as "pre-oxidation" and "post-disinfection." This is analogous to "primary disinfection" and "secondary disinfection" under the SWTR guidance.

Phase 1: before November 1983. Before 1983, GCWA, as much of the water industry, used free chlorine for primary and secondary disinfection. Bacterial quality was excellent and coliform bacteria were absent in routine monitoring. Plate counts were not analyzed at that time. The finished water produced during this period exhibited intermittent algae-related tastes and

Table 3.8

Disinfection strategies applied at GCWA

Phase	Period	Pre-oxidant/ primary disinfectant	Post-disinfectant/ secondary disinfectant
1	Before 11/83	Chlorine	Chlorine
2	11/83	Chloramines	Chloramines
3	12/83 - 4/84	Chloramines	Chlorine*
4	5/84 - 2/85	Chlorine dioxide	Chlorine
5	3/85 - 4/85	Chlorine dioxide	Chlorine dioxide
6	5/85 - 11/85	Chlorine dioxide	Chlorine dioxide/chlorine
7	12/85 - SWTR†	Chlorine dioxide	Chlorine dioxide/chloramines
8	SWTR - present	Chlorine dioxide‡/ chlorine dioxide§	Chloramines

*Breakpoint chlorination was used to achieve a free chlorine residual in the distribution system.

†Disinfection scheme changed after the promulgation of the SWTR.

‡Chlorine dioxide used intermittently as a pre-oxidant in the raw water.

§Chlorine dioxide used as the primary disinfectant following filtration.

odors. TTHM concentrations up to 350 µg/L, however, necessitated the investigation of other oxidant and disinfection strategies.

During the phases described below, there were changes in bacterial quality as measured by standard plate counts and bacterial type. At no time, however, did GCWA have a coliform violation. As discussed below, changes in bacterial populations reflected the response of the distribution system to a variety of disinfectants.

Phase 2: November 1983. In November 1983, GCWA implemented chloramines as both a pre-oxidant and post-disinfectant. A target residual of at least 2 mg/L at the customers' connections was established. Although TTHM concentrations were less than the 100 µg/L standard, bacteriological analyses in distribution-system samples yielded confluent growth in several plate cultures. In an attempt to improve the effectiveness of the distribution-system residual, the chlorine-to-ammonia weight ratio was modified from 3:1 to 7:1. Although this improved bacteriological performance to some degree, it also resulted in tastes and odors as a result of the formation of di- and trichloramine. At chlorine-to-ammonia mass ratios less than 5:1, chloramine stability was not a problem. At ratios higher than 5:1, stability was more of a problem. After this brief phase, GCWA never used a ratio higher than 5:1 when producing chloramines.

Phase 3: December 1983 - April 1984. Because lower winter temperatures suppressed THM formation, GCWA returned to free chlorine for post-disinfection while other alternatives were evaluated. The TNRCC (then the Texas Department of Health) was consulted regarding the use of chlorine dioxide as a pre-oxidant and post-disinfectant. Taste-and-odor complaints returned during the first week of May 1984, even though the chlorine dosage was increased and as much as 20 mg/L of powdered activated carbon (PAC) was added in the flocculating clarifiers.

Phase 4: May 1984 - February 1985. In an attempt to improve tastes and odors, chlorine dioxide was implemented as a pre-oxidant in May 1984. The generator at that time yielded a purity of approximately 80 percent, resulting in significant free chlorine being present in the feed stream. Nevertheless, THM production in the finished water was reduced. Further, no taste-and-odor complaints were received despite extremely high raw-water algae counts (up to 5000 blue-green algae/mL). Coliforms were not found in any distribution-system samples, and confluent growth was not observed in any standard plate counts.

As temperatures increased through the summer, however, TTHM concentrations increased in some samples above 100 μg/L. GCWA replaced the chlorine dioxide generator in January 1985 with one that could provide an average purity of greater than 98 percent, thereby reducing the amount of non-converted free chlorine from the generation process. For these colder periods, TTHMs were once again in compliance, and no taste-and-odor complaints were received. Nevertheless, GCWA recognized that TTHM compliance could again become an issue in the summer of 1985.

Phase 5: March 1985 - April 1985. GCWA received authorization from TNRCC in March 1985 to use chlorine dioxide as both a pre-oxidant and post-disinfectant. No other residual disinfectant was used during this phase. This authorization was granted with the stipulation that GCWA maintain a chlorine dioxide residual of less than 1.0 mg/L in the distribution system and increase microbiological water quality monitoring (e.g., plate counts).

Total coliforms were not measured in any samples from March through April 1985. Intermittent, elevated bacterial counts, however, were measured at three different points in the distribution system: (1) at the GCWA clearwell; (2) in Texas City, approximately two miles from the plant; and (3) in Bacliff, approximately five miles from the plant. In addition, it was difficult for GCWA to maintain a chlorine dioxide residual of 1.0 mg/L at the customer connections while at the same time providing a measurable chlorine dioxide residual throughout the customer distribution systems. The chlorine dioxide decay was relatively rapid. (A short study was done at the time, but because of the uncertainty in the analytical method used, chlorine decay data are not available.)

Figure 3.13 illustrates a comparison of average bacterial plate counts obtained at each monitoring location starting with the implementation of chlorine dioxide under Phase 4. It is evident that, initially, the use of chlorine dioxide increased standard plate counts above the desired maximum level of 500 colony forming units (cfu)/mL in distribution-system samples. The distribution of bacterial species also changed when chlorine dioxide disinfection began. Bacterial evaluation indicated that a shift from orange- and yellow-staining gram (-) rods to white-staining gram (+) rods, similar to slime-forming *Bacillus sp.*, occurred in the plant effluent. Figure 3.14 illustrates a comparison of bacterial species distribution for the phases in which chlorine dioxide was used for pre-oxidation or post-disinfection.

Figure 3.14 demonstrates that the microbial population found in the distribution system is a function of the disinfectant. Up to the time that this work was completed--and since that time--there has not been a significant amount of work relating the type of organisms as a function of disinfectant type or a relationship between gram (+) and gram (-) organisms and other water quality issues, such as taste and odor. What the data clearly demonstrate is that the population changes with disinfectant type and that both number and organism type are affected.

Phase 6: May 1985 - November 1985. The intermittent, elevated bacterial counts and the occurrence of unidentified white, gram (+) rods raised questions regarding the bacteriological quality of the finished water. Therefore, chlorine was used in addition to chlorine dioxide as a post-disinfectant until the bacteria could be identified.

The improvement in plate counts was almost immediate. The shift in bacterial species distribution continued as plate counts decreased, so that over 95 percent were yellow gram (-) rods and/or white gram (+) rods.

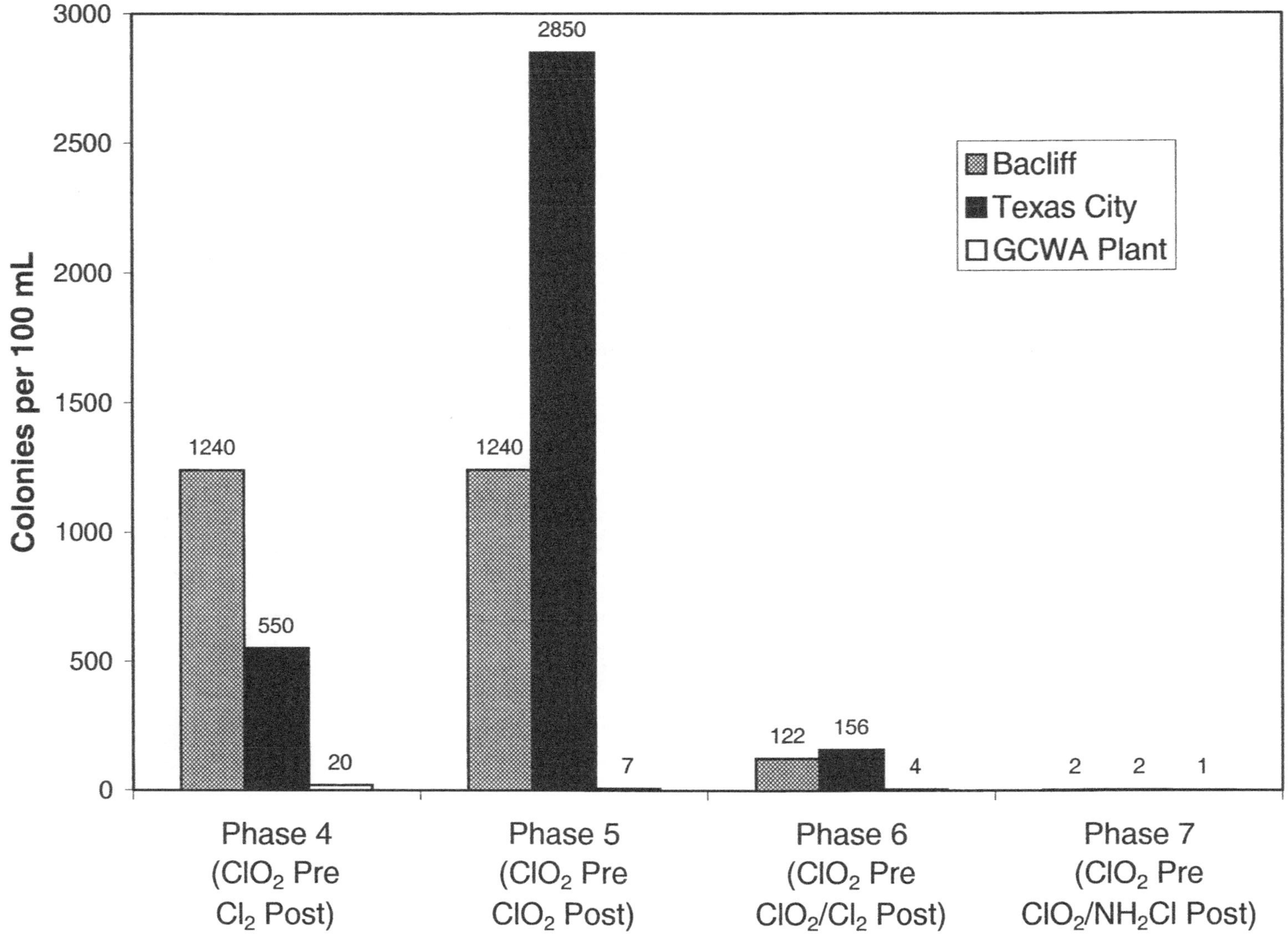

Figure 3.13 Bacterial counts versus disinfection alternatives at GCWA

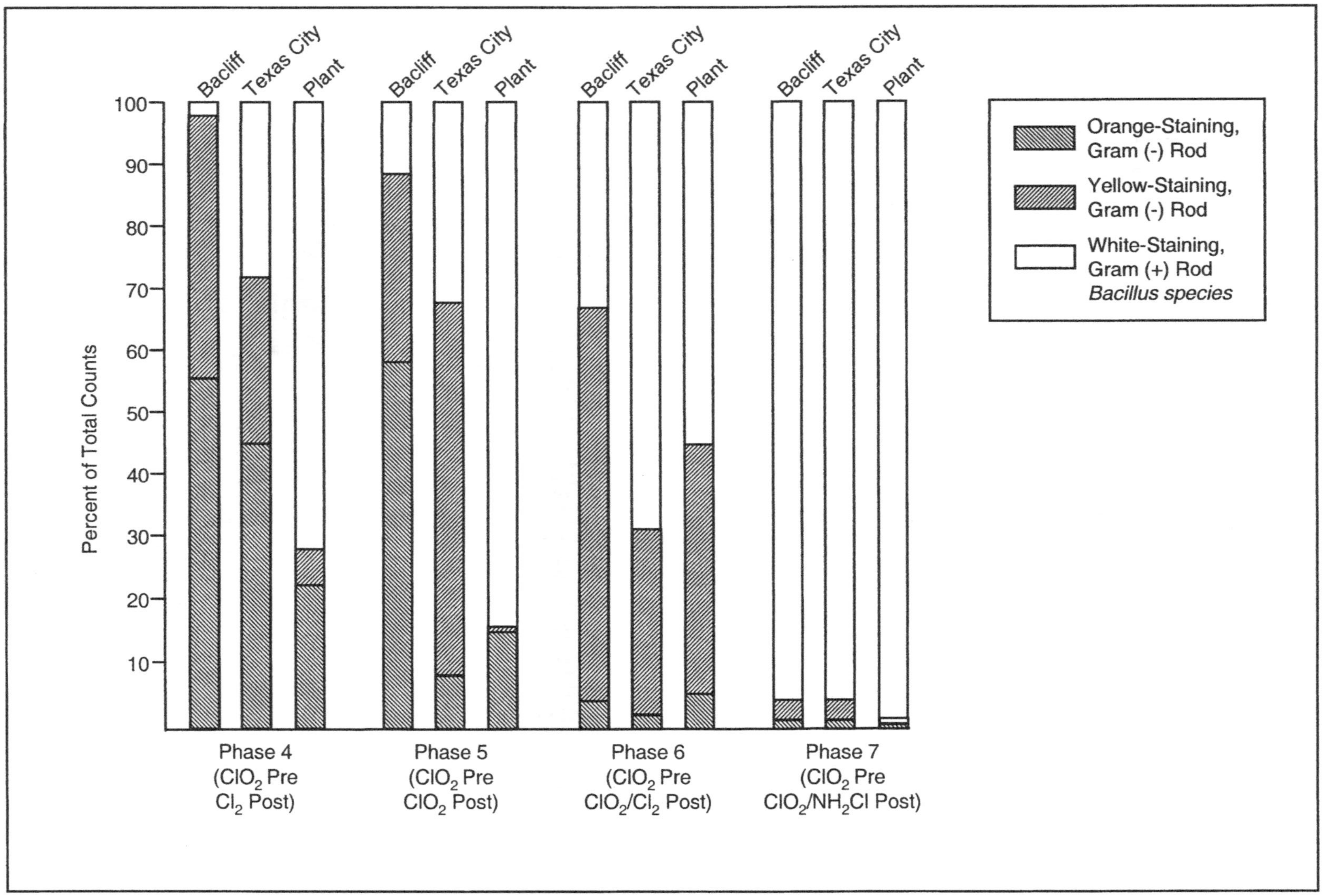

Figure 3.14 Bacterial species distribution versus disinfection alternatives at GCWA

Phase 7: December 1985 – SWTR. TTHM values approaching 100 μg/L indicated that the amount of free chlorine used at the WTP had to be minimized, despite its positive impact on distribution system bacteriological levels. In December 1985, GCWA implemented chloramines for post-disinfection. Chloramine residuals at customer connections were consistently between 2.0 and 2.5 mg/L.

After this modification, bacteriological counts in the distribution system improved further. Between that time and the present, plate counts have ranged between 1 and 30 cfu/mL. Of those colonies identified, over 95 percent were the white-staining, gram (+) *Bacillus sp.* variety. It is apparent that these organisms have preferential survivability when chlorine dioxide is used.

In retrospect, it is possible that the implementation of chlorine dioxide resulted in a "cleaning" of distribution pipe interiors, resulting in the release of bacteria embedded in biofilms. Further review of bacteriological sampling indicated that plate counts gradually increased with time further into the system. After bacterial counts peaked in a given location, they were reduced to lower levels than measured before the implementation of chlorine dioxide. It is believed that the strong disinfection capability of chlorine dioxide, together with a good biofilm control disinfectant such as chloramines, has allowed plate counts to continually remain low since December 1985.

Phase 8: SWTR – present. Following promulgation of the SWTR, GCWA discontinued the use of chlorine dioxide in the raw water, maintained chlorine dioxide as the primary disinfectant, following filtration, with chloramines as the secondary disinfectant. Chlorine dioxide was used intermittently as a pre-oxidant. This oxidation and disinfection strategy has remained to the present time, with TTHMs averaging between 50 and 70 μg/L together with complete SWTR compliance. The reduction from TTHM concentrations with free chlorine of over 300 μg/L to between 50 and 70 μg/L reflects a significant reduction (approximately 80 percent) in DBP formation. In many systems, free chlorine disinfection yields TTHM concentrations of 100 to 150 μg/L. It is difficult to determine whether the strategy employed by GCWA would result in 80 percent reductions in TTHM formation for other systems. Nevertheless, a well-operated and high-purity chlorine dioxide feed system, coupled with chloramines for secondary disinfection, should substantially reduce chlorinated DBP formation.

Parameters that Influenced Process Design

From a perspective of source-water quality, the most significant parameters controlling the process design were (1) the concentration of DBP precursors (e.g., TOC and UV) and (2) taste- and odor-causing compounds. These parameters led GCWA to investigate alternative disinfectants to either free chlorine and/or chloramines. Further, hardness and alkalinity led to the use of partial lime softening (to a pH of approximately 9), resulting in a need for the disinfectant to be compatible with the higher pH of the treated water. Finally, high bromide concentrations led to concerns about the formation of other DBPs (e.g., brominated DBPs) using potential disinfection strategies (e.g., ozone).

The most significant water quality criterion affecting design was the maximum allowable chlorine dioxide residual of 1.0 mg/L during the process-selection period. This limited the amount of chlorine dioxide that could be added and, essentially, prevented the use of chlorine dioxide as a secondary disinfectant.

When the use of chlorine dioxide was formally approved by TNRCC in 1986, TNRCC requested that GCWA follow USEPA guidance that the sum of chlorine dioxide, chlorite, and chlorate species be less than 1.0 mg/L (USEPA 1979). This essentially controlled the design dosage at the plant, because GCWA found that the maximum total chlorine dioxide dosage could not exceed 1.25 mg/L without violating this guidance. Initially, GCWA divided the dosage between raw and filtered water on a 50/50 basis. Since 1991, GCWA has added all of the chlorine dioxide to the filtered water to maximize *CT* credit. In the future, the chlorite MCL of 1.0 mg/L in the Stage 1 D/DBP will control the allowable chlorine dioxide dosage (USEPA 1998a).

The Decision-Making Process

The decision-making process employed by GCWA in their move to chlorine dioxide involved the logical individuals from the utility chain of command: General Manager, Assistant General Manager, Plant Manager, and Laboratory Manager. This was an appropriate technical contingent to have deployed on the issue because chlorine dioxide was an almost entirely new technology for potable water treatment at the time and posed a number of challenging technical questions. The technical challenges to introduce a new technology were well recognized by the staff.

The team spent years studying and preparing to introduce chlorine dioxide technology before the first trials. They conducted an extensive literature search and visited other facilities, including some in Europe where the technology had been more extensively applied in potable water treatment.

An additional challenge was presented by the need for a live, full-scale implementation. There was no pilot plant. A pilot plant would have been of only marginal use in any case because most of the technical issues related to distributed water quality. The major technical issues included:

- efficacy of disinfection;
- microbial side-effects in the distribution system habitat;
- chlorite production as a by-product of chlorine dioxide generation;
- taste-and-odor side effects.

The utility team's concerns in these areas were reinforced by concerns expressed by the state drinking water regulatory officials. The TNRCC was interested in seeing an extensive amount of monitoring accompanying implementation to track these concerns. TNRCC oversight was a major factor at every step of the decision process. TNRCC's public health focus was very consistent with the focus of GCWA staff, however, so this reinforced rather than conflicted with GCWA plans.

Another factor resulting from TNRCC oversight was the approach used to implement DBP compliance. TNRCC required GCWA (a wholesale authority) to perform compliance monitoring at all of its master meters and required each consecutive water system to perform compliance monitoring in their distribution system (all systems served having populations >10,000, so that THM monitoring was required by all consecutive systems). So two tests of compliance were involved.

Significant technical support was contributed to GCWA by Malcolm Pirnie, Inc., in evaluating the microbial and chlorite data. This support was provided by the engineering firm as a business investment in studying a new technology that might have broad application in Texas.

The GCWA Board was supportive throughout the process because of the need to act and the potential cost-effectiveness of the technology compared to alternative technologies. The Board exerted little management over the development process because of its highly technical content. GCWA has a relatively independent board consisting of one representative from Galveston, two municipal representatives, two industrial representatives (40 to 50 percent of GCWA demand is from industrial customers), and two representatives appointed by the County Commissioner's Court. Unlike some water authorities, the Board does not represent the wholesale customers responsible for distribution-system compliance on a one-to-one basis. Wholesale customers can recommend appointees to the Board, but can not directly appoint them.

Within these institutional arrangements, a bridge of trust and good working relations was necessary at the technical level to assure compliance through all affected networks. Although the multiple compliance burdens present the potential for differential impacts on different consecutive systems with differing residence times, the long-lasting chlorine dioxide/chloramines combination has proved to be the perfect technical solution for such an extensive set of networks in a regional authority. Some systems experienced problems using free chlorine at wells on their systems, but the conversion to chloramines is now nearly complete. Compliance and taste-and-odor issues have been worked through in all parts of the system.

Driving Forces

The advent of the original TTHM standard caused GCWA to recognize that DBP control would be a major issue. DBP formation is greater with high influent TOC, high bromide, high temperature, and long residence times during finished-water transmission and distribution. GCWA recognized that all of these characteristics were present in their system and fully understood that these factors made the system prone to by-product formation. Compared to the TTHM standard of 100 μg/L, levels measured in the system in the early 1980s were several hundred μg/L. GCWA also knew it would be necessary to come up with a unique solution. Ozone was ruled out because of the high TOC, high bromide levels, and a belief that ozone would exacerbate brominated DBP production (a bromate standard was not in place at this time).

In summary, the primary motivating force that drove the development of this new technology was the fact that GCWA was in a tight box, and because of the TTHM regulation, they had to move out of it. Because of these significant constraints, GCWA had to move outside the box to find a technical solution.

Constraints

The major constraint was technical knowledge of the new technology in a potable water application. GCWA surveyed the literature, visited other facilities, and enlisted the resources of an engineering partner to overcome this constraint.

Part of the technical knowledge constraint faced by GCWA consisted of the fact that the problem could not be fully addressed by pilot testing and required a live, full-scale performance test in distributed water. The end results in the distribution system were the great unknowns, and there was only one way to understand these effects--live full-scale implementation with significant monitoring.

The development was not otherwise constrained. The Board and TNRCC were supportive. Financial resources were not a constraint.

Technology Checklist - Anticipated Factors

Because the Assistant General Manager and Plant Manager both came from industrial backgrounds, chlorine dioxide technology was familiar and easily understood. Operators had a working knowledge of the technology and available monitoring methods. The technology capital and operating costs were within planning estimates.

Technology Checklist - Less Well Anticipated Factors

GCWA initially expected the generator's purity to be better than measured. Eighty percent purity was tolerable (purity defined as the concentration of chlorine dioxide divided by the sum of the concentrations of chlorine dioxide, chlorine, chlorite, and chlorate), but there was some contamination in the feed stream. This resulted in chlorite and chlorate concentrations that limited the total amount of chlorine dioxide that could be added when compared to a pure solution. For that reason, GCWA replaced its original chlorine dioxide generator after only eight months with a more efficient unit producing a higher purity product. This expense, although not extravagant (about $25,000), was more than originally budgeted or planned.

GCWA did not anticipate the relative insensitivity of residual measurements. Although available literature described drawbacks with monitoring techniques, the magnitude and extent of uncertainty was not clear until GCWA began full-scale testing. This resulted in GCWA modifying existing analytical techniques, including working with manufacturers and academicians to develop an ion chromatographic method for measuring chlorite and chlorate residuals. This method is now one of the standard methods approved for measuring chlorite and chlorate (USEPA 1996a, 1998a).

GCWA did not anticipate the impact of chlorine dioxide on distribution system microbial populations. As described previously under Process Selection, the chlorine dioxide cleansed the distribution system, initially increasing plate counts as biofilms released bacteria during oxidation of the biofilms by the chlorine dioxide. The nature of the bacterial population changed as well. Over a period of days, the plate counts increased in a given location and then decreased to levels lower than before chlorine dioxide was used (often to single digits). The increased plate counts moved in distance from the plant over time, indicating the cleansing impact of chlorine dioxide through the distribution system. Over a period of about four weeks, plate counts at all locations were lower than those measured before chlorine dioxide was implemented.

When GCWA began implementing chlorine dioxide in 1984, no other municipal water treatment plant in the state of Texas used chlorine dioxide for oxidation or disinfection. Consequently, TNRCC was initially hesitant to grant approval and wanted a significant amount of water quality data to support implementation. GCWA was not prepared initially for this effort but was able to respond. As a result of this demonstration, many other plants in Texas were able to convert according to GCWA's operating principles.

As stated earlier, the GCWA staff knew they were in a tight box with respect to DBP control as a result of having all of the conditions to favor by-product formation. This was part of the impetus for considering new technologies. Their subsequent journey through seven different configurations ultimately led them to an ideal combination for GCWA. The journey led to a significant breakthrough in understanding the dynamics of a chlorine dioxide/chloramine combination in extensive distribution networks, as well as an advance in analytical methods for chlorite and chlorate. The results provided an excellent template for systems facing similar constraints to achieve benefits in treatment and distributed water quality management. That success was made possible in large part by the scientific manner in which it was implemented, coupled with extensive monitoring.

Compliance Issues

The three most important compliance issues affecting the use of chlorine dioxide at GCWA were:

1. Maintaining compliance with the SWTR and recently promulgated IESWTR under all operating conditions.
2. Maintaining compliance with the current MCL for TTHMs of 100 μg/L, as well as the newly finalized MCLs of 80 μg/L for TTHMs and 60 μg/L for HAA5.
3. Currently limiting the use of chlorine dioxide so that the sum of the chlorine dioxide, chlorite, and chlorate concentrations are below 1 mg/L. The newly finalized Stage 1 D/DBP Rule will provide some relief to this regulatory constraint, allowing 1.0 mg/L for chlorite and 0.8 mg/L for chlorine dioxide, as measured at the first customer (USEPA 1998a).

GCWA is currently in the process of expanding their treatment plant from 25 to 50 mgd. Part of that expansion will improve baffling factors in the clearwell and ground storage to increase *CT* using free chlorine and chlorine dioxide. This will allow them to continue to meet SWTR inactivation requirements with an appropriate margin of safety. Because chlorine dioxide dosages can be increased under the newly promulgated Stage 1 D/DBP Rule, higher levels of disinfection can be achieved without adversely impacting TTHM formation. The strategy also will allow GCWA to comply with the new MCL for HAA5.

Through their own research and that of others (Aieta and Berg 1986), GCWA has determined that approximately 80 percent of their applied chlorine dioxide dosage is converted to either chlorite or chlorate, with 70 to 75 percent being converted to chlorite. Up to the present time, their total dosage has not exceeded 1.1 to 1.2 mg/L, because the total for chlorine dioxide, chlorite, and chlorate residuals measured in the distribution system could not exceed 1.0 mg/L. Under the Stage 1 D/DBP Rule to be enforced by the TNRCC, GCWA will be able to increase that dosage to the filtered water to between 1.25 and 1.4 mg/L, which will be controlled solely by the chlorite MCL of 1.0 mg/L. In the future, GCWA will evaluate the use of some chlorine dioxide as a pre-oxidant, because they potentially could convert to the use of ferrous sulfate as a coagulant. This may allow for some chlorite removal in the flocculating clarifiers (Griese et al. 1992, Iatrou and Knocke 1992).

Key Players

Key players involved in the decision-making process were the Assistant General Manager, Plant Manager, Laboratory Supervisor, and the GCWA Board of Directors. Staff from Malcolm Pirnie, Inc., worked with GCWA to inform TNRCC and ultimately to obtain approval for disinfection credit and full-scale implementation of chlorine dioxide. The Board fully supported the modification in disinfection strategy.

The five municipal wholesale customers at the time of the conversion were informed through frequent meetings with GCWA decision makers and by attendance at the Board meetings. GCWA also used the local newspaper to announce modifications to the disinfection strategy, including notifying hospitals, clinics, and other facilities that may have kidney dialysis.

Selection Issues and Future Strategies

GCWA has been pleased with the decision to use chlorine dioxide. It has been successful in meeting water quality goals and was a much less expensive option than ozone disinfection. In hindsight, this was fortunate because ozone would produce bromate concentrations above the 10 μg/L MCL in the Stage 1 D/DBP Rule, as a result of relatively high concentrations of bromide in the source water (up to 0.3 mg/L).

Improvements in generator yield and purity have allowed GCWA to more closely control their dosage, while minimizing undesirable by-products (e.g., chlorite, chlorate). In addition, the lower concentration of free chlorine in the feed stream resulting from more efficient generators has provided greater control for chlorinated by-products.

In the future, GCWA will continue to evaluate more efficient methods to produce chlorine dioxide. As a result of source-water characteristics, including TOC between 4 and

8 mg/L and bromide up to 0.3 mg/L, the next stage of D/DBP rule-making will likely present challenges that can not be solved by chlorine dioxide alone. With the concern for bromate production using ozone and the relatively alkaline water that makes pH depression difficult, GCWA may need to consider the use of GAC or membrane technology in the next 10 to 20 years. Regardless, chlorine dioxide will remain as an important strategy for microbial inactivation and reduction of taste- and odor-causing compounds.

Economic Analysis

The total cost of implementing the chlorine dioxide system was less than $50,000. The components included the chlorine dioxide generator, a shed to house the generator, a concrete pad to support the sodium chlorite storage tank and feed pumps, and the tank and pumps themselves. Free chlorine required for manufacturing chlorine dioxide was provided from the existing chlorine system through a new solution line.

These costs were funded out of the operations budget. A separate funding mechanism, such as bonds or modified rate structures, was not required. The cost of producing chlorine dioxide, calculated for a period between 1987 and 1991, was $0.019/1000 gallons produced for every mg/L added (typical dosages ranged between 0.75 and 1.25 mg/L).

Treatment Facilities

Water Supply

The Brazos River serves as the sole source of supply for GCWA. The watershed for the Brazos River extends from slightly within the New Mexico/Texas border southeast to the Gulf of Mexico. The drainage basin is about 600 miles long and ranges in width from one to 120 miles, including all or parts of 69 counties in Texas.

The GCWA raw water feed system consists of canals and a reservoir that convey water from the Brazos River to the treatment plant inlet. Two canals convey water from the Brazos River to the GCWA reservoir. At any given time, GCWA uses either canal, but raw water is not supplied to the GCWA reservoir by both canals simultaneously.

Water from the reservoir is discharged by gravity to the GCWA treatment plant. The plant intake is located along a canal approximately three miles from the reservoir.

Thomas S. Mackey WTP

The GCWA WTP has a design capacity of 18 mgd, was uprated to 25 mgd, and currently produces an average of 12 mgd. A process flow schematic was shown in Figure 3.12. A summary of existing design criteria is provided in Table 3.9.

Key features of the WTP design and operation include:

- on-site chlorine dioxide generation for disinfection and occasional pre-oxidation,
- ferric sulfate and cationic polymer for coagulation (partial lime softening to remove calcium at a pH of between 9.0 to 9.5 has been discontinued),
- up-flow solids contact flocculation/clarification,
- constant rate, dual-media filtration at a maximum filtration rate of 5 gpm/sf,
- thickening of spent filter backwash water and reactor/clarifier solids, with supernatant discharge to the canal downstream of the plant intake, and
- off-site land application of thickened solids.

The disinfection strategy uses a combination of free chlorine and chlorine dioxide to achieve *CT* through the plant clearwell. The free chlorine residual is quenched with ammonia to form a chloramine residual before the ground storage tank and distribution.

Instrumentation and Control

All mechanical plant functions can be controlled by a personal computer/programmable logic controller (PC/PLC). The PC is the interface with the PLC. The status of all operating equipment is conveyed through a SCADA system, and adjustments can be made either remotely or locally. In some instances, systems are controlled based on water quality monitoring. For example, coagulant feed used to be adjusted to maintain a set point with a streaming current

Table 3.9

Facility design criteria at Thomas S. Mackey WTP

Design parameters	Units	Existing design	TNRCC criteria
Flow			
Design peak flow	mgd	25	Peak annual daily flow
Design peak flow	gpm	17,350	
Raw-water pumping			
Number of pumps	No. of	3	
Type		Vertical turbine	
Capacity	gpm	2@10,000, 1@6,500	
Duty pumps at peak flow (firm capacity)	No. of (mgd)	2 (23.8)	
Rapid mixing			
Type		Low-lift pumps	
Number of units	No. of	2	
Maximum power	hp	150	
Detention time	sec	<0.1	
Energy input (20ºC)	1/sec	~5,000	
Volume	gal	<20	

(continued)

Table 3.9 (Continued)

Design parameters	Units	Existing design	TNRCC criteria
Flocculation			
Type		Upflow-solids contact	
Number of reactors	No. of	2	≥ 2
Design flow per reactor	mgd	12.5	
Detention time at peak flow	min	12.4	
Energy input (20ºC)	1/sec	13	
Maximum power	hp	10	
Volume per reactor	gal	101,000	
Clarification			
Type		Upflow-solids contact	
Number of reactors	No. of	2	≥ 2
Design flow per reactor	mgd	12.5	
Detention time at peak flow	h	1.5	$\geq 2^{*}$
Surface loading at peak flow	gpd/sf	2,100	
Weir length	ft	700	
Weir overflow rate	gpd/ft	17,900	15,000 to 20,000
Side water depth	ft	18	
Volume per reactor	gal	909,000	

(continued)

Table 3.9 (Continued)

Design parameters	Units	Existing design	TNRCC criteria
Filtration			
Type		Dual media	
Number of units	No. of	6	
Surface area per unit	sf	648	
Total surface area	sf	3,888	
Filter rate at peak flow	gpm/sf	4.5	≤ 5
Filter rate at peak and five units	gpm/sf	5.4	
Filter media depth	in.	33	≥ 24
GAC	in.	24	
Sand	in.	9	
Filter media support (gravel)	in.	12	
Backwash rise rate	in./min	27	
Backwash flow rate	gpm/sf (gpm)	21.6 (14,000)	12.5 to 19
Surface wash flow rate	gpm/sf (gpm)	0.93 (600)	
Bed expansion required	%		30 to 50
Freeboard required	in./min		>wash rate in inches vertical rise/min

(continued)

Table 3.9 (Continued)

Design parameters	Units	Existing design	TNRCC criteria
Backwash pumps			
Number of pumps	No. of	1	
Type		Vertical turbine	
Capacity	gpm	16,000	
Surface-wash pumps			
Number of pumps	No. of	2	
Type		Centrifugal	
Capacity (each)	gpm	600	
Filter backwash basin			
Volume	MG	0.5	
Clearwell basin			
Volume	MG	0.6	
Theoretical detention time at peak flow	min	35	
Clearwell pumps			
Number of pumps	No. of	3	
Type		Vertical turbine	
Capacity (each)	gpm	2@10,000, 1@6,500	
Duty pumps at peak flow (firm capacity)	No. of (mgd)	2 (20)	

(continued)

Table 3.9 (Continued)

Design parameters	Units	Existing design	TNRCC criteria
Ground storage tank			
Volume	MG	3	
Theoretical detention time at peak flow		Not applicable	
Distribution pumps			
Number of pumps	No. of	3	
Type		Vertical turbine	
Capacity (each)	gpm	2@10,000, 1@6,500	
Duty pumps at peak flow (firm capacity)	No. of (mgd)	2 (23.8)	

*Reactor clarifiers have been updated to 12.5 mgd each with TNRCC approval.

detector. The streaming current detector was used many years ago and discontinued approximately seven or eight years ago as it did not work quite well.

Challenges

Balancing Disinfection Needs and the Control of Chlorinated DBPs

The Brazos River is a challenging source water in terms of controlling chlorinated DBPs. The source is moderate to high in TOC (4 to 8 mg/L) and also is high in bromide (up to 0.3 mg/L). Consequently, TTHM formation under conditions of free chlorine in the distribution system can range up to 350 μg/L. In THMFP testing (7 days holding with a free chorine residual of approximately 1 mg/L after holding), THMFP concentrations can range between 800 and 1000 μg/L. These concerns initially led GCWA to a modified disinfection strategy using chlorine dioxide. Using this disinfection strategy, GCWA has been able to maintain TTHMs between 50 and 70 μg/L. HAA5 will not likely represent a compliance problem, given the relatively high source-water bromide concentration resulting in the preferential formation of brominated HAAs (Cowman and Singer 1996). This, coupled with the relatively high treatment pH, makes compliance with HAA5 less of an issue than TTHM compliance. Future regulatory efforts that may focus on HAAs not included in HAA5 may cause HAA formation to be more constraining.

Unfortunately, the very same issues that result in chlorinated DBP formation also result in a high demand for chlorine dioxide. The chlorine dioxide dosage is currently limited by TNRCC guidance to maintain the sum of chlorine dioxide, chlorite, and chlorate residuals less than 1.0 mg/L in the distribution system. Stage 1 of the D/DBP Rule (to be adopted by TNRCC) includes an MCL of 1.0 mg/L for chlorite, which will allow a slight increase in dosage (from about 1.0 mg/L to between 1.25 and 1.4 mg/L). Nevertheless, achieving the desired *CT* credit is difficult with the rapid decomposition of chlorine dioxide.

Consequently, GCWA achieves *CT* credit in four dedicated zones:

- zone 1: free chlorine disinfection in the clearwell,
- zone 2: chlorine dioxide in the clearwell,

- zone 3: chloramines in the ground storage tank, and
- zone 4: chlorine dioxide in the ground storage tank.

To maintain measurable residuals, chlorine dioxide is added to filtered water, before the clearwell. Free chlorine is added downstream of the chlorine dioxide, immediately before the clearwell. Ammonia is added between the clearwell and ground storage tank.

Using this disinfection scheme, GCWA achieved between 2 and 8 times the disinfection required under the SWTR, on a daily basis, in 1997 and 1998. Zone 4 was not required to achieve compliance. It is important to note, however, that the temperature varied between 10 and 30°C and the *CT* ratio (*CT* achieved/*CT* required) decreased as temperature decreased (as expected). Historically, temperatures can decrease to 5°C, resulting in a doubling of *CT* values for each disinfectant compared to 10°C. Under these circumstances, the credit for chorine dioxide in the storage tank (zone 4) may be required.

To maintain a reliable and measurable residual, it is critical to maximize the chlorine dioxide dosage without compromising the newly promulgated MCL for chlorite of 1.0 mg/L. The maximum residual disinfectant limit (MRDL) of 0.8 mg/L for chlorine dioxide (USEPA 1998a) will not be limiting (chlorine dioxide residuals leaving the treatment plant are 0.10 mg/L or less).

Aesthetic Concerns

The other major role of chlorine dioxide for GCWA is the control of tastes and odors in the finished water. Warm temperatures in Texas, coupled with the organic and nutrient content of the source water, promote algae growth. Reservoirs used to store water between the river and the treatment plant provide ample residence time to foster growth. Although chlorine dioxide together with PAC provided superior control compared to previous oxidation and disinfection strategies, GCWA upgraded their filters in 1991 to add GAC as a filtering medium. This allowed for a more proactive approach to adsorbing taste- and odor-causing compounds compared to PAC. Furthermore, GCWA baffled their source-water reservoirs in 1994, effectively eliminating dead zones and promoting flow-through patterns. These enhancements collectively have reduced taste-and-odor complaints, and a significant algal bloom has not been measured during the critical period (summer) since 1994.

System Performance

Water Quality

Data from January 1996 through November 1997 were reviewed to summarize raw- and treated-water quality and assess system performance. Additional data on DBP precursors, THMs, and HAAs were provided for March 26, 2001 in order to illustrate the effect of bromide on DBP speciation (Table 3.10). GCWA discontinued partial lime softening in 1994. The following discusses system performance on a parameter by parameter basis.

Turbidity. Figure 3.15 illustrates average daily raw-, settled-, and finished-water turbidities for the period of record in a "box and whisker" format.

Median raw water turbidity is about 35 NTU, with the 95th percentile at 80 NTU. Individual daily values, however, can exceed 300 NTU after a heavy rainfall. Nevertheless, settled-water turbidity averaged 1.3 NTU and finished-water turbidity was below 0.15 NTU (on average). The 95th percentile for the finished water during the period of record was 0.24 NTU.

GCWA uses a reservoir between the Brazos River and the plant to store water and provide some moderating influence to source-water quality. Although effective, water from the reservoir still flows through a canal to the plant intake. Although runoff along the canal is controlled, periods of high turbidity during rainfall events are still observed. These peaks, however, are lower than those measured in the Brazos River.

Hardness, pH, and alkalinity. GCWA's source water has moderate-to-high hardness and alkalinity, prompting GCWA to provide partial lime softening until 1994. The pH was increased in the treatment process to between 9.0 and 9.5 to remove calcium.

In 1994, GCWA discontinued their partial lime softening process. Since then, any reduction of hardness or alkalinity was a result of consumption or precipitation through the coagulation and settling process. Raw and treated-water pH, hardness, and alkalinity from January 1996 through November 1997 are presented in Figures 3.16 through 3.18, respectively. The median values for finished-water hardness and alkalinity were 185 and 120 mg/L,

Table 3.10
GCWA: DBP precursors, THM and HAA speciation on March 26, 2001

DBP precursors	Raw water
TOC, mg/L	5.66
UV, cm^{-1}	0.137
SUVA,* L/mg-m	2.42
Bromide, mg/L	0.18
DBPs, µg/L	**Finished water**
Chloroform	6
Bromodichloromethane	15
Dibromochloromethane	19
Bromoform	6
TTHMs	46
Monochloroacetic acid	ND
Monobromoacetic acid	ND
Dichloroacetic acid	9.7
Bromochloroacetic acid	11
Dibromoacetic acid	8.1
Trichloroacetic acid	3.8
Bromodichloroacetic acid	9.0
Dibromochloroacetic acid	7.5
Tribomoacetic acid	ND
HAA5	22
HAA9	49

*SUVA calculated assuming DOC equals TOC.

respectively. Treated-water pH was reduced (on average) from 8.2 to 7.6 with the addition of ferric sulfate in the coagulation process. GCWA added an orthophosphate to inhibit corrosion in the distribution system; therefore, the pH was not increased after treatment.

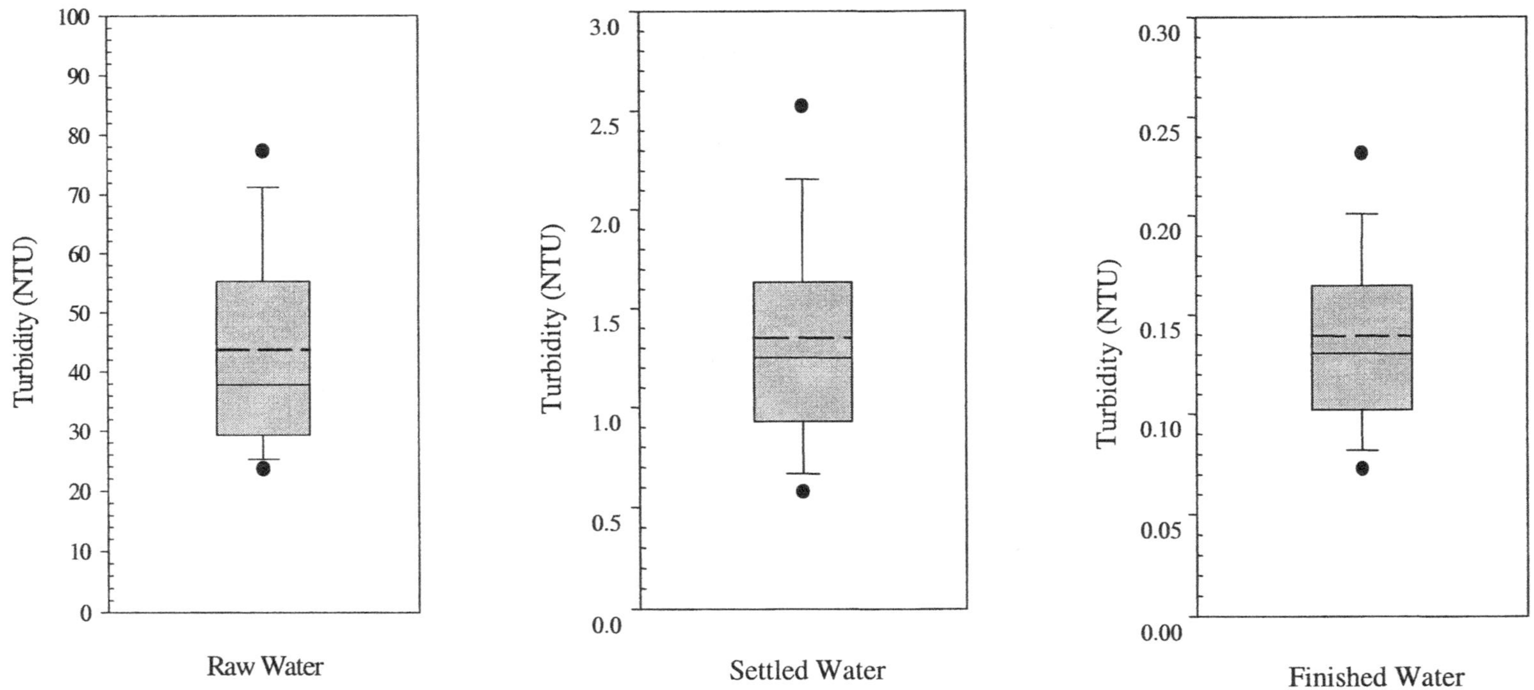

Figure 3.15 Raw-, settled-, and treated-water turbidity at Thomas S. Mackey WTP

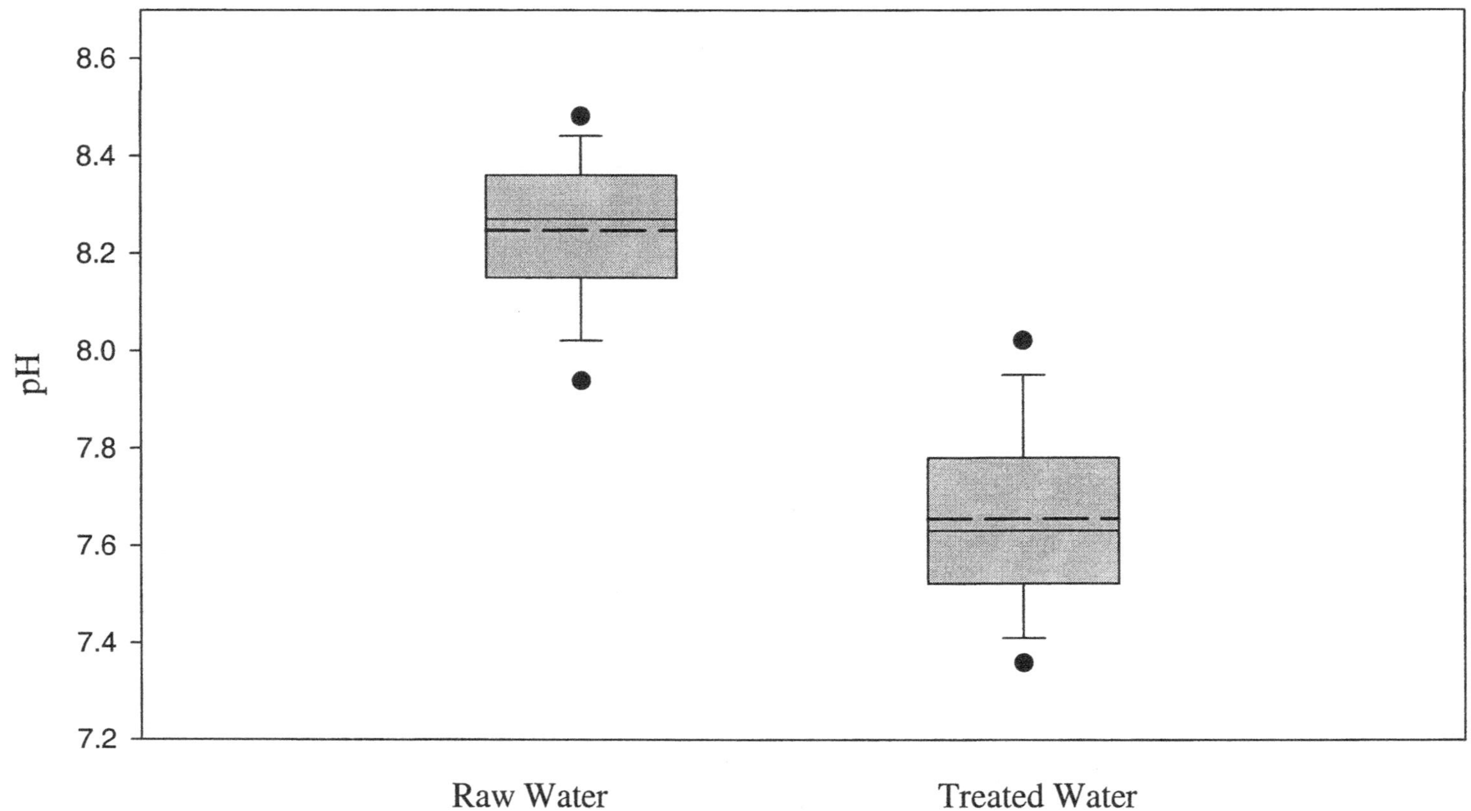

Figure 3.16 Raw- and treated-water pH at Thomas S. Mackey WTP

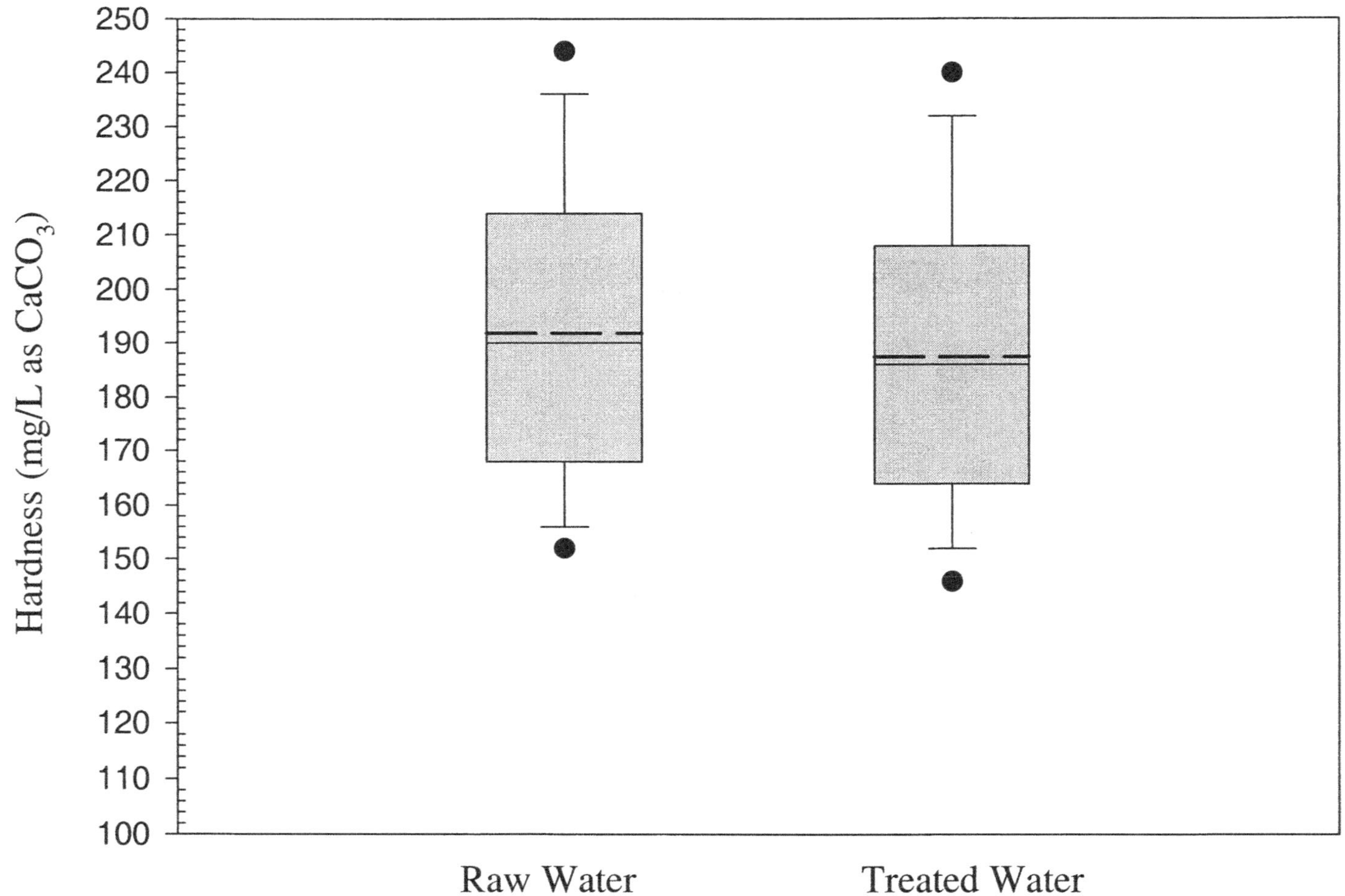

Figure 3.17 Raw- and treated-water hardness at Thomas S. Mackey WTP

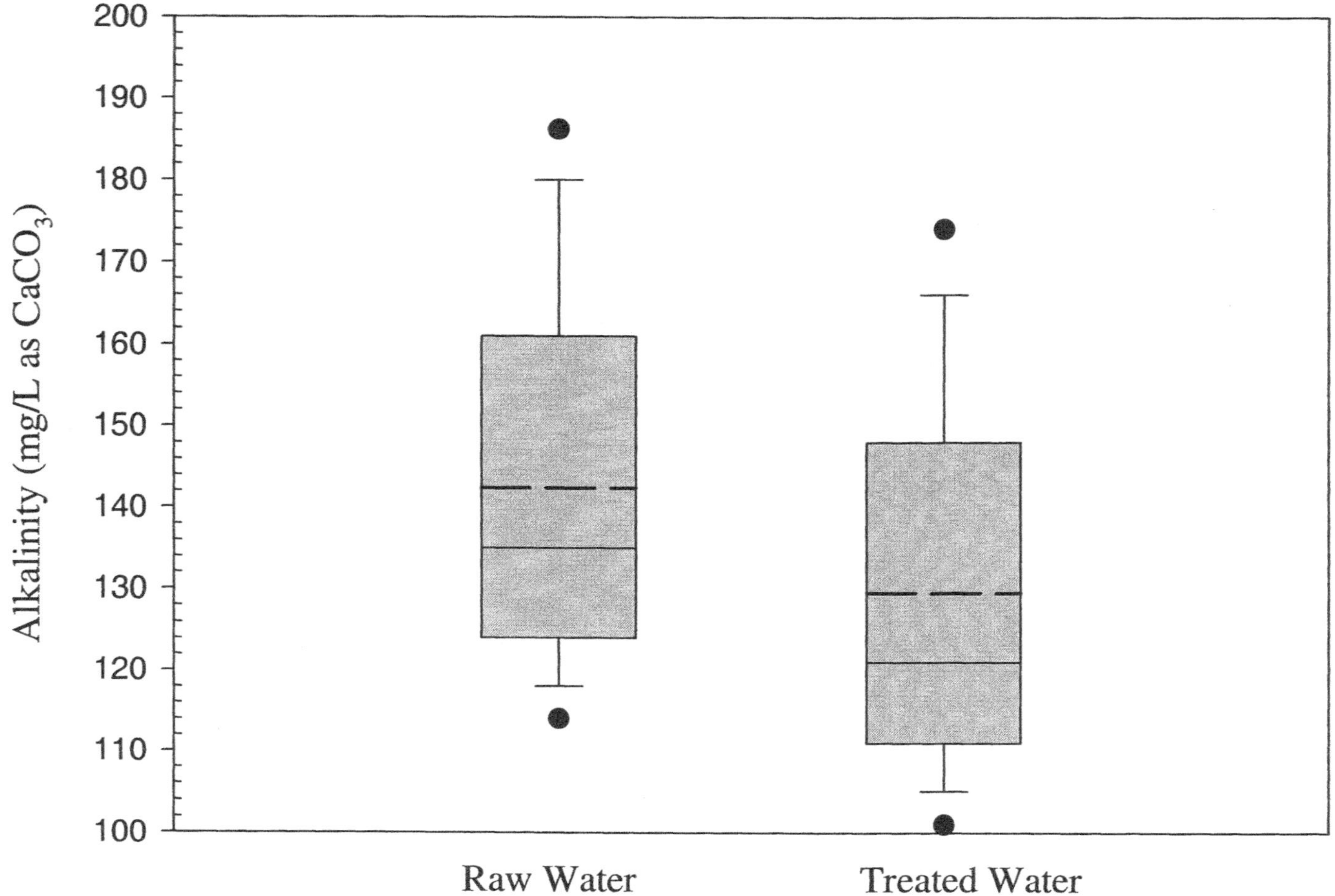

Figure 3.18 Raw- and treated-water alkalinity at Thomas S. Mackey WTP

TOC. Both partial lime softening and coagulation processes at GCWA have been effective at removing TOC. Under the newly promulgated Stage 1 D/DBP Rule, GCWA will be required to provide enhanced coagulation or enhanced precipitative softening (USEPA 1998a). As currently operated, enhanced coagulation will be implemented at GCWA.

Figure 3.19 illustrates raw- and treated-water TOC for the period of record. On a median basis, TOC was reduced from approximately 4.7 to 2.9 mg/L. Figure 3.20 shows percent TOC removal based on the data (1) in Figure 3.19 (without lime softening) and (2) from 1992 to May 31, 1994 when partial lime softening was practiced. It is evident that equivalent or slightly better TOC removal results can be achieved with coagulation compared to partial lime softening.

Based on raw-water TOC and alkalinity data presented previously, GCWA will be required to obtain either 25- or 35-percent TOC removal under enhanced coagulation. Most values fall in the 25-percent removal category (TOC >4.0 and ≤8.0 mg/L, alkalinity >120 mg/L or TOC >2.0 and ≤4.0 mg/L, alkalinity >60 mg/L and ≤120 mg/L). GCWA should be in compliance given the above historical TOC removal values under current operation, with only slight modifications on occasion if 35-percent removal is required.

Bromide. GCWA monitors bromide in the finished water on an irregular basis. Because bromide is expected to pass conservatively through the treatment process, finished-water bromide concentrations should reflect raw-water values. Figure 3.21 illustrates a box-and-whisker plot for treated-water bromide concentration for the period of record.

The median value for the period of record was 0.11 mg/L, with 5th and 95th percentile values of 0.05 and 0.23 mg/L, respectively. This median value places GCWA in the 75th percentile of systems nationwide, based on bromide concentration (Amy et al. 1994). Consequently, there is a high potential for brominated DBP formation.

THMs. GCWA analyzes for THMs in finished water on a quarterly basis, but wholesales their water to seven municipal customers. TTHM values in the customers' distribution system ranged between 50 and 70 μg/L, with a running annual average of approximately 55 μg/L. In a 35-utility survey conducted by MWDSC (Krasner et. al., 1989), THMs were sampled quarterly at GCWA for one year. Values for TTHMs ranged between 36 and 58 μg/L at the clearwell effluent, supporting the values measured in customer distribution systems. These values will be in compliance with Stage 1 of the D/DBP Rule. Because of the high level of bromide in this water, the major THMs formed are brominated species (Table 3.10).

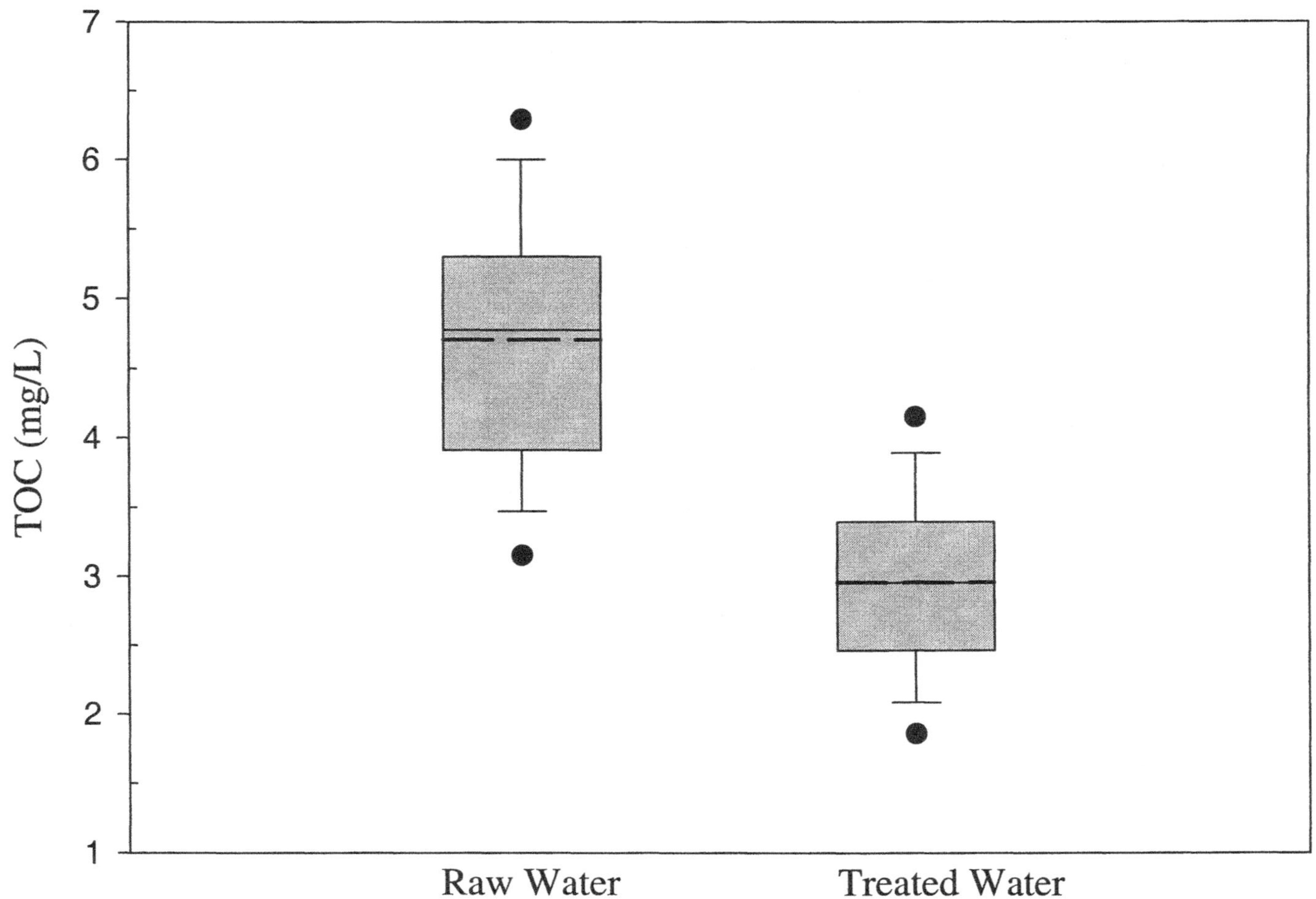

Figure 3.19 Raw- and treated-water TOC at Thomas S. Mackey WTP (without lime softening process)

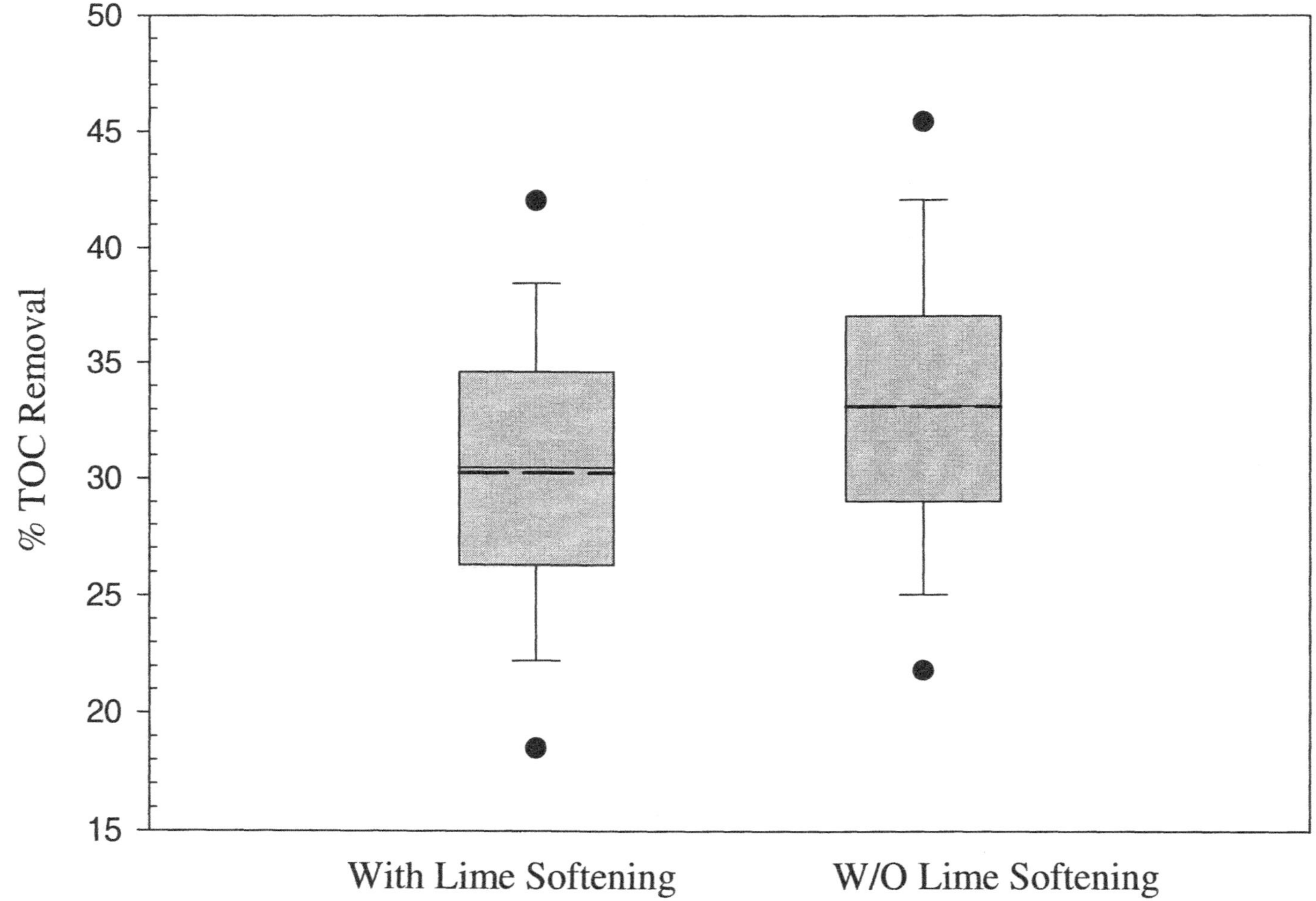

Figure 3.20 Percent TOC removal with and without (W/O) lime softening at Thomas S. Mackey WTP

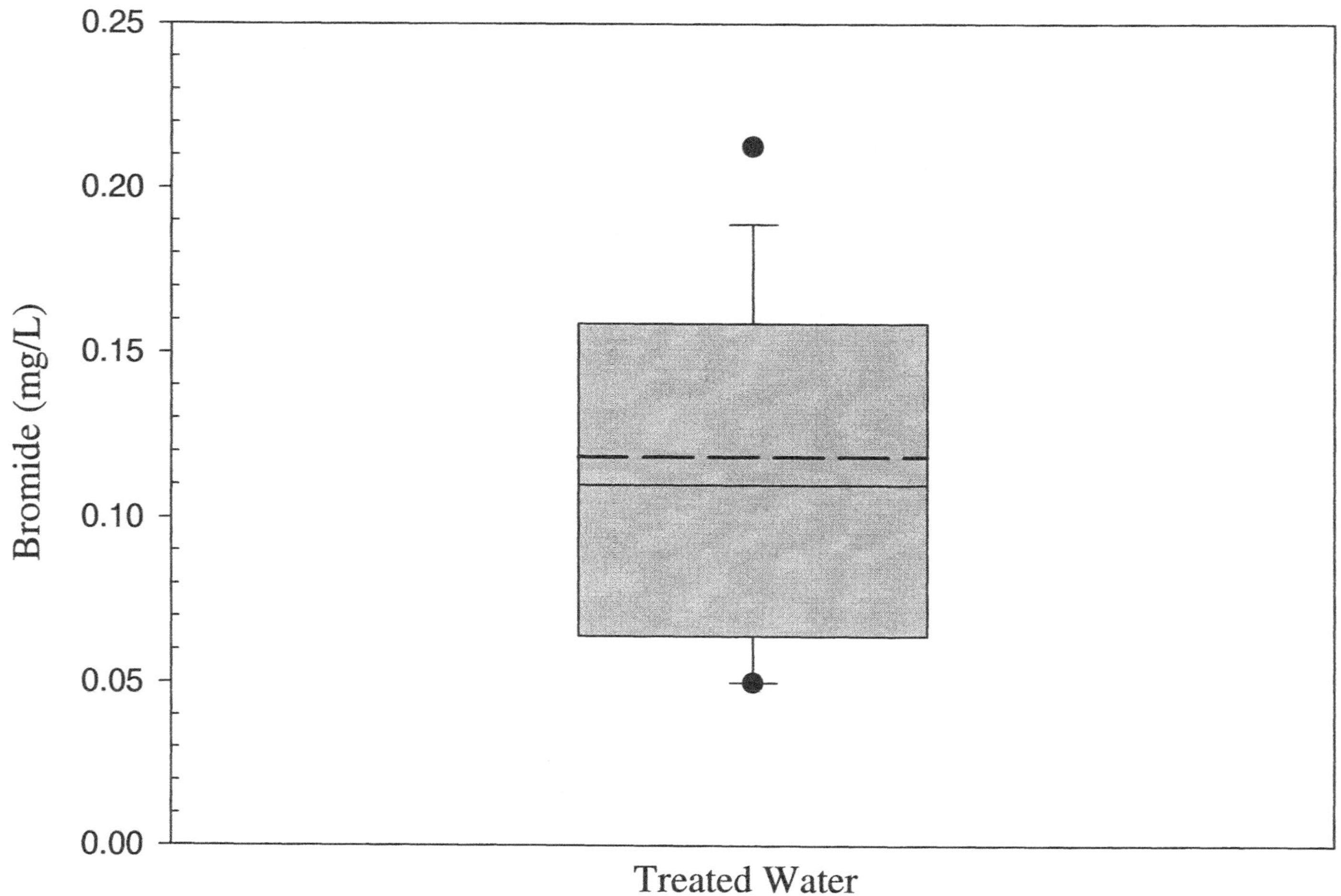

Figure 3.21 Treated-water bromide concentrations at Thomas S. Mackey WTP

HAAs. Neither GCWA nor their customers have conducted any routine monitoring for HAAs at this time (GCWA was not subject to ICR monitoring). In the 35-utility study, HAA5 concentrations were measured in the clearwell effluents between 13 and 16 µg/L. It is anticipated that HAA5 concentrations will be in compliance with Stage 1 of the D/DBP Rule, given the relatively high bromide content of the raw water. This raw-water bromide will result in the formation of brominated HAAs not included in the HAA5 value. Therefore, the 80 µg/L MCL for TTHMs will likely be the controlling parameter, as compared to 60 µg/L for HAA5. For a sample collected on March 26, 2001, the finished water contained 22 µg/L HAA5 and 49 µg/L HAA9. Because of the high level of bromide in this water, the major portion of the HAAs formed were the brominated species, many of which are not part of HAA5 (Table 3.10).

Chlorine dioxide. Chlorine dioxide decomposes rapidly after addition at GCWA (Hess 1985; Thompson et al. 1988, 1989), although no specific studies have been performed to measure decay rates. GCWA's typical chlorine dioxide dosage is 0.75 mg/L, with a maximum of 1.25 mg/L during taste-and-odor periods. A maximum dosage of 0.75 mg/L is added following filtration. Concentrations of chlorine dioxide entering the distribution system are almost always less than 0.05 mg/L.

Chlorite. Chlorite is formed at GCWA as a result of chlorine dioxide addition. To date, GCWA has limited the addition of chlorine dioxide to maintain less than 1.0 mg/L of the combined residuals for chlorine dioxide, chlorite, and chlorate.

GCWA analyzes for chlorite using either (1) ion chromatography—a standard method they developed together with input from chlorine dioxide manufacturers and academicians—or (2) amperometric titration. Figure 3.22 summarizes chlorite concentrations in the finished water for the period of record. The median chlorite value was 0.5 mg/L, with 5th and 95th percentile values of 0.2 and 0.9 mg/L, respectively. The 0.5 mg/L median is consistent with approximately 70-percent conversion of the average chlorine dioxide dosage of 0.75 mg/L.

Chlorate. Chlorate is also formed at GCWA either through a reaction during chlorine dioxide generation or through oxidation/reduction reactions involving chlorate. Once formed, chlorate is relatively stable. GCWA also analyzes for chlorate using either ion chromatography or amperometric titration, the latter of which has been shown to have significant analytical limitations. Figure 3.23 summarizes chlorate concentrations in the finished water for the period of record. The median value was 0.18 mg/L, with 5th and 95th percentile values of 0.08 and

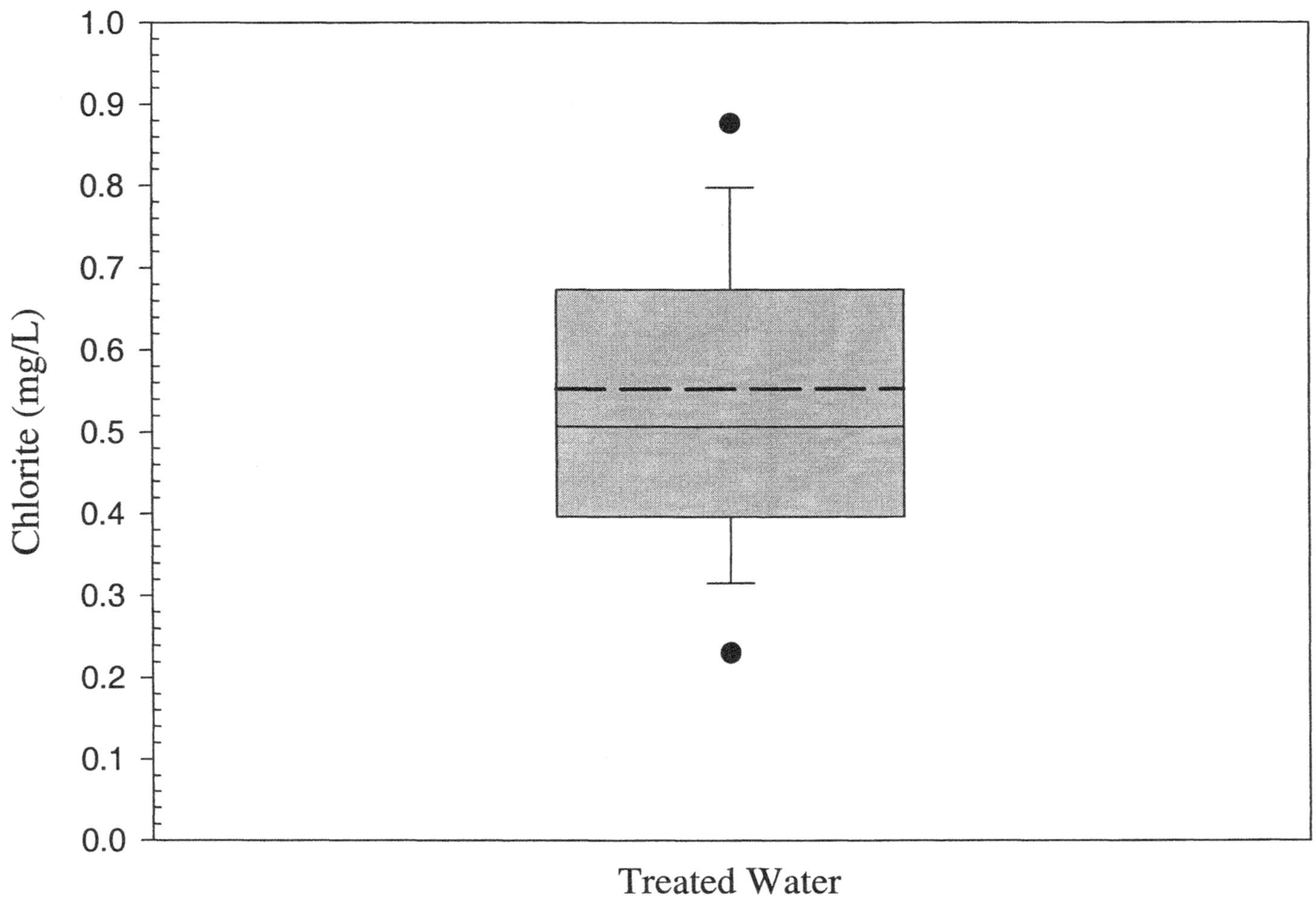

Figure 3.22 Treated-water chlorite concentrations at Thomas S. Mackey WTP

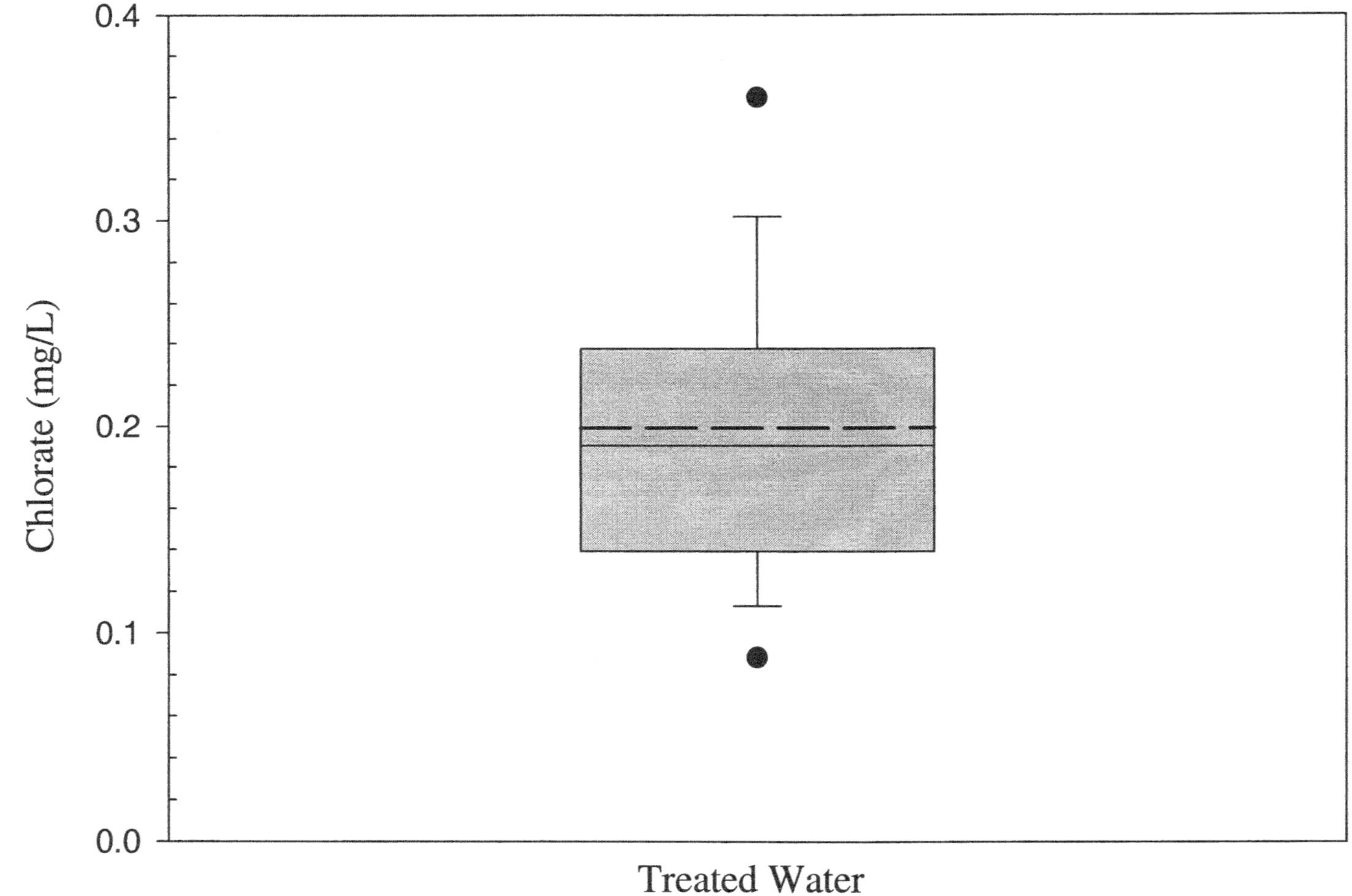

Figure 3.23 Treated-water chlorate concentrations at Thomas S. Mackey WTP

0.36 mg/L, respectively. These extreme values may not have much relevance given the accuracy and reproducibility of the amperometric analytical technique.

Summary of System Performance

GCWA currently complies with all existing treated-water regulations and is positioned to be in compliance with the newly finalized Stage 1 D/DBP Rule and IESWTR. The treated water continues to balance the microbial inactivation requirements and DBP production well. The Stage 1 MCL for chlorite and MRDL for chlorine dioxide will provide some relief on the chlorine dioxide dosage compared to GCWA's previous constraints (1.0 mg/L total residual for chlorine dioxide, chlorite, and chlorate). GCWA will also comply with the enhanced coagulation treatment technique.

The next stage of rule-making, however, may provide some challenges for GCWA. GCWA forms relatively high proportions of brominated DBPs as a result of the source-water bromide concentration. Therefore, limits on individual DBPs or a significant reduction in TTHM or HAA5 concentrations could set GCWA on a course to evaluate more expensive technologies. Potential regulation of additional HAAs beyond HAA5, which are brominated in nature, may also result in compliance difficulties. GCWA can form relatively high proportions of brominated HAAs that are not currently regulated. Because bromate will be a concern with ozone disinfection, GCWA likely will ultimately be required to evaluate membrane or GAC treatment in the long-term.

CHLORINE DIOXIDE: SKAGIT COUNTY PUBLIC UTILITY DISTRICT

Summary

The Public Utility District (PUD) of Skagit County operates the Judy Reservoir WTP. Before 1990, Skagit Co. PUD was an unfiltered-surface-water system. Water from Judy Reservoir was treated with chlorine only. In 1988-89, THM concentrations in the distribution system had ranged from 75 to 286 µg/L.

A WTP was built in 1989 and was placed on line in 1990 at the site of the unfiltered surface-water supply in order to meet the SWTR (which was promulgated in 1989) (USEPA 1989a) and the THM Rule (USEPA 1979) (however, THMs were the major concern then). As part of the WTP construction, the primary disinfectant was changed from chlorine to chlorine dioxide. The implementation strategy incrementally met changes in regulations one at a time.

Chloramines were introduced in 1991 for secondary disinfection to meet the THM Rule. The ammonia and caustic soda injection points were moved downstream of the chlorine addition point in 1994 to provide additional *CT* to meet the SWTR requirements. Additional changes were made in 1995-96 to comply with the LCR (USEPA 1991) and to meet anticipated requirements in the Stage 1 D/DBP Rule (USEPA 1994a). The Judy Reservoir WTP is a direct filtration plant and, thus, is exempt from the enhanced coagulation requirement in the Stage 1 D/DBP Rule (USEPA 1998a). However, Skagit Co. PUD enhanced their coagulation process in order to remove DBP precursors, facilitating meeting the DBP MCLs.

Each time a new regulatory compliance requirement was put forward for consideration, a new process was begun to re-optimize the chemistry of the plant. Thus, a global view of optimization was precluded by the incremental imperatives to comply with today's requirements. Cost was also an important driver that coaxed things to advance incrementally.

The current treatment and disinfection scenario at Skagit Co. PUD is allowing for simultaneous compliance with the *CT* requirements of the SWTR, corrosion-control requirements with the LCR, and the Stage 1 MCLs for THMs, HAAs, and chlorite (a by-product of chlorine dioxide) (Aieta and Berg 1986). In addition, the plant has been achieving good TOC removal.

Background

Case History

The PUD of Skagit County operates the Judy Reservoir WTP, which serves approximately 50,000 people. Built in 1989, the Judy Reservoir WTP—a direct filtration plant—had an initial design capacity of 12 mgd and utilizes the Judy Reservoir—fed by four streams—as the source water. The Judy Reservoir water has a moderate concentration of TOC (2.5-7.0 mg/L).

Before 1990, Skagit Co. PUD was an unfiltered-surface-water system. Water from Judy Reservoir was treated with chlorine only. In 1988-89, THM concentrations in the distribution system had ranged from 75 to 286 μg/L. A treatment plant was built to meet both the SWTR and THM Rule.

Initially, the WTP operated with chlorine dioxide and chlorine for primary and secondary disinfection, respectively. However, THMs still exceeded the MCL. In 1991, the secondary disinfectant was changed to chloramines, which resulted in THM concentrations in the distribution system (in 1991-92) of 3 to 15 μg/L.

However, primary disinfection with chlorine dioxide was inadequate to meet *CT* while still meeting the proposed chlorite MCL, thus a free chlorine contact time was necessary to comply with the *CT* requirements before chloramines were formed. In 1994, the ammonia and caustic soda injection points were moved downstream of the chlorine addition point to provide additional *CT* to meet the SWTR requirements. This change in disinfection practices resulted in increasing the THMs to >40 μg/L.

In 1995-96, additional changes were made to comply with the LCR and to meet anticipated requirements in the Stage 1 D/DBP Rule. Although the LCR was promulgated in 1991 (USEPA 1991), changes in the treatment process in order to comply with the LCR were not immediately implemented. Carbon dioxide was added ahead of the flocculation basins to lower the pH. In 1995, the finished-water pH increased from 7.3 to 9.0. The main reasons for these treatment and pH changes were corrosion control and enhancing coagulation (TOC removal); increased *CT* was an added bonus. Lowering the pH in the plant improved inactivation as chlorine has a different (higher) regulatory inactivation credit based on (lower) pH (USEPA 1989a). Increasing the finished-water pH improved corrosion control. Because chloramines were used for secondary disinfection, an increase in the finished-water pH level did not result in a significant increase in THM formation. (Alkalinity, however, was doubled.)

Process Selection

Because the turbidity levels in water from Judy Reservoir are quite low (0.2 to 3 NTU), a direct filtration plant was built. Bench-scale testing with optimized coagulation indicated that high concentrations of THMs would be formed with free chlorine, whereas the use of chlorine dioxide and chloramines produced far less THMs. Although optimized coagulation at a conventional plant with sedimentation basins could improve precursor removal, the choice of using alternative disinfectants at a direct-filtration plant could produce even lower THM concentrations than optimized coagulation using free chlorine. Moreover, full conventional treatment is not required for source waters that are quite low in turbidity.

After the WTP was built, modifications in the treatment and disinfection practices were needed to meet *CT* requirements in the SWTR and corrosion-control requirements in the LCR. Therefore, various tests—bench- and full-scale—were conducted in order to define optimal disinfection strategies for *CT* compliance, improve corrosion control, and minimize DBP formation. Many of these tests were conducted after the D/DBP Rule was proposed in 1994.

Full-scale testing was conducted to evaluate the use of carbon dioxide (CO_2) addition ahead of the flocculation basins and moving the point of chlorine (Cl_2) addition until after the flocculation basins (Table 3.11). For corrosion control, the pH was adjusted to ~9.0 simultaneously with ammonia addition, which was relocated or delayed. Because the pH was not brought up to 9 until chloramines were formed, the pH depression—in addition to adding *CT* and promoting enhanced coagulation—reduced THM formation, as THM formation is lower at a reduced pH. Also, moving the point of chlorine addition lowered the THM formation. Thus, the THM improvements shown in Table 3.11 are a result of delayed chlorination (Skagit Co., however, had chlorine dioxide pre-disinfection with low chlorine levels), improved precursor removal from the reduced pH of coagulation, and lower THM formation at a reduced pH. By minimizing THM formation prior to the finished-water reservoirs, Skagit Co. PUD was able to produce THM concentrations well below the Stage 1 MCL of 80 μg/L.

Note that lowering the pH can increase trichloroacetic acid formation, although dichloroacetic acid formation should not be affected (Stevens et al. 1989). Although HAAs were not evaluated in this test, they were evaluated in subsequent testing (see below).

Table 3.11

The effect of carbon dioxide addition and moving the point of chlorination on THM formation at Judy Reservoir WTP

	THM concentration (μg/L) at each sample location*			
Treatment scenario	Flocculation basin effluent	Filter Effluent	Plant effluent	Reservoir effluent
No CO_2, pre-Cl_2	8.1	35	37	59
+ CO_2, delayed Cl_2	1.6	17	26	40

*The data represent single data points.

In 1995, a series of bench-scale tests was conducted to evaluate different disinfection treatment options for their potential impacts on DBP formation, corrosion control, and disinfectant residual decay. Table 3.12 shows the results of increasing the chlorine dose to the filter effluent water at two pH levels. For example, increasing the pH to between 8.0 and 8.5 and allowing two hours of free chlorine contact time prior to ammonia addition (i.e., case 3) caused an increase in THM formation, but did not significantly affect HAA formation (the HAA concentrations for case 2 and 3 were the same within analytical variability) (Table 3.12). These tests demonstrated that pH adjustment should occur simultaneously or after ammonia addition to minimize DBP formation. Testing also indicated that the current chlorine dioxide practices would not exceed the chlorite MCL of 1.0 mg/L (which had been proposed in 1994) (USEPA 1994a).

Parameters that Influenced Process Design

The parameters and issues that influenced process design were TOC, DBPs, *CT*, and corrosion control.

TOC removal: The D/DBP Rule requires surface-water systems with conventional treatment to utilize enhanced coagulation to achieve designated TOC removal levels (USEPA 1998a). The Judy Reservoir WTP is a direct filtration plant and, thus, is exempt from this requirement (USEPA 1998a). However, Skagit Co. PUD enhanced their coagulation process in order to remove DBP precursors, facilitating meeting the DBP MCLs.

Table 3.12

The effect of different disinfection scenarios on DBP formation at Judy Reservoir WTP

	DBP formation* for tests evaluated		
DBP	Control†	Case 2‡	Case 3§
THMs (μg/L)	21	32	52
HAAs (μg/L)	45	52	55
Chlorite (mg/L)	0.8	0.5	0.5

*2-h contact time; TOC concentrations in samples tested varied from 3.0 to 4.6 mg/L.

†Control: Existing chlorine dioxide/chlorine/chloramine treatment, pH ~6.7-7.

‡Case 2: Increased chlorine dose to filter effluent water, pH 6.7-7.

§Case 3: Increased chlorine dose to filter effluent water, pH 8-8.5.

DBP control: In addition to enhanced coagulation, which included a change in the primary coagulant from alum or sulfated polyaluminum chloride to aluminum chlorohydrate, optimization of the disinfection process and pH adjustment was selected for DBP control.

CT requirements: One of Skagit Co. PUD's goals was to define optimal disinfection strategies for *CT* compliance while minimizing DBP formation.

Corrosion control: Another one of the goals was to raise the treated-water pH and bicarbonate alkalinity at some point in the process train for corrosion control without significantly increasing DBP formation.

The Decision-Making Process

Skagit Co. PUD implemented treatment modifications at Judy Reservoir WTP in response to increasingly stringent regulations. In fact, the WTP was initially placed on line in 1990 at the site of an unfiltered surface-water supply in order to meet the SWTR (which was promulgated in 1989) and the THM Rule (however, THMs were the major concern then). As part of the WTP construction, the primary disinfectant was changed from chlorine to chlorine dioxide. The implementation strategy incrementally met changes in regulations one at a time. Chloramines were introduced in 1991 for secondary disinfection to meet the THM Rule. The ammonia and caustic soda injection points were moved downstream of the chlorine addition point in 1994 to provide additional *CT* to meet the SWTR requirements. Additional changes were made in 1995-96 to comply with the LCR and to meet anticipated requirements in the Stage 1 D/DBP Rule. The coagulation process was modified even though, as a direct-filtration plant, they are exempt from the enhanced coagulation requirements. However, the modified coagulation process is helping in meeting the Stage 1 DBP MCLs.

Driving Forces

The major driving forces in implementing treatment modifications at the Judy Reservoir WTP were the SWTR, the THM Rule, the LCR, and the D/DBP Rule (Figure 3.24). Just as regulations were incrementally promulgated, one at a time, the decision process responded in similar fashion. Each time a new regulatory compliance requirement was put forward for consideration, a new process was begun to re-optimize the chemistry of the plant. Thus, a global view of optimization was precluded by the incremental imperatives to comply with today's requirements. Some "back-tracking" resulted in moving the point of ammonia injection and increasing the THM levels, where the current configuration had to "give back some ground that had previously been won." In meeting the lead and corrosion-control requirements, it was only possible to hold onto current THM gains through some clever pH and alkalinity re-balancing.

Looking forward, the onward march of the regulations may shake things up once again at Skagit Co. PUD if tougher requirements for *Cryptosporidium*, THMs, and HAAs emerge. An incremental response to such requirements may not be as likely, as a technology shift to ozone may be required. Ironically, ozone was considered as a preferred option at the outset, but was by-passed due its higher cost. So, cost was also an important driver that caused things to advance incrementally (Figure 3.24). At the outset of this compliance process, Skagit Co. PUD was faced with the $10 million cost of a filtration plant as the first step. That represented a large adjustment in rates for a city of this size moving from a very low-cost unfiltered status. Ozone would have added perhaps another $3 million to that total, which made it much more expensive than the alternative of chlorine dioxide.

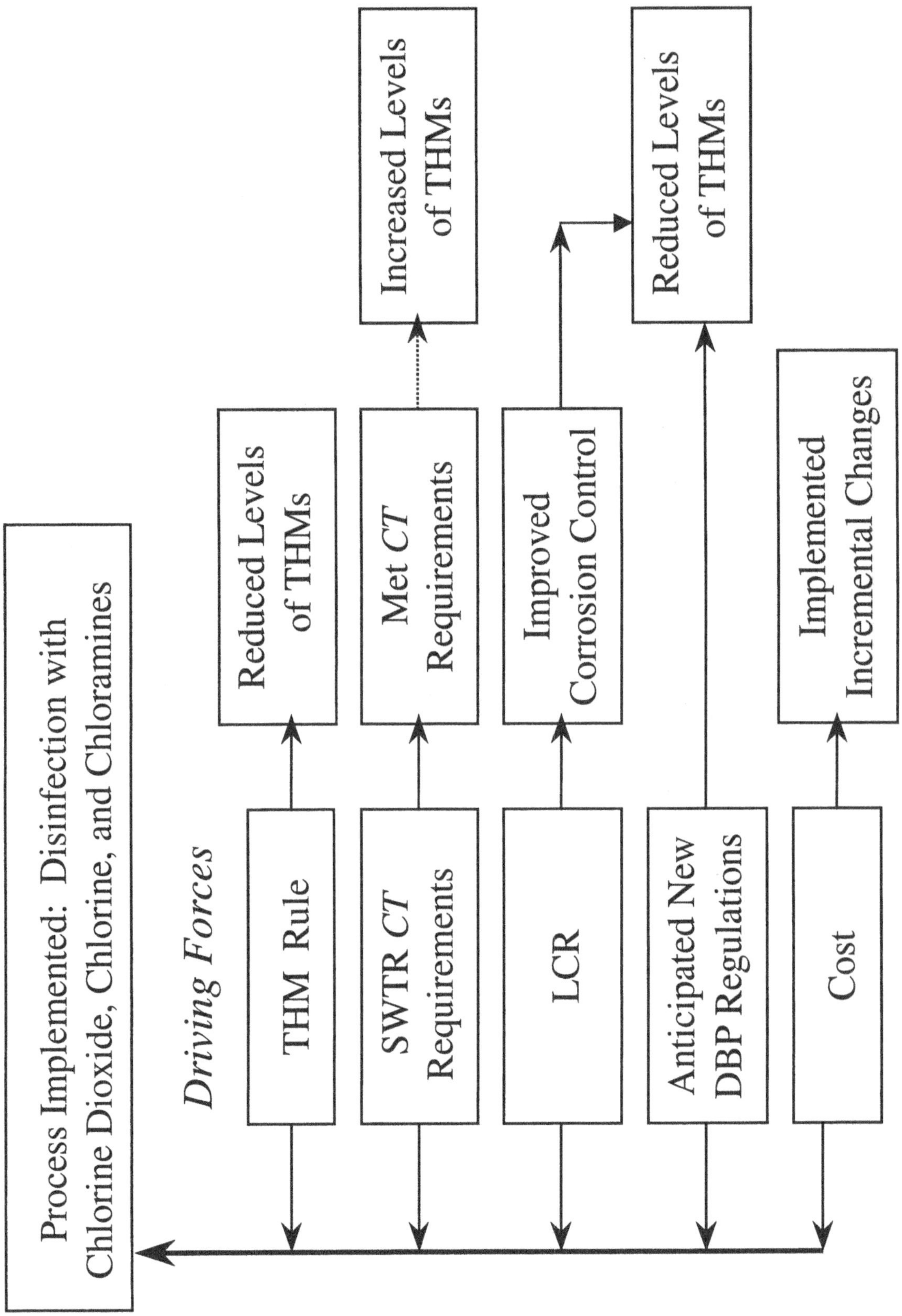

Figure 3.24 The decision-making process at Skagit Co. PUD: chlorine dioxide and chloramines for DBP control

Constraints

Because of the reactivity of the TOC in Judy Reservoir to form THMs, free chlorine contact time needs to be limited. However, chlorine dioxide alone—when using a dose that keeps chlorite formation below the 1.0 mg/L MCL—was not providing adequate *CT* for meeting the requirements of the SWTR. Because ~70 to 80 percent of chlorine dioxide is converted to chlorite (Aieta and Berg 1986), the chlorine dioxide dose needs to be <1.25-1.4 mg/L. A careful balancing of disinfection and DBP control had to be developed in order to comply with both *CT* requirements and DBP MCLs.

Typically, an elevated pH in the distribution system for corrosion control can be a constraint when chlorine is used as the secondary disinfectant. Because Skagit Co. PUD switched to chloramines, this constraint no longer applies. The monochloramine residual is actually more stable as pH increases, thus pH adjustment can be beneficial for secondary disinfection. In addition, increasing the pH of chloraminated water did not result in significant increases in THM formation, which would have occurred with chlorine.

The history of incremental changes has been one of constantly weaving in and out of the competing constraints on plant chemistry imposed by the successive regulatory waves. With further microbial and DBP requirements, there may be a need for ozone or another advanced technology (e.g., membranes), because the current technology will be over-constrained at that point. Fortunately, the Judy Reservoir supply is characterized by relatively low bromide levels, making this option unconstrained on the bromate side of the equation. Ozone may be the best for Skagit (low bromide), but may not be right for others.

Technology Checklist—Anticipated Factors

They correctly anticipated that chlorine dioxide and chloramines would produce lower concentrations of THMs. They correctly anticipated that enhanced coagulation with its efficient removal of DBP precursors and moving the point of chlorine addition would allow Judy Reservoir WTP to meet the Stage 1 THM MCL.

Looking back on the path followed to attain and re-attain compliance at Skagit Co. PUD through successive waves of regulation, it can be concluded that the incremental approach taken was capable of meeting each new challenge through adaptations. However, these changes were not anticipated beforehand, because the response mode was incremental. Moreover, most plants are responding to regulations that are competitive in process practice, e.g., pH up for corrosion control versus pH down for THM control. The same issue applies to chlorine dose and the point of addition regarding *CT* and DBP formation.

Another issue was the selection of a direct filtration plant rather than a full conventional process. It looks like this choice remains very appropriate to this rather typical Northwest surface supply and will not require revisiting as regulations continue to move forward.

Technology Checklist—Less Well Anticipated Factors

With the initial disinfection scenario of chlorine dioxide and chlorine for primary and secondary disinfection, respectively, they initially had some taste-and-odor problems in certain parts of the distribution system. This problem was eliminated with the switch to chloramines for secondary disinfection. In other AwwaRF research, it has been found that taste-and-odor problems can be generated in the chlorinated distribution system of systems that use chlorine dioxide (Dietrich and Hoehn 1991).

With the disinfection scenario of chlorine dioxide and chloramines for primary and secondary disinfection, respectively, they were not able to meet the *CT* requirements of the SWTR. This was resolved by moving the ammonia and caustic soda injection points downstream of the chlorine addition point. However, the use of free chlorine alone to provide additional *CT* to meet the SWTR requirements resulted in significantly increasing the formation of THMs during the late spring and summer months when water temperature increases.

Typically, many systems expect that raising the distribution-system pH for corrosion control will result in an increase in THM formation. However, there was not a significant increase in THM formation with increasing distribution-system pH at Skagit Co. PUD because the secondary disinfectant was chloramines.

Compliance Issues

The current treatment and disinfection scenario at Skagit Co. PUD is allowing for simultaneous compliance with the *CT* requirements of the SWTR, corrosion-control requirements with the LCR, and the Stage 1 MCLs for THMs, HAAs, and chlorite. Although the level of THMs has been reduced, HAAs have not been as significantly reduced. For example, in 1998-99, the average value of the THMs in the distribution system was 33 μg/L (the average value in the distribution system in each monthly sampling period was typically 24 to 39 μg/L). In this time period, the average value of the HAAs in the distribution system was 39 μg/L (the average value in the distribution system in each monthly sampling period was typically 27 to 52 μg/L). Because the coagulation pH was lower with carbon dioxide addition and disinfection with free chlorine occurred before final pH adjustment, chlorination at a lower pH would also have some effect on lowering THM formation and potentially raising (or having little effect on) HAA formation.

Key Players

The decision-making process involved a variety of WTP staff and consultants. For example, CH2M-Hill was involved with their chloride dioxide and chloramine control work in 1990. Economic and Engineering Services conducted a corrosion-control study in 1995.

Demographic information from 1990 was obtained for the County of Skagit. The median household income (in 1989) was $28,389. The median cost of housing was $80,800. The median age of the population was approximately 34 years.

The ultimate decision on technology choices was made by the utility staff.

Selection Issues and Future Strategies

Skagit Co. PUD has been monitoring THMs and HAAs at several locations in the distribution system. Although they have increased *CT* and increased the distribution-system pH to slightly below 9.0 (closer to 8.6), they are able to attain compliance with Stage 1 of the

D/DBP Rule. However, changes in treatment may be required depending on what happens with the Stage 2 Rule.

Economic Analysis

Although Skagit Co. PUD has made several modifications to the treatment process over time, the major capital cost was the initial cost of building the filtration plant. In terms of annual operating costs, in addition to the cost of adding carbon dioxide for pH depression, there has been an increase in the cost for chemicals to raise the pH. For example, caustic expenditures were $15,640 in 1994 and $32,300 in 1995.

Treatment Facilities

Water Supply

Skagit Co. PUD obtains water from Judy Reservoir. The reservoir is fed by four similar streams: Gilligan Creek, Mundt Creek, Turner Creek, and Salmon Creek typically represent 50, 35, 10, and 5 percent of the water in the reservoir, respectively.

Typical water quality parameters for the plant influent are presented in Table 3.13. The influent to the plant typically has color in the range of 15 to 35 pcu, with average-to-moderate TOC (2.5-7.0 mg/L, average = 4.3 mg/L) and UV absorbance in the range of 0.14 to 0.17 cm^{-1}. The water has little to no bromide (e.g., in 1988-89, the bromide was <0.01 to 0.01 mg/L). The water is high in THM formation potential (240-800 μg/L, average = 550 μg/L). The water is low in alkalinity and hardness (average values of 14 and 16 mg/L as $CaCO_3$, respectively). The water is low in turbidity (0.2-3 NTU, average = 0.5 NTU) and microbial pathogens.

Judy Reservoir WTP: Before Treatment Modifications

Judy Reservoir WTP is a 12-mgd direct filtration plant (Figure 3.25), which uses chlorine dioxide (ClO_2) and chlorine for primary disinfection and chloramines for secondary disinfection. It typically produces a flow of 8 mgd. Water treatment processing systems include a static

Table 3.13

Typical raw water quality at Judy Reservoir WTP

Parameter	Minimum	Average	Maximum
TOC (mg/L)	2.5	4.3	7.0
UV (cm^{-1})	0.138		0.166
Temperature (°C)	0.5	10	22
pH	6.8	7.2	7.5
Alkalinity (mg/L)		14	16
Hardness (mg/L)		16	17
Turbidity (NTU)	0.2	0.5	3
Total coliforms (#/100 mL)	ND	10*	1500
Giardia (# cysts/100 L)	ND	ND	1 unconfirmed
Cryptosporidium (# oocysts/100L)	ND	ND	ND

*Arithmetic average value.

mixer, flocculation basins, and dual-media (coal, sand) filters. As a direct-filtration plant, there are no sedimentation basins.

Before 1990, Skagit Co. PUD was an unfiltered surface-water system. Water from Judy Reservoir was treated with chlorine only. When the WTP was first constructed, it was operated with chlorine dioxide and chlorine for primary and secondary disinfection, respectively. In 1991, the secondary disinfectant was changed to chloramines. In 1994, the ammonia (NH_3) and caustic soda injection points were moved downstream of the chlorine addition point. These treatment modifications were in response to the SWTR and the THM Rule and represent how the Judy Reservoir WTP was operated prior to additional modifications to meet the Stage 1 D/DBP Rule.

Raw water (pH varies between 6.8 and 7.5) was treated with alum (6-16 mg/L, average = 10 mg/L) or sulfated polyaluminum chloride and cationic polymer (2.0-4.0 mg/L, average = 3.0 mg/L). The pH of the filtered water was raised with sodium hydroxide (NaOH) to 7.3-8.0 in 1994 (and to 8.0-8.5 in 1995) for corrosion control.

Primary disinfection/oxidation were attained by applying chlorine dioxide and chlorine to the raw water. After filtration—in addition to caustic (for pH adjustment)—ammonia was added

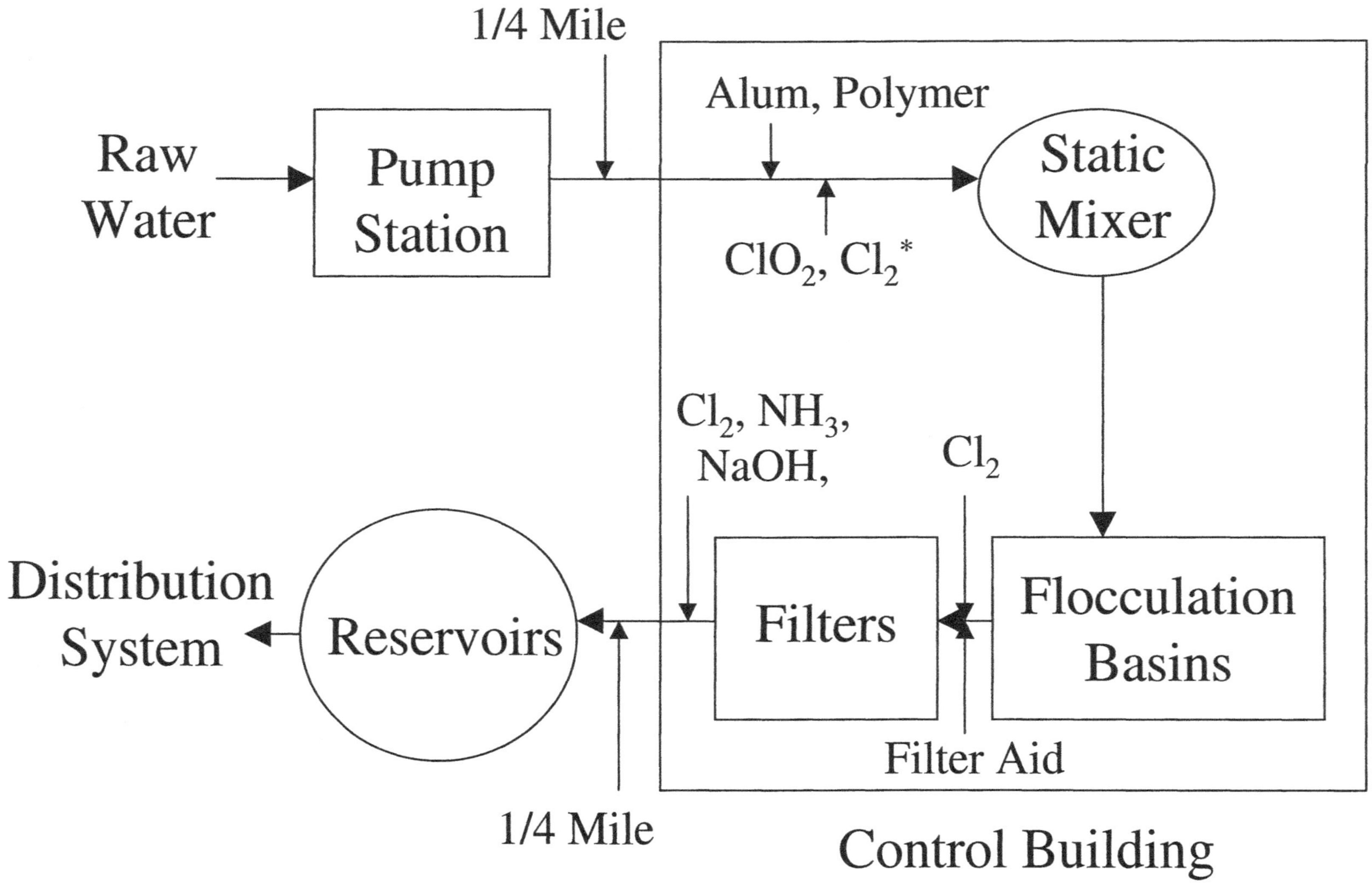

*Present in the chlorine dioxide feed solution

Figure 3.25 Judy Reservoir WTP: before modifications (1991)

to form chloramines. The finished water was then put in treated-water storage in a covered finished-water reservoir before distribution.

Judy Reservoir WTP: After Treatment Modifications

Figure 3.26 shows a schematic of the current treatment plant after modifications. Raw water is now treated with aluminum chlorohydrate (1.0-3.0 mg/L, average = 2.0 mg/L) and cationic polymer as the primary coagulants. Carbon dioxide is added to depress the coagulation pH to 6.4-6.6 (average = 6.5). The pH of the filtered water is not raised with sodium hydroxide until the finished-water reservoir influent and effluent. The pH is raised to 8.5-8.8 (starting in 1995).

Primary disinfection is attained by applying chlorine dioxide and free chlorine to the raw water and additional chlorine to the (water after flocculation or) filtered water. (There is a free chlorine residual in the chlorine dioxide feed solution. Conventional chlorine dioxide generators are not 100-percent efficient. Nonetheless, small amounts of free chlorine will only produce minimal concentrations of DBPs.) Between the filters and the finished-water reservoir (a ¼-mile distance), the filtered water is disinfected with free chlorine at a suppressed pH of 6.3. The travel (contact) time in the ¼-mile pipe varies with plant flow.

Ammonia to form chloramines and caustic are not added until the finished-water reservoir influent or effluent. These chemicals can be added at either location in order to provide operational flexibility. The location of ammonia addition is based on *CT* considerations (e.g., it can be added at the finished-water-reservoir effluent during the winter because extra contact time is needed with colder temperature). This is important because it allows the system to increase contact time in the winter for inactivation and to reduce contact time in the summer for DBP control.

Implementational and Operational Issues

With the initial disinfection scenario of chlorine dioxide and chlorine for primary and secondary disinfection, respectively, Skagit Co. PUD had some taste-and-odor problems in

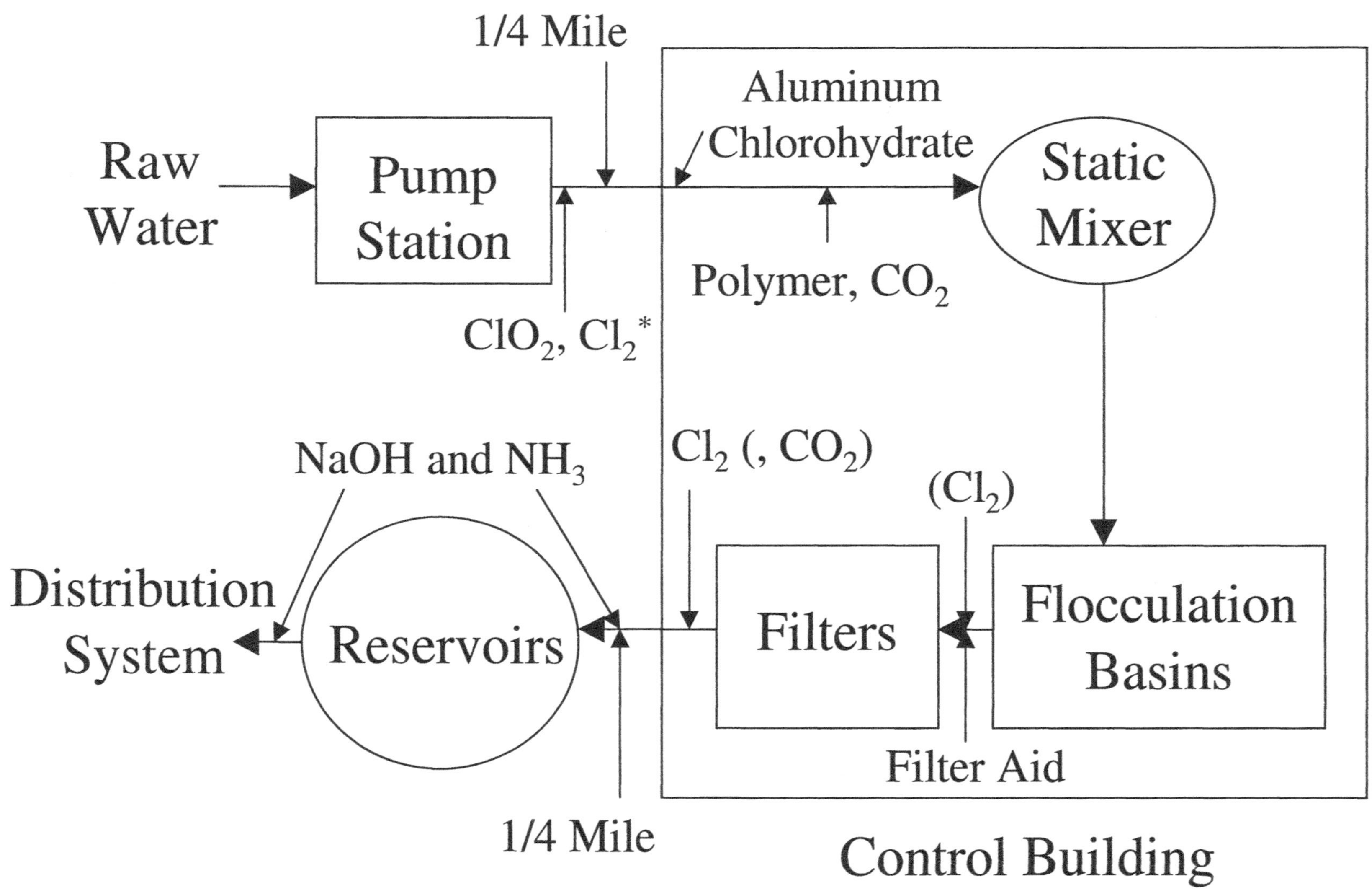

*Present in the chlorine dioxide feed solution

Figure 3.26 Judy Reservoir WTP: after modifications (current)

certain parts of the distribution system. At that time, chlorine was being added in the distribution system at booster stations. The taste-and-odor problem was eliminated, in part, with the switch to chloramines for secondary disinfection. In addition, the chlorine pumps in the distribution system were turned off. In other AwwaRF research, it has been found that taste-and-odor problems can be generated in the chlorinated distribution system of systems that use chlorine dioxide (Dietrich and Hoehn 1991). In that AwwaRF study, the researchers found that chlorite from the chlorine dioxide could react with free chlorine (not chloramines) in the distribution system to reform chlorine dioxide. The reformed chlorine dioxide was found to react with new carpeting in the homes of consumers to produce very objectionable odors.

The filtration and disinfection processes and the maintenance of a detectable chloramine residual in the distribution system have been effective in preventing coliform problems. There is no excess ammonia in the treated water leaving the WTP.

The only significant problem that was experienced in implementing these modifications was an accidental ammonia overdose (January 1995) due to a mechanical failure. In addition, chlorine was used in the distribution system in December 1997 while a leaky, fiberglass ammonia storage tank was being repaired. The only significant "trade-off" that was experienced in implementing these modifications was that the SWTR *CT* requirements required changes in the disinfection practices that resulted in THMs increasing from ~5 μg/L to ~40 μg/L.

Caustic storage was not an issue, even though there was an increased need due to carbon dioxide addition. The increased distribution-system pH (e.g., 8.5-8.8) did not present a problem. Many utilities in the U.S.—especially softening plants—distribute water with a pH ~9.

Nitrification has not been a problem in the distribution system since going to chloramines. There were no problems from the public when they switched to chloramines.

System Performance

Water Quality

TOC: The influent TOC has typically varied from 2.5 to 7.0 mg/L, with an average of 4.3 mg/L. In general, the character of the TOC (UV, THMFP) was the same before and after the modifications. TOC removal before modifications ranged from 5 to 30 percent, with an average value of 15 percent. TOC removal after modifications—i.e., different coagulant and pH depression with carbon dioxide—has typically been from 26 to 50 percent (average = 36 percent) (Figure 3.27). In Figure 3.27, the increase in TOC removal starting in October 1995 was probably due to a combination of lower-pH coagulation and the treatment of a higher TOC in the influent water, as TOC removal tends to increase with increasing influent TOC concentration. Although TOC in the influent increased in concentration over the time period in Figure 3.27, TOC variations can vary from year to year.

Turbidity: The raw water had turbidity levels ranging from 0.2 to 3 NTU, with an average of 0.5 NTU. The finished water had an average turbidity value of 0.04 NTU (range = 0.02-0.07 NTU).

Alkalinity and pH: The raw-water pH has ranged from 6.8 to 7.5 (average = 7.2). The finished-water pH has been increased (in 1993 to 1994 to 1995) from 7.3-8.0 to 8.0-8.5 to 8.5-8.8 as part of their corrosion-control program. The raw-water alkalinity was 14 mg/L as $CaCO_3$ (on average). Finished-water alkalinity has ranged from 21 to 28 mg/L (average = 26 mg/L). Due to the low alkalinity of the water, the addition of chlorine can depress the pH (i.e., poor buffering capacity).

THMs: In 1988-89—when Skagit Co. PUD was an unfiltered surface-water system with chlorine disinfection—THM concentrations in the distribution system had ranged from 75 to 286 μg/L. With the construction of the WTP—which was initially operated with chlorine dioxide and chlorine for primary and secondary disinfection, respectively—THMs still exceeded the MCL. In 1992—after the secondary disinfectant was changed to chloramines—THM concentrations in the distribution system averaged from 5.7 to 13.6 μg/L. In 1993—after the ammonia and caustic soda injection points were moved downstream of the chlorine addition point—THMs were increased to >40 μg/L in the distribution system.

In 1996, carbon dioxide was added ahead of the flocculation basins and the finished-water pH was increased to 8.5-8.8. In 1996, THM concentrations were typically <60 μg/L (often <50 μg/L) (Figure 3.28). However, in 1998-99, the average value of the THMs in the distribution system was 33 μg/L (the average value in the distribution system in each monthly sampling period was typically 24 to 39 μg/L) (Figure 3.28). In the 1996-99 time period, the only significant increase in THM formation in the distribution system was observed in December 1997 when chloramines were not used.

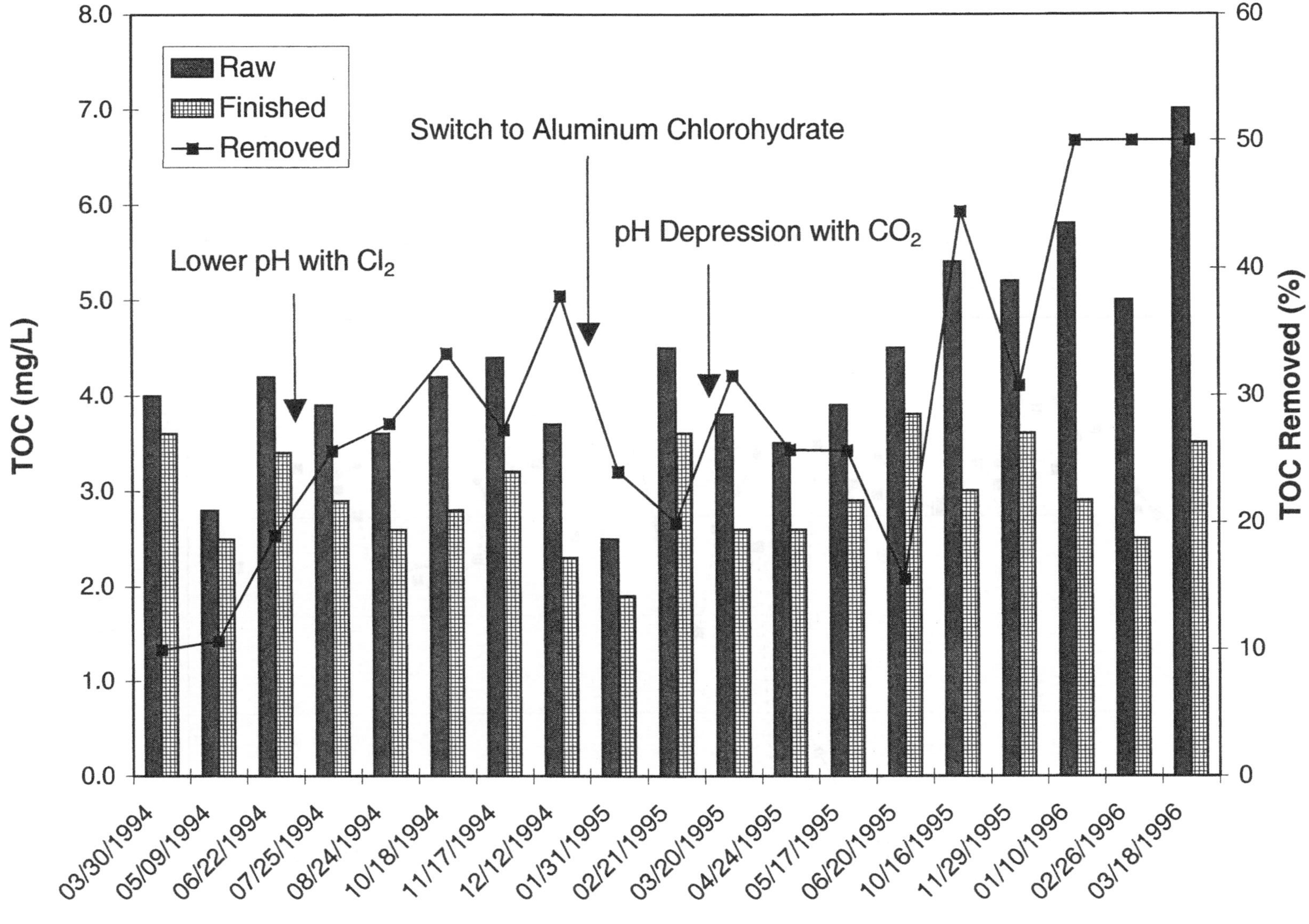

Figure 3.27 Effect of treatment modifications on TOC removal at Judy Reservoir WTP

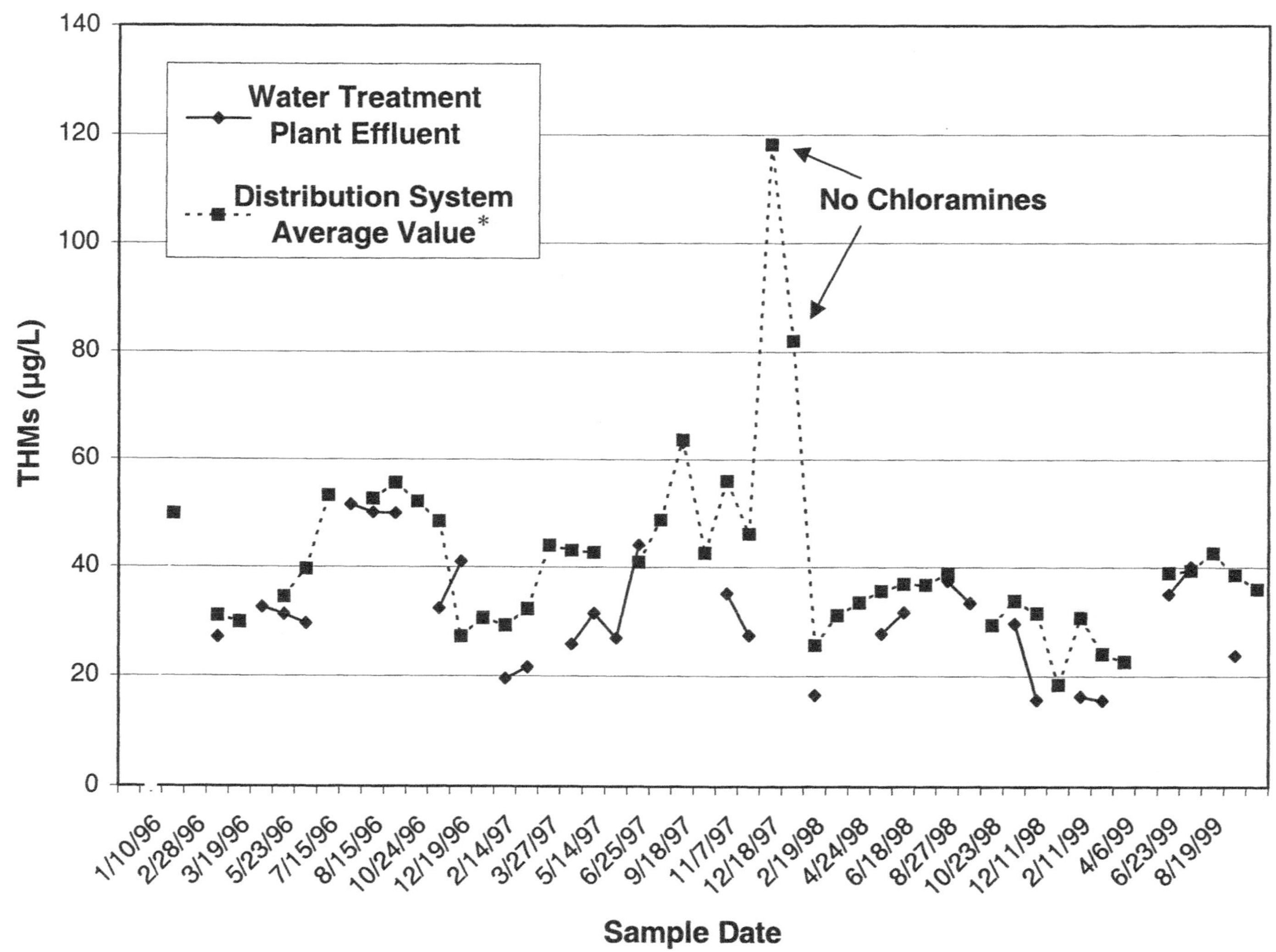

*Average of all distribution-system samples collected on a particular date

Figure 3.28 THM occurrence at Judy Reservoir WTP and in the distribution system, 1996-1999

HAAs: In 1988-89—when the finished-water (at the beginning of the distribution system) had THM concentrations of 24-44 µg/L—the finished-water HAA concentrations were 46-57 µg/L. (At that time, Skagit Co. PUD was part of a 35-utility DBP study that included HAA monitoring [Krasner et al. 1989].) There were no HAA data collected in the distribution system at that time, but HAAs were probably quite high because the THMs in the distribution system were greater than the 100 µg/L MCL. After treatment modifications, HAAs had initially typically ranged from ~40 to ~60 µg/L, with an average of ~50 µg/L in the distribution system (Figure 3.29).

In 1998-99, the average value of the HAAs in the distribution system was 39 µg/L (the average value in the distribution system in each monthly sampling period was typically 27 to 52 µg/L) (Figure 3.29). In this low-bromide water, dichloro- and trichloroacetic acid were the major HAAs produced. Thus, the sums of HAAs with either five or six species—which includes all of the species with chlorine only—have been quite similar.

Chlorite and chlorate: In the finished-water, chlorite has ranged from 0.4 to 0.5 mg/L and chlorate has averaged 0.2 mg/L.

Summary of System Performance

The current treatment and disinfection scenario at Skagit Co. PUD is allowing for simultaneous compliance with the *CT* requirements of the SWTR, corrosion-control requirements with the LCR, and the Stage 1 MCLs for THMs, HAAs, and chlorite. In addition, the plant has been achieving good TOC removal.

OZONE: ANN ARBOR WATER UTILITIES

Summary

The City of Ann Arbor operates a two-stage lime softening plant (50-mgd design flow), which treats a blend of surface water and groundwater. The influent water is hard with moderate-to-high concentrations of TOC. The driving forces in the decision to switch from chloramines to ozone for primary disinfection included: (1) Ozonation would meet the SWTR

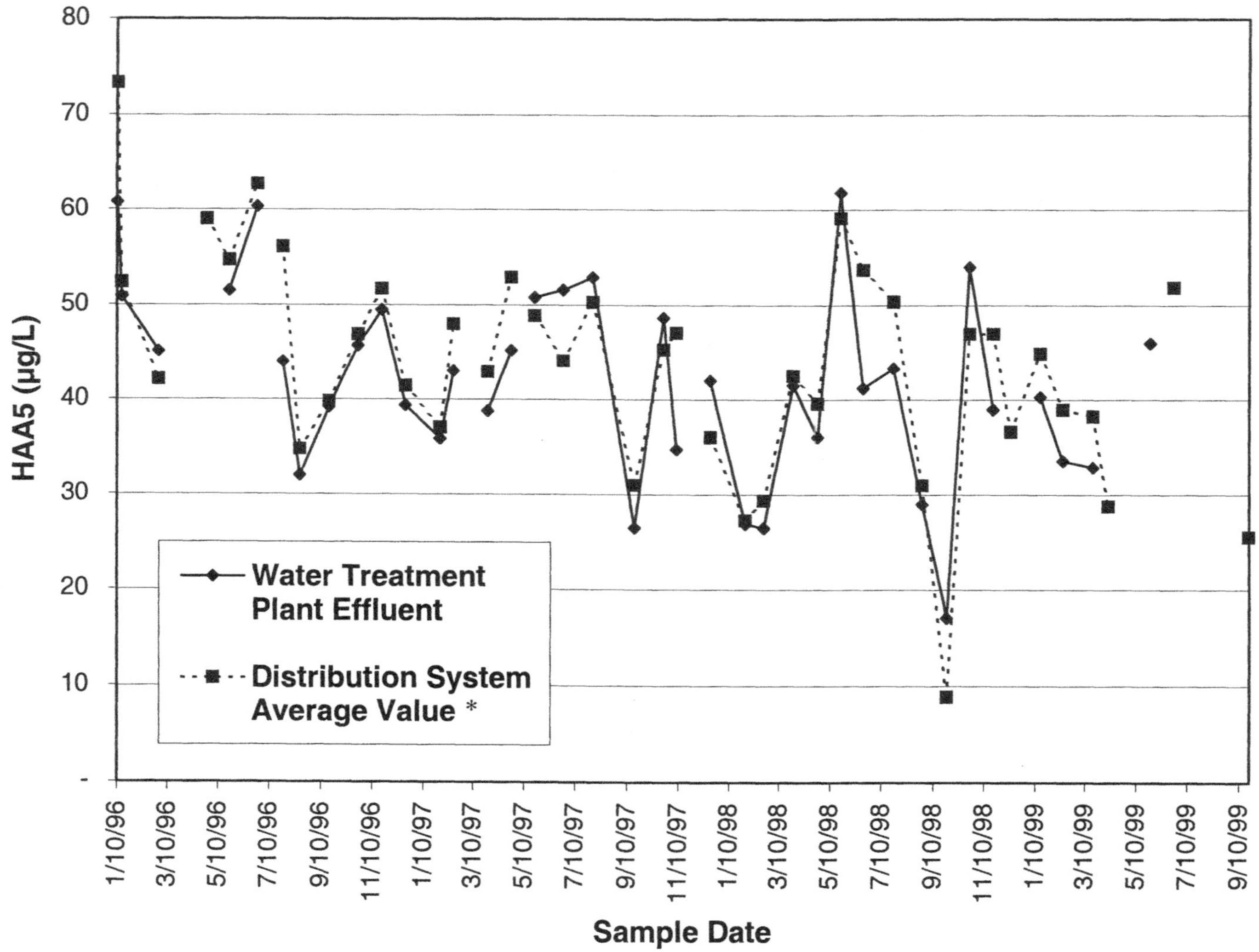

*Average of all distribution-system samples collected on a particular date

Figure 3.29 HAA5 occurrence at Judy Reservoir WTP and in the distribution system, 1996-1999

CT requirements for viruses at low temperatures. (2) Ozonation was expected to satisfy future anticipated DBP regulations. (3) Ozonation was expected to improve taste and odor. (4) Ozone was the most desirable alternative based on cost and space constraints.

The process design was influenced by pathogens, DBPs, tastes and odors, and water temperature. Ozonation of softened water before filtration was selected because the disinfection criteria could be met at the lowest overall cost. Ozonation at pH 8.0 (rather than at the softened water pH) was chosen to reduce the amount of ozone required, control the formation of some DBPs, and lower costs. The final design included the option to add hydrogen peroxide in the event ozone alone was not able to control taste-and-odor episodes. Biological filtration (GAC/sand filters) and the addition of chloramines to the filtered water further addressed the issues of DBPs and microorganisms. Extreme seasonal variations in temperature between winter and summer influenced the treatment design and operating parameters at Ann Arbor. The ozone retrofit cost approximately $7.5 million. Annual costs for electricity, chemicals, and natural gas have increased due to the modifications.

Some of the key facts that were highlighted during this case study are: (1) This decision process was being undertaken at a time (1989-90) when ozone was just beginning to dawn as an accepted technology in the U.S. and when many engineers were projecting that the combination of microbial and DBP regulations would push a large number of plants to ozone. (2) Extensive pilot-plant studies of ozonation were carried out to address major areas of uncertainty, which included: seasonal swings in temperature and raw water quality, optimizing pH control, ozone dosage, and points of ozone application. (3) Ozonation resulted in low levels of THMs and HAAs, allowed Ann Arbor to meet the virus and *Giardia* requirements as specified in the SWTR, and improved taste and odor in a cost-effective manner. (4) One of the unanticipated issues in the Ann Arbor case study was an increased number of interferences in the total coliform testing due to the emergence of a non-coliform bacteria of some variety. This resulted in a need for many unnecessary confirmation samples. Changes in bacterial fauna may be an issue to be expected when utilities move to ozone. Ann Arbor has been optimizing the use of biological filtration to control AOC to limit regrowth.

The treatment strategy significantly changed the operational needs of the plant and additional staff were needed. Provisions for training the staff had been included in the construction contract. It took the treatment plant a period of three to five months to operate like an "integral unit."

Before treatment modifications, the TTHMs ranged from 14 to 39 μg/L and HAA5 ranged from 4 to 21 μg/L during the case-study period. This low level of DBP formation was due to the use of chloramines as the primary and secondary disinfectant. Chloramines were not preformed; chlorine and ammonia were added sequentially. The short free chlorine contact time before ammonia addition minimized DBP formation, but was inadequate for obtaining virus *CT* credit throughout the year.

After treatment modifications, with ozonation and chloramination, the TTHM levels ranged from 1.4 to 13 μg/L and HAA5 ranged from 1.5 to 15 μg/L. The bromate levels ranged from 2 to 8 μg/L for the sampling period studied. The levels of HPC bacteria in the distribution system increased (on average) from 105 to 237 cfu/mL after the introduction of ozonation. Four distribution locations, all at long detention time, accounted for the majority of this increase. The closest locations actually have been showing a decrease. The interference rate on the total coliform test for distribution samples increased from an annual average of 1.4 to 6.7 percent.

The Ann Arbor case study demonstrated how ozonation could successfully be used to meet its diverse multiple water quality objectives while operating within cost and space constraints. Primary disinfection with ozonation produced a high quality effluent water that met both the disinfection criteria and the DBP goals of the plant. However, current levels of bromate (less than the Stage 1 MCL) are in a range that could be problematic under some Stage 2 regulatory scenarios. In addition, ozonation introduced new challenges in maintaining and monitoring the microbial quality of the water in the distribution system.

Background

Case History

The City of Ann Arbor operates the Ann Arbor WTP and serves approximately 115,000 people. The influent to the plant consists of a blend of Huron River water (typically 85 percent) and well water (15 percent). The treatment plant is a two-stage lime-softening plant, which used chloramines for primary and secondary disinfection since 1981 to comply with the THM rule. In addition, they had been using GAC in their filters for the removal of tastes and odors. The treatment plant has a typical flow of 20 mgd with a design capacity of 50 mgd. The influent water is hard with moderate-to-high concentrations of TOC and low levels of bromide.

With promulgation of the SWTR in 1989 (USEPA 1989a), the disinfection (chloramination) practice was not sufficient—the treatment was not effective at meeting the *CT* requirements for viruses at low temperatures—thus mandating the need for treatment modifications. A future anticipated DBP rule was also taken into consideration while treatment modifications were being considered. Other problems experienced by the plant, prior to modifications, included nitrification episodes (prior to 1990) (Skadsen 1993) and taste-and-odor problems due to seasonal (summer) algae spikes in the river. (Note that the seasonal algal taste-and-odor problems were effectively controlled by the use of GAC with a three-year change out cycle. If they ran the GAC longer, they had taste-and-odor problems.)

After an evaluation of treatment alternatives and completion of a pilot-plant study, ozonation was selected to replace chloramines for primary disinfection. Ozone was chosen as the most desirable alternative based on cost, expected future regulations, and space constraints. The ozone retrofit cost approximately $7.5 million and ozonation was on-line in 1996. Ozonation resolved the *CT* compliance issues and taste-and-odor problems, however this treatment modification raised some downstream water quality issues. After ozonation was installed in the treatment plant, increases in HPC bacteria and interferences in coliform testing were detected at distant locations in the distribution system. The closest locations actually have been showing a decrease.

Routine monitoring for water quality parameters, DBPs, and microbial parameters was conducted before and after the treatment modifications were introduced. Fiscal year 1994/1995 (July to June) was chosen for the "before treatment modifications" data analysis and the calendar year 1997 was chosen for the "after treatment modifications" data analysis.

Process Selection

Ann Arbor conducted an "in house" evaluation of its process alternatives. Ultimately a consultant was hired to do a pilot-plant study; they also re-evaluated Ann Arbor's options.

Alternative treatment strategies that were identified for complying with both the SWTR and existing and anticipated future rules governing DBPs included:

- Increase the size of the reservoir and meet the disinfection criteria using monochloramine. Increasing the size of the reservoir was considered in order to increase contact time, no change in disinfectant concentration was considered.
- Baffle the reservoir, add chlorine to the reservoir inlet, provide multiple points for the addition of ammonia, and meet the disinfection criteria with free chlorine.
- Use additional groundwater blending to dilute the Huron River TOC.
- Use chlorine dioxide as the primary disinfectant and continue use of monochloramine as a distribution-system disinfectant.
- Use ozone as a primary disinfectant and continue use of monochloramine as a distribution-system disinfectant.

Ozone was chosen as the preferred primary disinfectant as it provides the best disinfection and overall water quality and was expected to achieve the lowest level of regulated (both current and anticipated) DBPs. In addition, ozone was expected to improve taste and odor, reduce the level of chlorine required, and extend the useful life of the GAC filter media.

An ozone pilot-plant study was conducted by Ann Arbor and CH2M Hill in 1990/1991 for a cost of $130,000. The study included two seasonal test periods (September/October and March) and examined several treatment processes. The main objective of the pilot-plant tests was to determine a cost-effective method of using ozone to meet drinking water objectives and improve water quality. Additional studies were performed at full-scale in February-March and August-September 1997 to further optimize the ozonation process (Skadsen, Myers, and Bellamy 1997).

Point of ozone addition studies: Tests were conducted to evaluate the best location to add ozone in the treatment process, the amount of ozone required, and the effect of ozone on other treatment processes and downstream water quality.

Alternative points for ozone application included: (1) the raw river water, (2) softened water (combination of river and groundwater) before filtration, and (3) both the raw river and softened water. The study found that the amount of ozone required for disinfection of raw river water (8 mg/L) was twice as much as that required for disinfection of softened water (4 mg/L) due to the high ozone demand of the source water without precursor removal. Therefore, application of ozone to the softened water before filtration was recommended because the disinfection criteria could be met at the lowest overall cost.

CT studies: The SWTR requires 3-log *Giardia* and 4-log virus inactivation and/or removal (USEPA 1989a). Because Ann Arbor uses sedimentation and filtration as part of their treatment process, they receive a credit of 2.5- and 2.0-log removal for *Giardia* and viruses, respectively. Disinfection must provide the remaining treatment (0.5-log *Giardia* and 2-log virus inactivation). The *CT* values (pH 7 to 9) necessary for Ann Arbor to achieve *Giardia* and virus inactivation with ozone are 0.5 mg-min/L in the summer (minimum temperature = 10°C) and 0.77 mg-min/L in the winter (minimum temperature = 3°C). In summer, both water flow rates and organic content are at their greatest. Therefore, the summer conditions govern the maximum daily ozone demands (Table 3.14). Even though the *CT* requirements are greater in winter, the winter flows are so low that winter conditions govern the minimum daily ozone demands (Table 3.14).

Table 3.14

Estimated ozone daily doses and demands based on pilot-plant studies at Ann Arbor (pH = 8.0)

Condition	Flow (mgd)	Dose (mg/L)	Demand (lb/day)
Maximum daily	50	3.4	1,575
Average daily	20	3.0	556
Minimum daily	10	3.2	293

Ozonation pH studies: The ozone dose required to establish a given ozone residual decreases with decreasing pH. Ozone decay rates also decrease with decreasing pH and TOC concentrations. Therefore, pH adjustment of the softened water prior to ozone addition was necessary to reduce the amount of ozone required and to control costs. In addition, it was predicted that lowering the pH (and, thus, lowering the ozone dose) would allow for control of some DBPs. Pilot-plant tests were conducted at pH levels of 7.5, 8.0, 8.3, and 8.8. Full-scale tests were performed at pH levels of 7.5, 8.0, 8.5, and 9.3. A pH of 8.0 was chosen as the optimum ozonation pH when considering both capital costs (ozone and pH-adjustment facilities) and operating costs (Skadsen, Myers, and Bellamy 1997). It was decided not to decrease the pH further because that would increase chemical costs more than the cost savings realized by lower ozone use. This was not due to concerns over corrosion in the distribution system as they raise the pH before distribution to handle these and other distribution water quality concerns.

Ozone/hydrogen peroxide studies: A combination of ozone and hydrogen peroxide was tested to see if taste-and-odor control and organic compound removal were improved over ozone treatment alone. The addition of hydrogen peroxide increased the ozone demand and accelerated the ozone decay rate, necessitating higher ozone dosages. The downstream effects of ozone/hydrogen peroxide treatment were similar to those when ozone was used alone. Hence, there was no perceived advantage to using ozone in conjunction with hydrogen peroxide. In the final design, the option to add hydrogen peroxide later was included but not installed.

Type of ozone system studies: Several alternative ozone systems—air feed, oxygen feed generated onsite, oxygen feed using purchased liquid oxygen (LOX)—were evaluated to meet the ozone demands. In addition, low- and medium-frequency ozone generators were considered.

The system chosen was LOX-based, using medium-frequency ozone generators, as it had the lowest operator maintenance requirements, required the least space, had the lowest initial capital cost, and had an overall noise level lower than an air-feed system.

GAC studies: Ann Arbor had been using GAC in their filters for the removal of tastes and odors. Experience at other plants indicates that ozone can increase the life of GAC for taste-and-odor removal, thus reducing costs for GAC regeneration. In terms of the filtration process, pilot-plant studies indicated that biological activity on the filters did not significantly increase filter head loss or effluent turbidity. Although ozonation increases the AOC level of the water, GAC filtration (in the pilot-plant studies) reduced the AOC by an average of 76 percent (to values equal to or lower than those without ozone). However, the ozonated GAC filter effluent averaged five times higher HPC levels than unozonated GAC filter effluents in pilot tests. A short free chlorine contact time (1 to 2 minutes) prior to ammonia addition was required to effectively lower the bacteria levels.

Parameters that Influenced Process Design

The parameters that influenced process design were pathogens, DBPs, tastes and odors, and water temperature.

Pathogens: The SWTR requires 99.9 percent removal and/or inactivation of *Giardia* cysts and 99.99 percent removal and/or inactivation of viruses (USEPA 1989a). Chloramination at Ann Arbor was not effective at meeting the virus requirement. Ozonation was primarily selected to meet the SWTR requirements. An added benefit of ozonation is that it may also offer future protection against *Cryptosporidium* depending on *CT* requirements and what the currently designed contactors can provide.

DBPs: One of the major goals of implementing ozonation at Ann Arbor was to ensure that the disinfection changes maintained the low levels of DBPs that they were obtaining with the chloramination disinfection scheme.

Tastes and odors: Algae present in the Huron River during summer has led to taste-and-odor episodes in Ann Arbor's water. Ozonation was also selected to better resolve this problem.

Water temperature: Extreme seasonal variations in temperature between the winter and summer have influenced the treatment design and operating parameters at Ann Arbor. The rate of ozone decay was 30 to 45 percent slower in the winter (pH = 8) compared to summer (average water temperature = 25°C), which may have been due to colder water temperatures (6-8°C) slowing down ozone decay.

The Decision-Making Process

The promulgation of the SWTR in 1989 with its *CT* requirements (USEPA 1989a) provided the impetus for treatment modification talks at the Ann Arbor Treatment Plant. The decision-making process, which took place in 1989-90 was an informal "in house" consensus process and involved a number of key individuals from the treatment plant. Cost and feasibility issues were evaluated on paper and ozone was selected for primary disinfection.

The decision emerged as "handwriting on the wall" from consideration of the imperative need to comply with the SWTR contrasted against the lack of viable alternatives. The construction of additional contact basins for detention time to maintain the existing chloramine strategy was ruled out due to space constraints. The addition of chlorine to the reservoir inlet was never seriously considered by the plant staff due to past experiences with THMs and free chlorine. Consideration was given to additional groundwater blending, but the Huron River TOC is so reactive that a dilution strategy would have required almost a complete switch to groundwater during some seasons of the year. In addition, for the groundwater alternative, they considered softening costs (it is much harder than the river), transportation costs (i.e., new pipe line[s]), and the unknown availability of groundwater sources sufficient to produce the required flow rates. Chlorine dioxide was considered operationally risky and possibly not clear of regulatory concerns. UV and membranes were not considered mainstream options. This decision process was being undertaken at a time when ozone was just beginning to dawn as an accepted technology in the U.S. and when many engineers were projecting that the combination of microbial and DBP regulations would push a large number of plants to ozone (Ferguson, Gramith, and McGuire 1991). Two ozone plants were already operational in Michigan.

Interestingly, these options were evaluated and a tentative selection of ozone was made at the plant staff level without working through any more of the details. The "handwriting on the wall" was plain enough. After reaching an early decision at the plant level, pilot-plant studies were performed to work out the details and costs of ozone application. Major areas of uncertainty to be addressed by the pilot-plant work included: seasonal swings in temperature and raw water quality, optimizing pH control, ozone dosage, and points of ozone application. All of these questions had to be answered to assess the cost of applying ozone.

The pilot-plant work, conducted in 1991, evaluated a worst-case scenario for microbial inactivation that assumed a 2.0-log level could be eventually required. Although *Cryptosporidium* was recognized by plant staff as a concern, the 2.0-log assumption was unrelated to any specific organism. However, *CT* requirements for viruses at low temperatures were a driving force. The pilot study was designed to evaluate TTHMs as a group. Aldehydes were also measured because they were known to be by-products that might be an eventual concern (Jacangelo et al. 1989). However, bromate and HAAs were not on everyone's radar screen at this point in time and were not covered in the pilot-plant work. Bromate was identified as a possible concern, but they lacked sufficient information on its formation, concentration, testing methodology, and potential regulations to make any sort of informed decision.

Once the pilot-plant work was done, a firm proposal was put forward for a $15 million expenditure for installation of ozone and renovation of the WTP. But the executive management of the utility did not put the proposal for the ozone facility bond issue before the city council right away. In fact a year and a half went by and council action was finally triggered by SWTR compliance issues that resulted in a formal consent order. Over this period, the utility management was attempting to implement a gradual rate increase program to fund the project rather than having to increase rates in a single sudden change. A public hearing held on the proposal to implement ozone as part of the response to the consent order attracted one attendee from the general public.

Driving Forces

The single most important driving force to install ozone was that the plant would meet the *CT* requirements for viruses and *Giardia* as laid out in the SWTR (USEPA 1989a) and at the same time be in compliance with present and future proposed DBP regulations (Figure 3.30). An added benefit of ozone is that it may provide future protection against *Cryptosporidium*.

A second benefit of ozonation was that it would resolve any taste-and-odor issues present in the Huron River water. Because of taste and odor problems, Ann Arbor had already installed GAC in all filters in 1989. The use of GAC in concert with ozone was envisioned to be a good combination for meeting their treatment objectives.

One of the major driving forces in the decision to go to ozone was cost and space. There simply was no room to expand detention time enough to stay with the chloramine strategy. Ozone implementation cost less than increasing the size of the finished-water reservoir to meet disinfection criteria using chloramines. Space constraints were one of the drivers in the decision to use LOX. Even fitting in the ozone plant was difficult.

Constraints

One of the major constraints faced by the decision makers was balancing the SWTR and anticipated DBP rule and enhanced SWTR, which greatly constrained treatment choices.

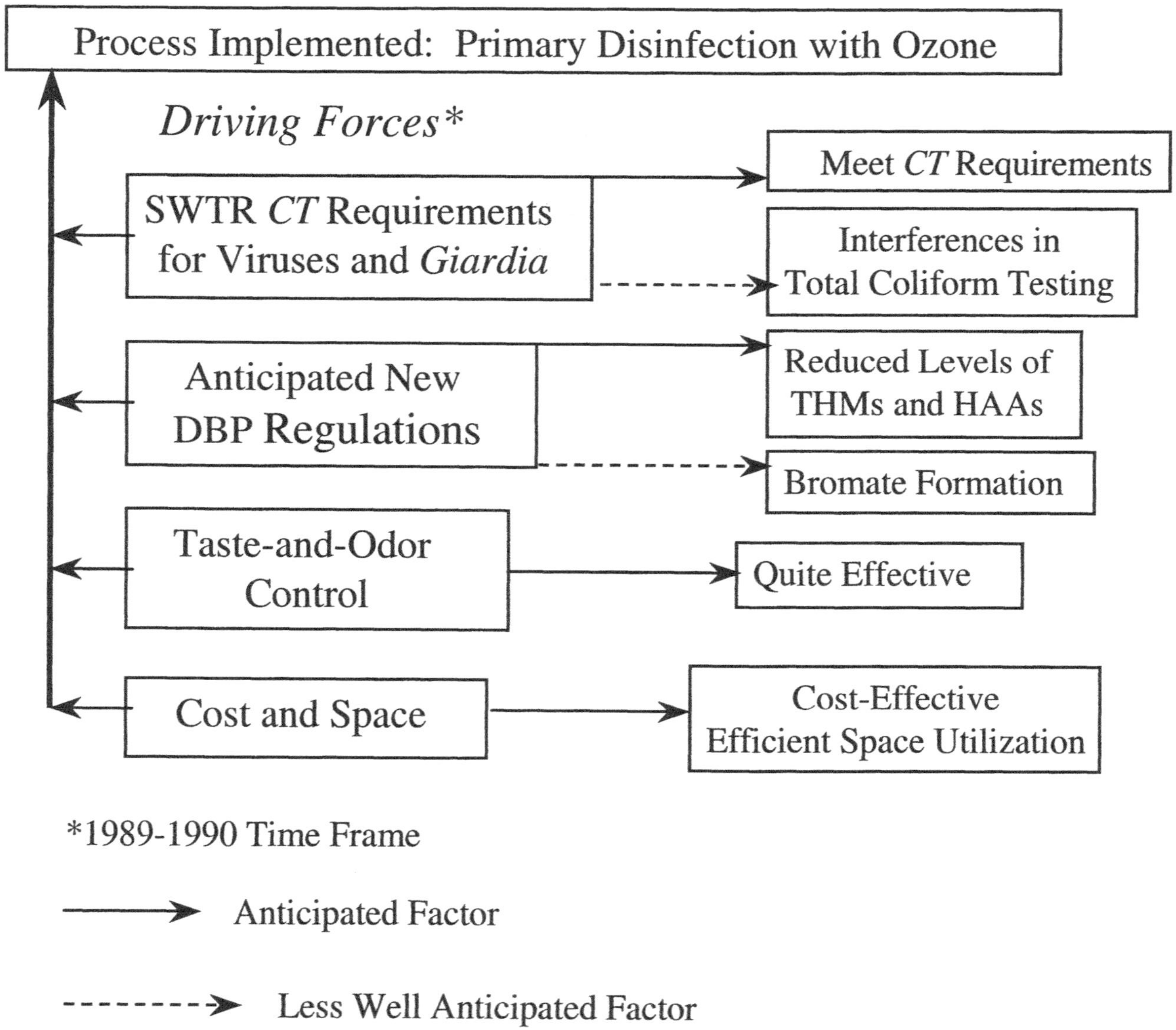

Figure 3.30 The decision-making process at Ann Arbor: switched from chloramines to ozone for primary disinfection

Another major constraint was the lack of physical space at the plant site. The constraints on financial resources played a role in not expediting implementation.

Technology Checklist—Anticipated Factors

The City of Ann Arbor correctly anticipated that ozone application would result in low levels of chlorination DBPs. THM concentrations were, in fact, lower than predicted. The pilot-plant work predicted that ozone would produce DBP levels comparable to those of the chloramines strategy, but full-scale results are producing much lower levels. They had also correctly predicted the types of organic ozonation DBPs (aldehydes) they were likely to find.

They had correctly anticipated the fact that ozonation would enable them to comply with all future anticipated DBP regulations with the exception of bromate, which remains a central concern. Current levels of bromate are in a range that could put this plant in jeopardy under some regulatory scenarios.

Ann Arbor correctly anticipated that ozonation would provide them with adequate pathogen control, resolve their virus *CT* issues and allow them to be in compliance with the SWTR and the recently promulgated interim enhanced SWTR (USEPA 1998b). They designed redundancy and capability to provide a high enough ozone dose for future *Cryptosporidium* regulations.

Although ozone improved the water taste, it was subtle. However, ozonation did provide Ann Arbor with a comfort level that they would no longer risk any taste-and-odor breakthrough with the GAC. In addition, full-scale experiences have not indicated any concerns over increases in GAC filter-effluent HPC bacteria with or without ozone.

They correctly predicted the operating costs to run an ozone system. They designed dual electrical feed points and are planning to add a generator backup.

Technology Checklist—Less Well Anticipated Factors

One of the unanticipated issues in the Ann Arbor case study was an increased number of interferences in the total coliform testing. They currently utilize a presence/absence media that has been exhibiting turbid interference reactions due to the emergence of a non-coliform bacteria of some variety. This resulted in a need for many unnecessary confirmation samples. The same effect is seen in the membrane filter method. It is hypothesized that there is simply one or more microorganisms that are causing this interference. Switching to the colilert method is one strategy they are considering. It is concluded that changes in bacterial fauna are an issue to be expected in moving to ozone. This has been highlighted in other AwwaRF studies. Interestingly, the interference rate has declined signficantly in the last year.

A number of process-control issues were not well anticipated. In addition, they did not anticipate leaks, the lack of adequate water quality for cooling or sludge accumulation.

Compliance Issues

Implementation of ozonation resolved compliance problems with the SWTR, in particular, the virus *CT* requirements (Figure 3.31). With ozonation, the Ann Arbor Water Plant is producing THMs and HAAs below the previously proposed placeholder Stage 2 MCLs (USEPA 1994a). Current levels of bromate are below the Stage 1 MCL (USEPA 1998a), but may not meet a future Stage 2 MCL under some regulatory scenarios. In addition, high levels of HPC bacteria have been found at distant locations in the distribution system and an increase in interferences has been seen in the coliform testing. However, the closest locations actually have been showing a decrease.

Key Players

The decision-making process involved a variety of water treatment plant staff in a team process. Consultants from CH2M Hill conducted the pilot-plant study and prepared the preliminary design and cost estimates. The city council accepted staff recommendations. Various efforts were made to inform the public about the treatment modifications through a public meeting, newspaper articles, and radio interviews. The public was supportive and accepting of the rate increase.

Demographic information from 1990 were obtained. The median household income was \$34,594, whereas the average household income was \$44,916. The median cost of housing was \$111,996. The median age of the population was 28.5 years. Approximately 60 percent of the

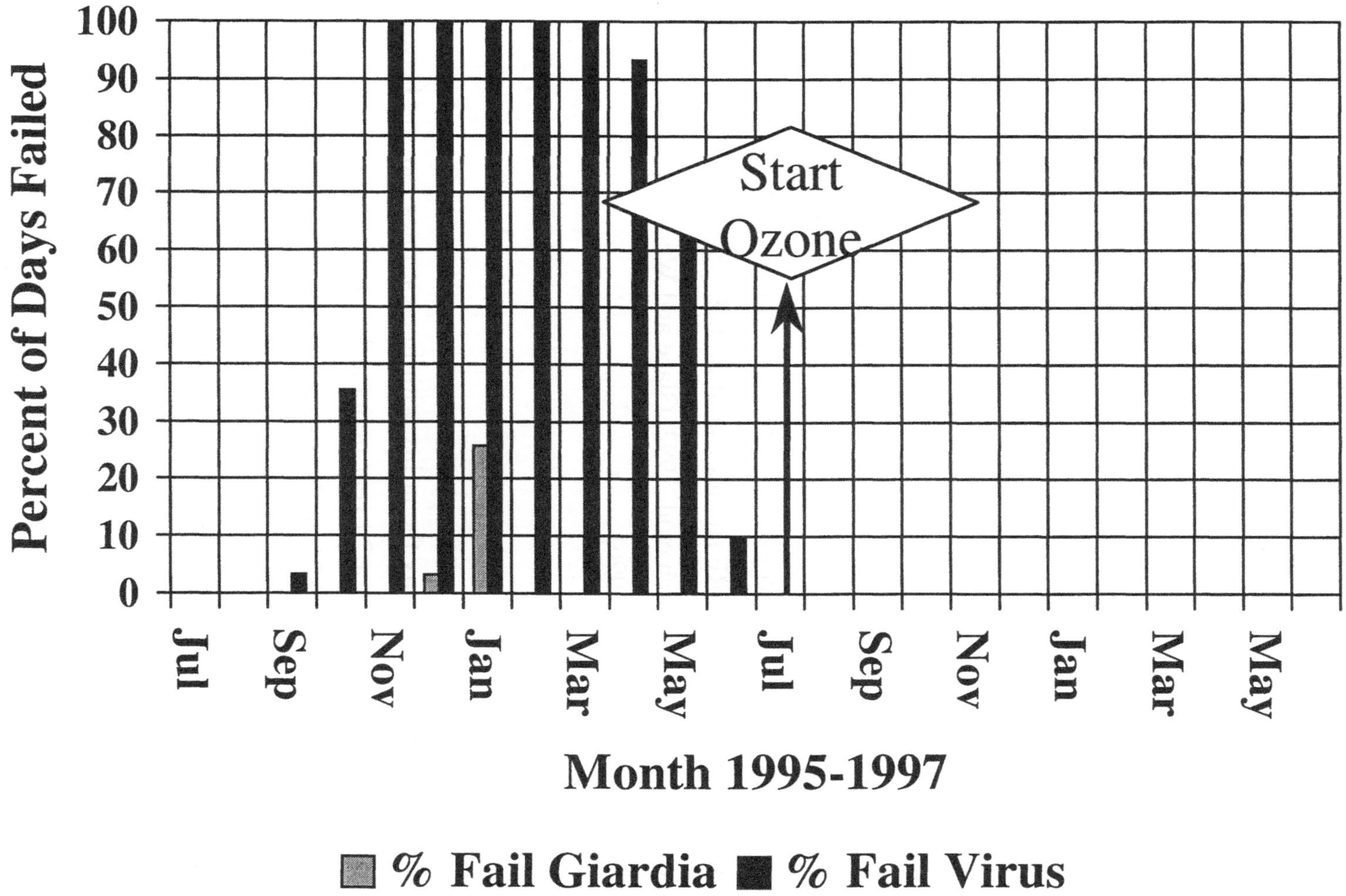

Figure 3.31 Percent *CT* failure before and after ozone addition at Ann Arbor

population was between the ages of 18 and 44, and 81 percent of the population was Caucasian. For the population 25 years or older, 59 percent had obtained a bachelor's degree or higher.

Selection Issues and Future Strategies

The City of Ann Arbor has been satisfied with the results of the ozonation process. The only thing they would have done differently would have been to make some minor modifications in the design.

Ann Arbor has been conducting experiments on feedpoints (in-line injection versus diffusers) to optimize the system. In addition, experiments on the effect of pH on ozone application were conducted. Ann Arbor has been developing a model for the determination of *CT* credit. Other research include an examination of the impacts of ozonation on distribution-system water quality, use of off-gas recycle and impact on ozone feed, and methods (chemical addition) to improve secondary settling to reduce the sludge impact on the ozone system.

The staff found a number of AwwaRF studies useful in their post decision-making phase of implementation (Dyksen et al. 1992, Finch et al. 1992, Glaze and Weinberg 1993, Finch et al. 1994, Bellamy 1995, Becker and O'Melia 1996).

Economic Analysis

Water revenue bonds were floated by the utility with interest rates ranging between 5.4 and 7.4 percent with payments spread over a 20-year period. The water rates were also slowly increased to fund the project. A total of $17.1 million were issued of which $15.9 million was attributable to the ozone/WTP renovation. The ozone portion of the project (ozone generators, ozone contactor, LOX tank and vaporizers, lift station, CO_2 tank, CO_2 contactor, cooling system, destruct system, modifications to the electrical system) accounted for approximately $7.5 million and the remaining money was used for plant improvements.

Annual costs for electricity, chemicals, and natural gas have increased due to the modifications (Table 3.15). An additional cost is a labor cost ($70,000 x 1.67) due to the hiring of two additional staff members (an electrician and an instrument technician).

Table 3.15

Annual costs at the Ann Arbor WTP before and after modifications

	Before modifications ($/MG)*	After modifications ($/MG)†
Electricity	53	73
Chemicals	73	91
Natural gas and miscellaneous	4.0	28

*1994-95: 5895 MG delivered.

†1996-97: 6102 MG delivered.

The annual electrical consumption at the water treatment plant increased from 4,867,920 kW•h to 7,235,200 kW•h with the implementation of ozonation. The annual natural gas usage at the water treatment plant increased from 57,584 HCuFT to 374,050 HCuFT after the treatment modifications. The main reason for the natural gas increase was the amount of new space added (administration, conference room, ozone building, sodium hydroxide storage, laboratory, maintenance building). Lime is one of the major chemical cost items; however, its usage and cost did not change (annual averages of $40/MG and $43/MG). Likewise, post-chlorine usage and cost did not change (annual average of $13/MG); whereas, pre-chloramination (annual average cost of $2.7/MG) was eliminated with the implementation of ozonation. The implementation of ozonation introduced chemical costs for liquid oxygen (annual average of $8.5/MG) and sodium hydroxide (for pH adjustment after ozonation) (annual average of $8.8/MG). Other chemical costs (ammonia, phosphate, coagulant, liquid carbon dioxide) were relatively comparable before and after the modifications.

The ozone dose required to reach the same *CT* value increases with increasing ozonation pH because ozone demand and decay rates increase with pH (e.g., at approximately 9°C, 2.6 to 5.8 mg/L of ozone was required to achieve a *CT* of approximately 2 mg-min/L at pH 7.5 to 9.3) (Skadsen, Myers, and Bellamy 1997). The sodium hydroxide dose required to raise the pH to 9.3 after ozonation increased as the ozonation pH decreased (e.g., 1.3 to 3.4 mg/L of sodium hydroxide was required for ozonation pH levels of 8.5 to 7.5). The cost of chemicals (ozone, carbon dioxide, and sodium hydroxide) for ozonation pH experiments at 9 and 20°C are shown in Figures 3.32 and 3.33, respectively (Skadsen, Myers, and Bellamy 1997). (Some of the chemical dosages did not follow the predicted trends, most likely due to the impact of the

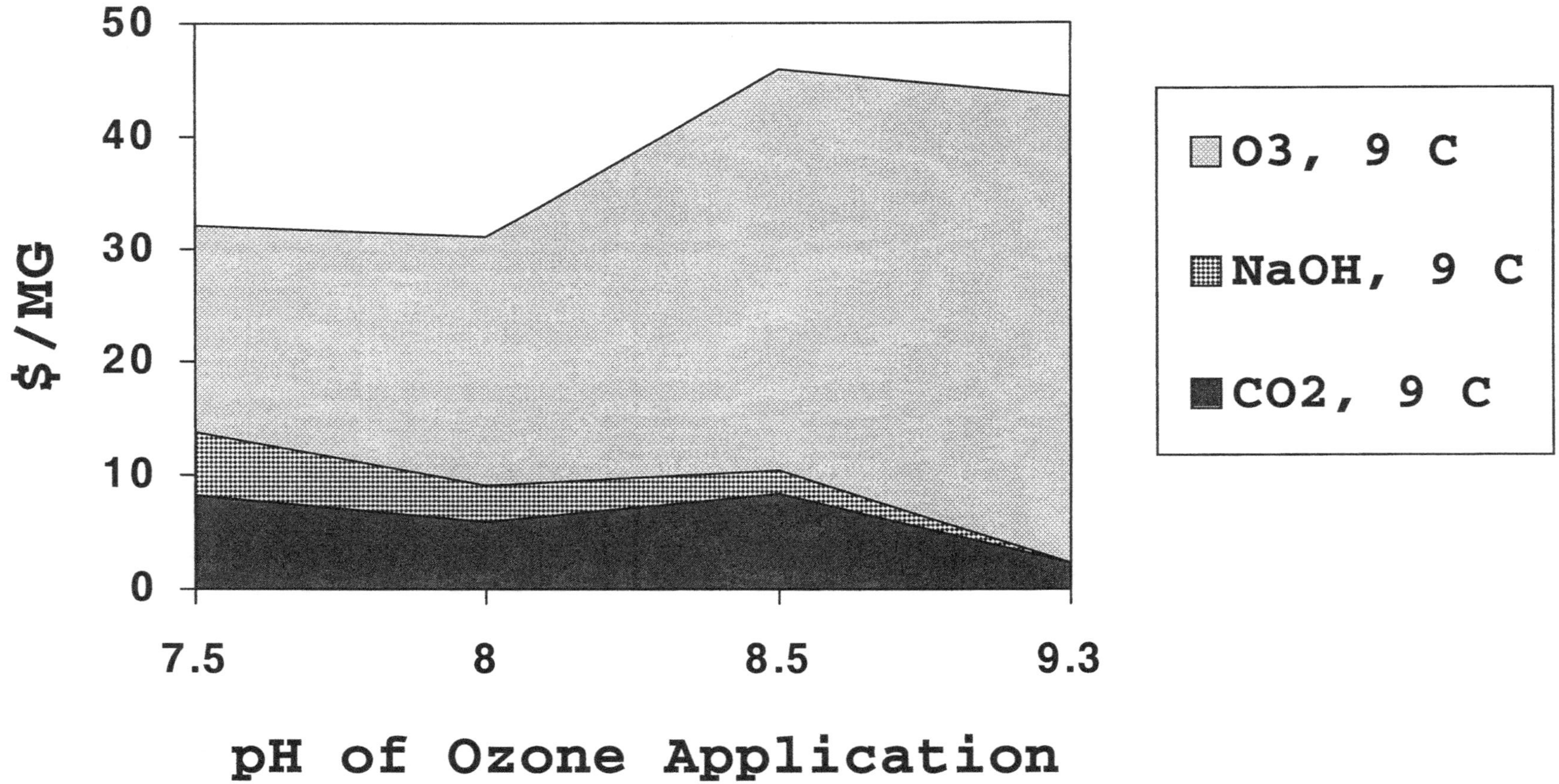

Figure 3.32 Chemical costs at Ann Arbor: pH effects at 9°C

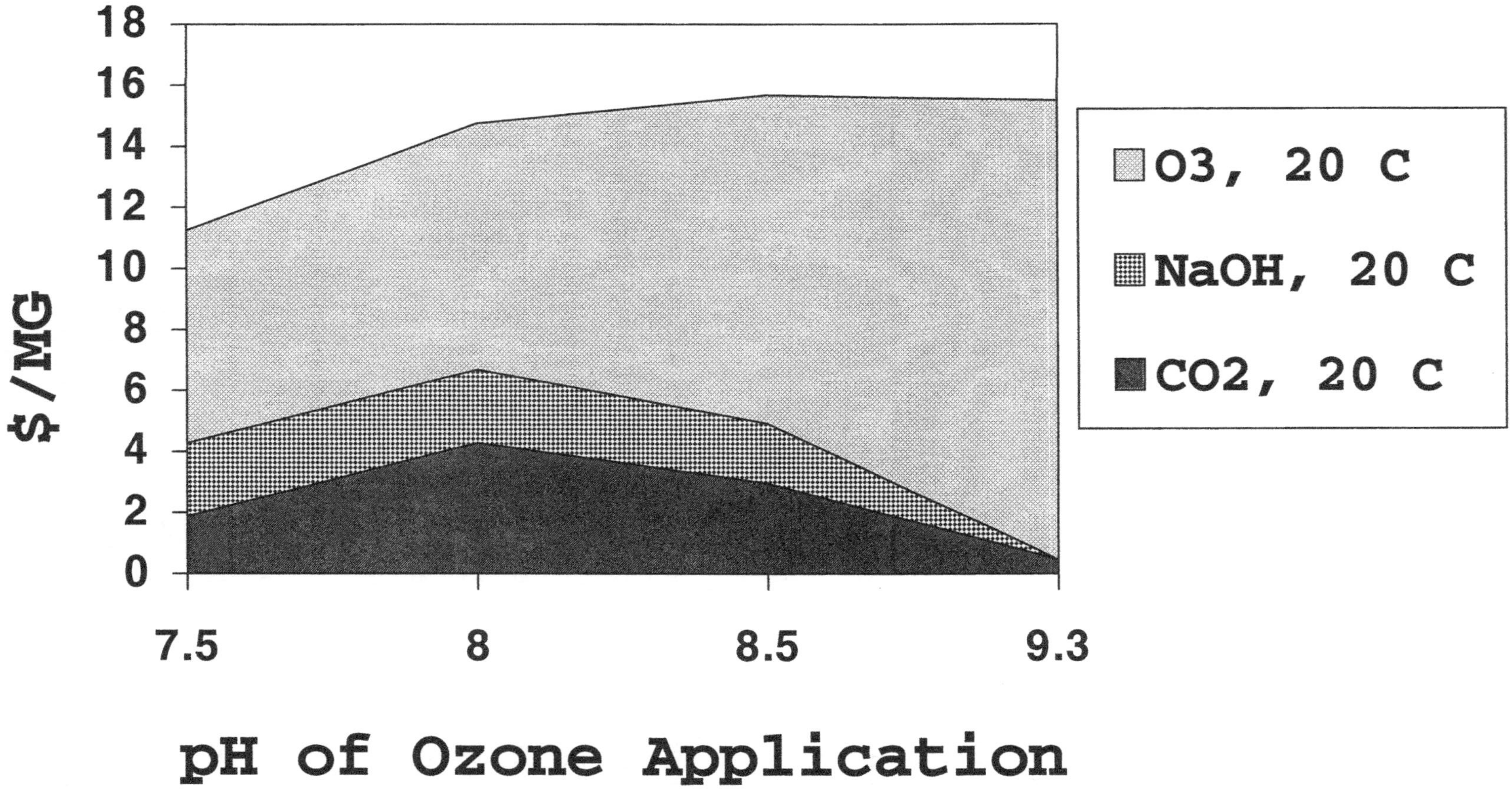

Figure 3.33 Chemical costs at Ann Arbor: pH effects at 20°C

changing amount of well water being used as split treatment in the process during this time.) Ozone made up the largest portion of the chemical costs. Considering full-scale plant operation variations, the chemical costs at ozonation pH 7.5 and 8.0 were not significantly different. At an ozonation pH of 8.5 or higher, the increasing cost of ozone was more significant than the reduction in the use of other chemicals.

Treatment Facilities

Water Supply

The City of Ann Arbor's water supply is obtained from an impoundment on the Huron River and from groundwater wells. Well water constitutes 0 to 30 percent of the total plant flow with river water providing the rest. Typical water quality parameters for raw river water and well water are presented in Table 3.16.

Ann Arbor WTP: Before Treatment Modifications

The treatment plant is a 50-mgd two-stage lime-softening plant (Figure 3.34), which used chloramines for primary and secondary disinfection. It typically operates at a flow of 20 mgd. The influent (flow-weighted river and well water) to the plant is hard (285-344 mg/L, average = 314 mg/L), with moderate-to-high levels of TOC (5.1-8.1 mg/L, average = 6.0 mg/L).

Raw river water (pH = 7.9-8.2) was preliminarily disinfected with chlorine followed by ammonia addition to form chloramines (very minimal, if any, free chlorine contact time at the river). The water was softened with lime (127-210 mg/L, average = 187 mg/L) at a pH of 11.3 (range = 11.0-11.4). During the summer months (April through November), well water was blended with softened river water from the first-stage clarifier effluent (split treatment). The blended water was recarbonated to a pH of ~10. It then entered a second-stage softening/flocculation/clarification process. A cationic polymer (0.37-1.09 mg/L, average = 0.62 mg/L) was added to enhance settling and removal of sludge. The water was subsequently recarbonated to a pH of 9.3 before filtration (GAC/sand filters). In fiscal year 1994-95, 2.2-29 mg/L of carbon dioxide (average = 8.9 mg/L) was used at the plant. Sodium hexametaphosphate

Table 3.16

Ann Arbor water quality: comparison of before and after ozone treatment

Water quality	Before modifications*				After modifications†			
parameters	River	Well	Influent‡	Effluent	River	Well	Influent	Effluent
TOC, mg/L								
Minimum	5.5	1.0	5.1	2.1	5.7 (4)§	1.2	4.5	2.1 (1.9)
Average	6.9	2.2	6.0	3.0	6.5	1.5	5.5	2.7
Maximum	8.7	5.9	8.1	3.5 (3.7)	7.9 (10)	2.0	7.0	3.4 (3.6)
TOC removal,** %								
Minimum				46				40
Average				51				51
Maximum				62				59
UV, cm^{-1}								
Minimum					0.176	0.026	0.142	0.045
Average					0.198	0.037	0.172	0.051
Maximum					0.228	0.058	0.209	0.057
SUVA, L/mg-m								
Minimum					2.7	2.0	2.7	1.5
Average					3.1	2.4	3.1	1.9
Maximum					3.3	2.9	3.3	2.2
Bromide, mg/L								
Minimum					<0.02	0.03	0.03	
Average					0.06	0.10	0.07	
Maximum					0.08	0.13	0.08	
THMFP,†† μg/L								
Minimum					(150)			
Average					484			
Maximum					(1050)			

(continued)

Table 3.16 (Continued)

Water quality parameters	Before modifications River	Well	Influent	Effluent	After modifications River	Well	Influent	Effluent
Temperature, °C								
Minimum	4.1			4.3 (4)	7.9 (3)	12		4.7 (4)
Average	12.6			12.8	14.3	15.7		13.6
Maximum	22.3			21.1 (23)	22.5 (24)	21.2		22.8 (25)
pH								
Minimum	7.9			9.3 (9.1)	7.9	7.2		9.2 (9.0)
Average	8.1			9.4	8.2	7.4		9.3
Maximum	8.2			9.7	8.3	7.6		9.3 (9.7)
Alkalinity, mg/L as $CaCO_3$								
Minimum	205	302	218	28 (20)	187 (180)	278	216	21 (18)
Average	215	314	234	39	209	308	227	34
Maximum	228	335	250	48 (55)	229 (240)	328	252	54 (60)
Total hardness, mg/L as $CaCO_3$								
Minimum	255	438	285	96 (90)	253 (250)	461	295	103 (95)
Average	272	502	314	121	272	514	315	131
Maximum	293	593	344	141 (160)	290 (290)	560	346	159 (180)
Calcium hardness, mg/L as $CaCO_3$								
Minimum	(155)	(320)		(63)	(190)			(60)
Average	205	383		93	210			92
Maximum	(235)	(423)		(113)	(225)			(125)
Ammonia, mg/L as NH_3-N								
Minimum	0.16	0.09			0.02 (<0.01)	0.06		
Average	0.26	0.12			0.06	0.09		
Maximum	0.46	0.16			0.15 (0.2)	0.11		

(continued)

Table 3.16 (Continued)

Water quality Parameters	Before modifications River	Well	Influent	Effluent	After modifications River	Well	Influent	Effluent
Turbidity, NTU								
Minimum	1.3			0.04 (0.02)	1.1 (1.0)			0.06 (0.02)
Average	3.8			0.06	3.6			0.07
Maximum	7.0			0.07 (0.2)	10.1 (20)			0.08 (0.2)
HPC bacteria, cfu/mL			Filter Effluent				Filter Effluent	
Minimum	7,010	0	(<1)	3	14,716 (<100)		(<1)	15
Average	12,708	0	2,080	200	29,129		2,650	151
Maximum	21,243	0	(75,700)	958	45,014		(19,950)	328
					(>570,000)			
Total coliforms, #/100 mL								
Minimum	62	0			203 (<10)	0		0
Average	781	0			836	0		0
Maximum	2,890	0			4,120 (58,700)	0		0
Fecal coliforms, #/100 mL								
Minimum	10				0 (<1)			
Average	88				30			
Maximum	356				86 (300)			
Giardia[‡‡]		# cysts/100 gal				# cysts/100 L		
Minimum	ND[§§]			ND	8 or <10			<0.5
Average	10			ND	ND			ND
Maximum	71			ND	20 or <63			<1
Cryptosporidium[‡‡]		# oocysts/100 gal				# oocysts/100 L		
Minimum	ND[§§]			ND	<2			<0.5
Average	114			ND	ND			ND
Maximum	1,739			ND	63			<1

(continued)

Table 3.16 (Continued)

Water quality	Before modifications				After modifications			
Parameters	River	Well	Influent	Effluent	River	Well	Influent	Effluent
Viruses, pfu/100 L								
Minimum					<1			<0.1
Average					ND			<0.1
Maximum					3.2			<0.1

*Data collected between July 1994 and June 1995; based on monthly (average) data.

†Data collected in calendar year 1997; based on monthly (average) data.

‡Influent water quality = flowed-weight average of river and well water quality.

§Minimum and maximum in parentheses represent minimum and maximum of all measurements, not limited to monthly average data.

**Percent removal of TOC in flow-weighted influent.

††THM formation potential, Cl_2 dose = 25-50 mg/L, Cl_2 residual = 2-5 mg/L, pH = 9.2, temperature = 22°C, incubation time = 7 days.

‡‡Units for older *Giardia* and *Cryptosporidium* data were in gallons, not in liters.

§§For older *Giardia* and *Cryptosporidium* data, detection limits varied from 3 to 284/100 gal.

(target feed rate = 1.0 mg/L) was added before filtration to reduce calcium carbonate precipitation on the filter media. After filtration, sodium hypochlorite and ammonia were added. The finished water was distributed at a pH of 9.4 (on average).

During the winter months (generally December through March), chemical application points were similar to those in summer months. However, split treatment was not practiced and well water was blended with river water prior to the first-stage lime-softening process to raise the water temperature and improve sludge dewatering.

The chloramine dosage was 4.1-6.2 mg/L (average = 5.6 mg/L). *Giardia* log inactivation ranged from 0.5 to 1.1 logs.

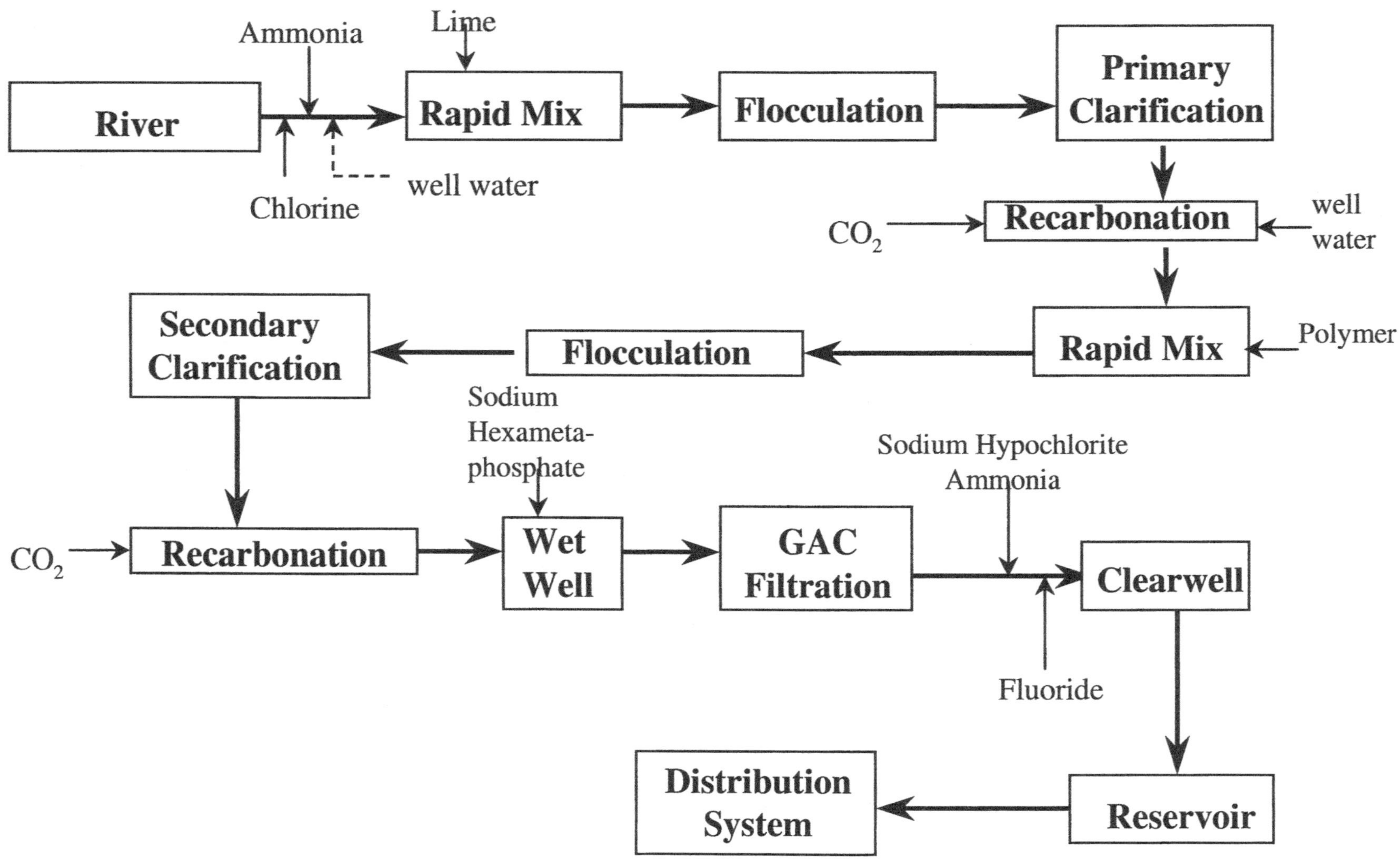

Figure 3.34 Ann Arbor WTP: before modifications

Ann Arbor WTP: After Treatment Modifications

The influent water quality (e.g., TOC, hardness) was similar to that prior to modifications (Table 3.16). SUVA—which is an indicator of humic content—was moderate (2.7-3.3 L/mg-m).

Figure 3.35 shows a schematic of the treatment plant after modifications. Ozonation of softened water was implemented for primary disinfection. Prechlorination was eliminated. The first point of disinfectant (ozone) addition is after secondary clarification. The lime-softening treatment conditions and chemical dosages practiced in the plant prior to the modifications have (in general) remained the same. The softened water is ozonated at a pH of 8.0. In fiscal year 1996-97, 3.5-42 mg/L of carbon dioxide (average = 20 mg/L) was used at the plant.

There are eight ozone contact cells. Six of the cells have contact times of 2.1 minutes (cell number 3: 4.2 minutes; cell number 8: not used). The contact time for the overall contactor is 16.8 minutes. The ozone system is operated at a 6- to 10-percent gas concentration. An off-gas recycle system applies ozone to the first cell, which reduces the demand in subsequent cells but does not produce an ozone residual. A 70/30 percent split of ozone gas can be fed into cells 2 and 3, respectively. Alternatively, they can feed 100 percent into cell 1 or 100 percent into cell 2. The goals are to achieve a 0.1 mg/L residual or greater in the first contact cell and to maintain sufficient residuals in subsequent cells to meet the *CT* target. The residuals in the effluents of cells 2, 3, and 4 have been 0.28-0.45 mg/L (average = 0.34 mg/L), 0.14-0.35 mg/L (average = 0.23 mg/L), and 0.05-0.22 mg/L (average = 0.17 mg/L), respectively.

After ozonation, the pH of the finished water is adjusted with NaOH for corrosion control (average finished-water pH = 9.3). In fiscal year 1996-97, NaOH usage ranged from 1.0 to 32 mg/L (average = 5.3 mg/L). Then the ozonated water undergoes biofiltration in GAC/sand filters. EBCT for the GAC filter has been 3.75 minutes in the summer and 7.5 minutes in the winter. No ozone residual is carried on top of the filters. The filter effluent is disinfected with an average chloramine dose of 3.5 mg/L for approximately three hours in the reservoir, which helps inactivate HPC bacteria shed from the filters (Table 3.16). Free chlorine contact time is variable, but typically short. Some portion of the filter effluent has no free chlorine contact time (those filters right by the clearwell). Others have some contact time depending on the distance from the filter effluent to the point of chlorination. Maximum contact time would be 200 feet of 30-inch pipe (probably a few minutes).

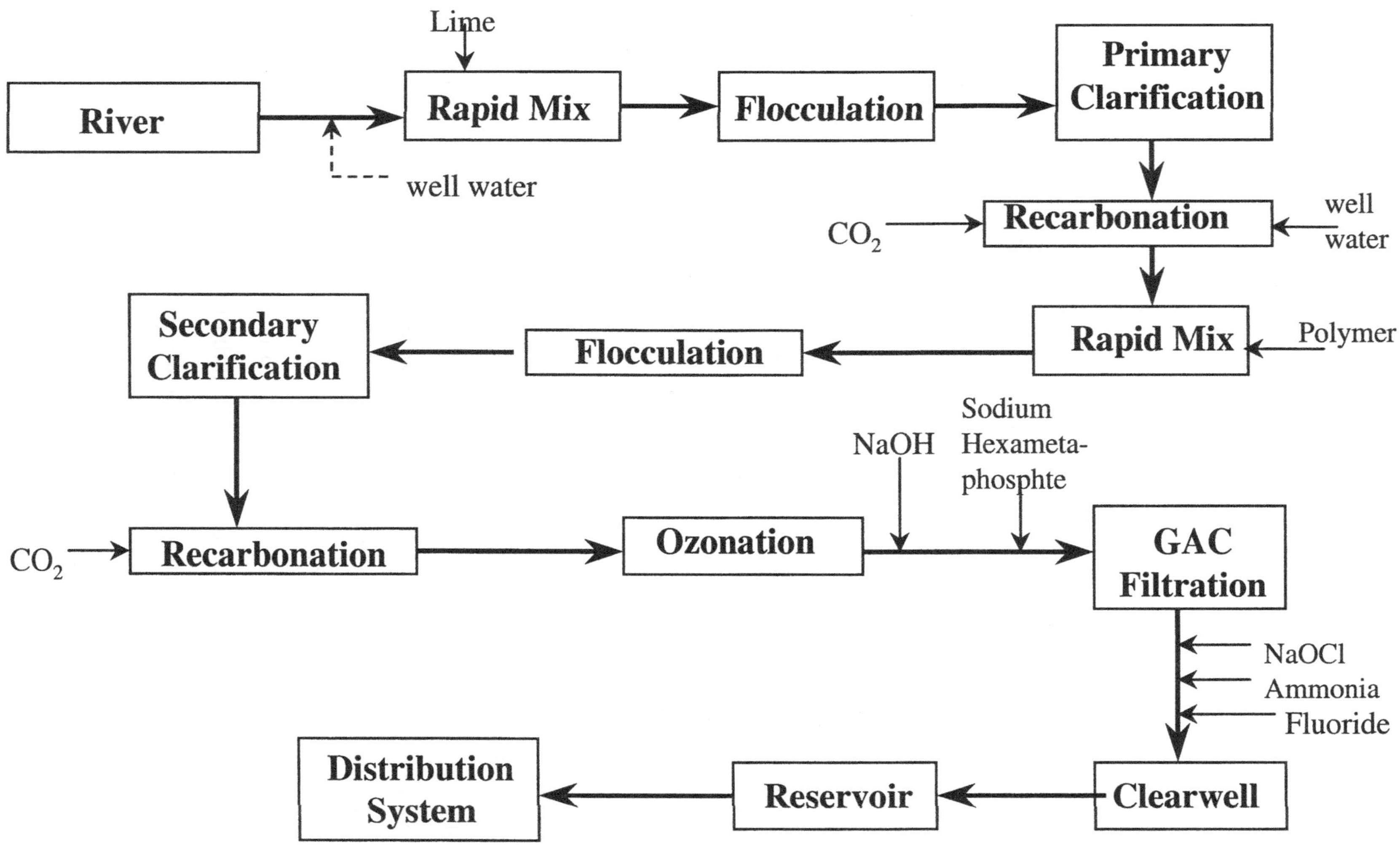

Figure 3.35 Ann Arbor WTP: after modifications

Operations

Instrumentation and Control

Before the implementation of ozonation, all the information required to calculate *CT* values were fed into a SCADA system. These values—flows, volumes, chlorine residuals—were real-time data. Calculations were made within the SCADA software, and real-time *CT* values were displayed for operations. The operators could make adjustments accordingly.

After the modifications, operators input data consisting of ozone residuals, concentrations, flows, etc., into an Excel spreadsheet to calculate *CT* values. Operating parameters are based on virus *CT* required. The *CT* achieved ranges from 100 to 285 percent, with an average of 177 percent being achieved. In the very near future, they will have on-line continuous *CT* reporting.

In order to ensure compliance, the design incorporated redundancy. Dual electrical feed points were designed and there are plans for adding a backup generator.

Filter Operations

The filters consist of 18 inches of GAC and 6 inches of sand. The GAC (Calgon F300) used is bituminous and is sized at 0.8 to 1 mm, with a uniformity coefficient $\leq$2.1. The filtration rates varies from 0.76 to 3.0 gpm/sf (average = 1.5 gpm/sf). The EBCT of the GAC is 3.7 min at design flow and 7.4 min at typical flow. The filters are backwashed with chloraminated water (3 mg/L), which may negatively impact the biological activity on the filters. The criterion for backwashing the filters include a turbidity level of 0.2 NTU (0.3 NTU is an alarm), number of hours in service (usually 80), and headloss (which has not turned out to be a controlling criterion). (Ann Arbor's backwash is typically based on hours in service. However, if a filter begins to exhibit problems, as identified by an effluent turbidity of 0.2 NTU or greater, that filter is removed from use regardless of the number of hours and is then backwashed [this is an unusual occurrence]. The turbidity applied to the filters is quite variable, typically ranging from 5 to 20 NTU [because they are a lime-softening plant]). The filters are backwashed until visually clean. (Visually clean is a typical way of assessing the adequacy of backwashing. However,

Ann Arbor is experimenting with backwash turbiditimeters as a means to standardize this process.) The GAC in the filters is usually replaced or regenerated after three years of service.

Challenges

HPC Bacteria

One of the consequences of ozone implementation has been an increase in HPC bacteria in certain parts of the distribution system. HPC increases were noted at distant locations. The closest locations actually have been showing a decrease. Maybe with the more aggressive disinfection scheme (ozone/chloramines versus chloramines/chloramines), it takes the bacteria longer to recover and grow in the distribution system. It is postulated that the use of ozone has increased the AOC levels in the distribution system, thereby promoting bacterial growth, which is being reflected in higher HPC numbers. This has also caused interference in coliform measurements. Ann Arbor was in the process of conducting additional microbiological monitoring and tests to resolve this issue; however, the interference problem with total coliform testing went away on its own.

Implementation and Operational Issues

Operator Training and Start-Up

Provisions for training the plant staff had been included in the construction contract. In addition, ongoing training is needed. It took a period of two to three months for the operational staff to be "comfortable" with the new technology and three to five months for the treatment plant to operate like an "integral unit". The treatment strategy significantly changed the operational needs of the plant, and additional staff (mechanics, instrumentation technician) were needed. An electrician and an instrumentation technician were added. Some maintenance staff time is required, but is still being assessed.

Design and Operational Issues

Several problems were encountered during implementation and operation. Some of the issues were:

- Dealing with plant flow fluctuations: flow pacing could not handle (filter press effluent, filter backwash).
- Sludge accumulation over diffusers: caused fluctuating ozone residuals, which resulted in *CT* compliance difficulties. Assessing methods (chemical addition) to improve secondary settling to reduce sludge impact on the ozone system.
- LOX vaporizers do not always defrost well in winters: caused the system to shut down due to low gas flow.
- There was no backup provided for the cooling system: process water used for cooling and lubrication was too dirty and caused operational problems.
- General contractor significantly underbid project: was not able to perform at those costs; created an adversarial environment between all parties.
- The chemical cost estimates for caustic soda were too low.
- Regulatory agency required design modifications to provide for 2- (from 1-) day tanks for NaOH (after pre-design was complete): two tanks were constructed, but each was too small to accept a full NaOH delivery.
- Size and depth of piping and construction made for difficult operations: had to shut down significant parts of the plant to deal with the problem, which affected the ability to deliver water, thus creating a scheduling challenge.
- There were leaks in the ozone system, which had to be fixed.
- Design of the backup free chlorination feedpoint: was put in the river water feedline; however, it had to be moved to achieve backup *CT*. At the time, they knew that a backup system was necessary, however they were not aware of the need to disinfect after precursor removal nor the fact that they would be disallowed *CT* credit at high pH. The decision had been based on information available at the time; only later information (regulations) made the location incorrect.

- The system had to be temporarily shutdown due to outdoor ozone ambient levels, dust, and paint fumes.

System Performance

Water Quality

The water quality data from 1994/1995 (July - June) represents the treatment plant "before treatment modifications" and the "after treatment modifications" water quality data was collected in 1997 (January - December) and were based on monthly averages (Table 3.16 and 3.17). The influent to the plant is a flow-weighted average of surface water and groundwater. The water quality parameters (e.g., TOC, temperature, pH, alkalinity, hardness, turbidity) of the influent to the plant were similar before and after treatment modifications.

TOC: The influent TOC, before modifications, varied from 5.1 to 8.1 mg/L, with an average of 6.0 mg/L. The TOC values of the river (range = 5.5 to 8.7 mg/L, average = 6.9 mg/L) heavily influenced the influent values. Well water had low values of TOC, with an average value of 2.2 mg/L (range = 1.0 to 5.9 mg/L). Treatment with lime softening resulted in plant effluent values ranging from 2.1 to 3.5 mg/L, with an average of 3.0 mg/L. TOC removal—compared to the flow-weighted influent—ranged from 46 to 62 percent, with an average value of 51 percent. Good TOC removal is typically achieved during softening at pH levels in which magnesium hardness is removed (e.g., $\geq$ 10.7) (Thompson et al. 1997), which Ann Arbor achieved.

The influent TOC to the plant, after modifications, ranged from 4.5 to 7.0 mg/L, with an average value of 5.5 mg/L. The plant-effluent TOC ranged from 2.1 to 3.4 mg/L, with an average of 2.7 mg/L. TOC removal ranged from 40 to 59 percent, with an average of 51 percent, which is similar to the TOC removal achieved before the implementation of ozonation.

The TOC removal requirements for softening plants treating water with an influent TOC of >4.0-8.0 and >8.0 mg/L are 25 and 30 percent, respectively (USEPA 1998a). Ann Arbor greatly exceeds these requirements. Higher TOC removal is most beneficial at this plant, as it significantly lowers the ozone dosage requirements.

Table 3.17

Ann Arbor DBPs: comparison of before and after ozone treatment

DBPs	Before modifications* Finished water	After modifications† Finished water
TTHMs, µg/L		
Minimum	14	1.4
Average	24	7.2
Maximum	39	13
HAA5, µg/L		
Minimum	4.2	1.5
Average	16	5.0
Maximum	21	15
HAA6,‡ µg/L		
Minimum	4.2	1.5
Average	18	5.6
Maximum	23	16
TOX, µg Cl^-/L		
Minimum	<50	110
Average	137	146
Maximum	150	215
CNCl, µg/L		
Minimum	1.4	<0.5
Average	2.7	1.4
Maximum	5.2	2.6
Bromate, µg/L		
Minimum		2
Average§		3 or 4
Maximum		8

(continued)

Table 3.17 (Continued)

DBPs	Before modifications Finished water	After modifications Finished water
Bromate/bromide,[**] % (molar basis)		
Minimum		1.8%
Average		2.9%
Maximum		5.3%
Formaldehyde, μg/L		
Minimum		4.4
Average		11
Maximum		18
Sum of aldehydes,[††] μg/L		
Minimum		4.4
Average		19
Maximum		30
AOC, μg C/L		
Minimum		40
Average		129
Maximum		210

[*]Monthly THM data collected between July 1994 and June 1995; other DBP data collected quarterly between October 1995 and May 1996.

[†]Data collected in calendar year 1997; based on monthly or quarterly data.

[‡]HAA6 = HAA5 + bromochloroacetic acid.

[§]NDs set at 0 or ½ DL, respectively.

[**]Percentage of bromide converted to bromate (on a molar basis), assuming ozonation was the only source of bromate.

[††]Acetaldehyde, butanal, formaldehyde, glyoxal, propanal, methyl glyoxal, and pentanal.

UV: UV data were available for the raw and finished water for the ozonation plant. The influent had an average UV absorbance of 0.17 cm^{-1} (range = 0.14-0.21 cm^{-1}). The raw-water UV was reduced—via softening and ozonation—to a range of 0.05 to 0.06 cm^{-1}.

Total hardness: Total hardness levels in the influent, before modifications, ranged from 285 to 344 mg/L as $CaCO_3$. Hardness levels in the well water averaged 502 mg/L (range = 438-593 mg/L). The influent levels, after modifications, showed a similar range in hardness (Table 3.16). The plant effluent, before and after modifications, had average total hardness levels of 121 and 131 mg/L, respectively. Lime softening adequately lowered the hardness level in this water.

Temperature: Extreme seasonal differences in temperature have played a major role in process treatment at Ann Arbor. Average monthly river water temperatures in 1997 ranged from 7.9 to 22°C, with an average of 14°C. (A river temperature of 7.9°C as a monthly average in winter was real, but was unusually high that year.) The lowest river temperature in the winter during a sampling interval was 3°C. Well water temperatures are a fairly constant 14°C. The well water temperatures measured at the plant (after transporting through miles of piping and into the building) had monthly averages ranging annually from 12 to 21°C, with an average of 16°C. Ann Arbor raises the temperature of the influent water in the winter by mixing a larger proportion of the groundwater with the surface water (24-29 percent versus 10-20 percent)—as well as introducing the well water at the head of the plant—to bring the temperature to a level where treatment processes like softening, coagulation, flocculation, disinfection, etc., are more effective.

Bromide: Influent bromide at the ozonation plant ranged from 27 to 80 μg/L, with a mean of 67 μg/L. The groundwater is higher in bromide, with levels varying from 30 to 130 μg/L (average value of 100 μg/L). The levels of bromide from the Huron river reflect those of the plant influent with a range of <20 to 76 μg/L (average value of 59 μg/L). Bromide can play a significant role in compliance issues due to the oxidation by ozone of bromide to bromate, which is now a regulated DBP (USEPA 1998a). Bromide can also be an issue because of brominated THMs and HAAs, however it is currently not a problem at Ann Arbor because the free chlorine contact time is short.

Bromate: The bromate levels (Table 3.17) in the finished water of the ozonation plant ranged from 2 to 8 μg/L. The annual average was 3 or 4 μg/L, depending on whether NDs were set equal to zero or ½ the DL. Assuming that ozonation was the only source of bromate, the percentage of bromide converted to bromate, on a molar basis, ranged from 1.8 to 5.3 percent (average = 2.9 percent). At its current level of bromate formation, Ann Arbor meets the Stage 1 bromate MCL of 10 μg/L.

Typically, bromate levels in the finished water were somewhat higher than in the ozone contactor effluent. For example, in July through November 1997, the difference ranged from 0.6 to 2.6 μg/L (average = 1.4 μg/L). Postdisinfection with chloramines uses hypochlorite and ammonia. In a USEPA study of 14 systems using hypochlorination, bromate was detected in more than half of the hypochlorite solutions in concentrations that ranged from 4 to 51 mg/L (Bolyard, Fair, and Hautman 1992). For example, subsequent analysis of a hypochlorite solution used at Ann Arbor in May 1999 had 38 mg/L of bromate. If one is using 14-15 percent bleach (the latter solution was 14.6 percent), which is approximately 180 g/L available chlorine, then for a chlorine dose of 5 mg/L in which the bleach has 38 mg/L of bromate, the finished water will have a 1.1 μg/L bromate contribution from the hypochlorite.

TTHMs: Table 3.17 shows the TTHM concentration before and after treatment modifications at Ann Arbor. Before ozonation, the finished-water TTHM levels ranged from 14 to 39 μg/L, with an annual average of 24 μg/L. This low level of TTHM formation was due to the use of chloramines as the primary and secondary disinfectant. After the installation of ozonation, the TTHM levels ranged from 1.4 to 13 μg/L, with an annual average of 7.2 μg/L. The use of ozonation further reduced TTHM formation, enabling the plant to meet the Stage 1 TTHM MCL of 80 μg/L (USEPA 1998a), as well as the previously proposed placeholder Stage 2 TTHM MCL of 40 μg/L (USEPA 1994a).

HAAs: Table 3.17 shows how the use of ozonation also reduced HAA formation. Before treatment modifications, the plant had an annual average of 16 μg/L for the sum of HAA5 (range = 4-21 μg/L). After the installation of ozone, the HAA5 annual average was 5.0 μg/L (range = 1.5-15 μg/L). The HAA5 concentrations are below the Stage 1 MCL of 60 μg/L (USEPA 1998a) and the previously proposed placeholder Stage 2 MCL of 30 μg/L (USEPA 1994a).

TOX: Before treatment modifications, the TOX levels in the finished water ranged from <50 to 150 μg Cl^-/L (average = 137 μg Cl^-/L). Installation of ozone did not reduce the TOX levels in the effluent (range = 100 to 215 μg Cl^-/L), average = 146 μg Cl^-/L.

CNCl: The chloramine by-product cyanogen chloride (CNCl) had somewhat lower formation after the installation of ozonation (average value lowered from 2.7 to 1.4 μg/L)

Aldehydes: The following seven aldehydes were analyzed for: acetaldehyde, butanal, formaldehyde, glyoxal, propanal, methyl glyoxal, and pentanal. After ozonation and biofiltration, the sum of these aldehydes ranged from 4 to 30 μg/L, with an average concentration of 19 μg/L. Formaldehyde was the most predominant aldehyde formed.

In full-scale tests in summer (at an ozonation pH of 8), >80 percent of the aldehydes produced by ozonation were removed during biofiltration (e.g., >50 μg/L produced by ozonation reduced to <10 μg/L with biofiltration), whereas in winter there was no removal (e.g., 27 and 29 μg/L in the ozonated and filtered water, respectively). Aldehydes can be readily removed in a filter with biological activity, which suggests poor biological activity on the filters in the winter months.

Ann Arbor is currently experimenting with pilot filters. One of their tasks is to optimize biological activity in the winter. However, these experiments are finding that their filters (at temperatures of 7-9°C) are removing aldehydes very effectively.

AOC: After ozonation and biofiltration, AOC ranged from 40 to 210 μg C/L (average = 129 μg C/L). In full-scale tests in summer (at an ozonation pH of 8), approximately 40 percent of the AOC produced by ozonation was removed during biofiltration, whereas in winter there was no removal. As with the aldehydes, a certain percentage of the AOC can be removed in a filter with biological activity, which again suggests poor biological activity on the filters in the winter months.

Bacteria: HPC bacteria and coliforms were found in relatively high numbers in the Huron River (Table 3.16). HPC bacteria in the influent before treatment modifications averaged 12,708 cfu/mL (range of monthly averages = 7,010 to 21,243 cfu/mL) and after treatment modifications averaged 29,129 cfu/mL (range = 14,716- 45,014 cfu/mL). The average and maximum levels of HPC bacteria in the plant effluent before treatment modification was 200 and 958 cfu/mL, respectively, and after treatment modification was 151 and 328 cfu/mL, respectively. HPC bacteria levels (before and after ozonation; i.e., 1995/96 versus 1996/97) in the distribution system are depicted in Figure 3.36. HPC levels increased (on average) from 105 to 237 cfu/mL after the introduction of ozonation (Skadsen, Myers, and Bellamy 1997). Four distribution locations, all at long detention times, accounted for the majority of this increase.

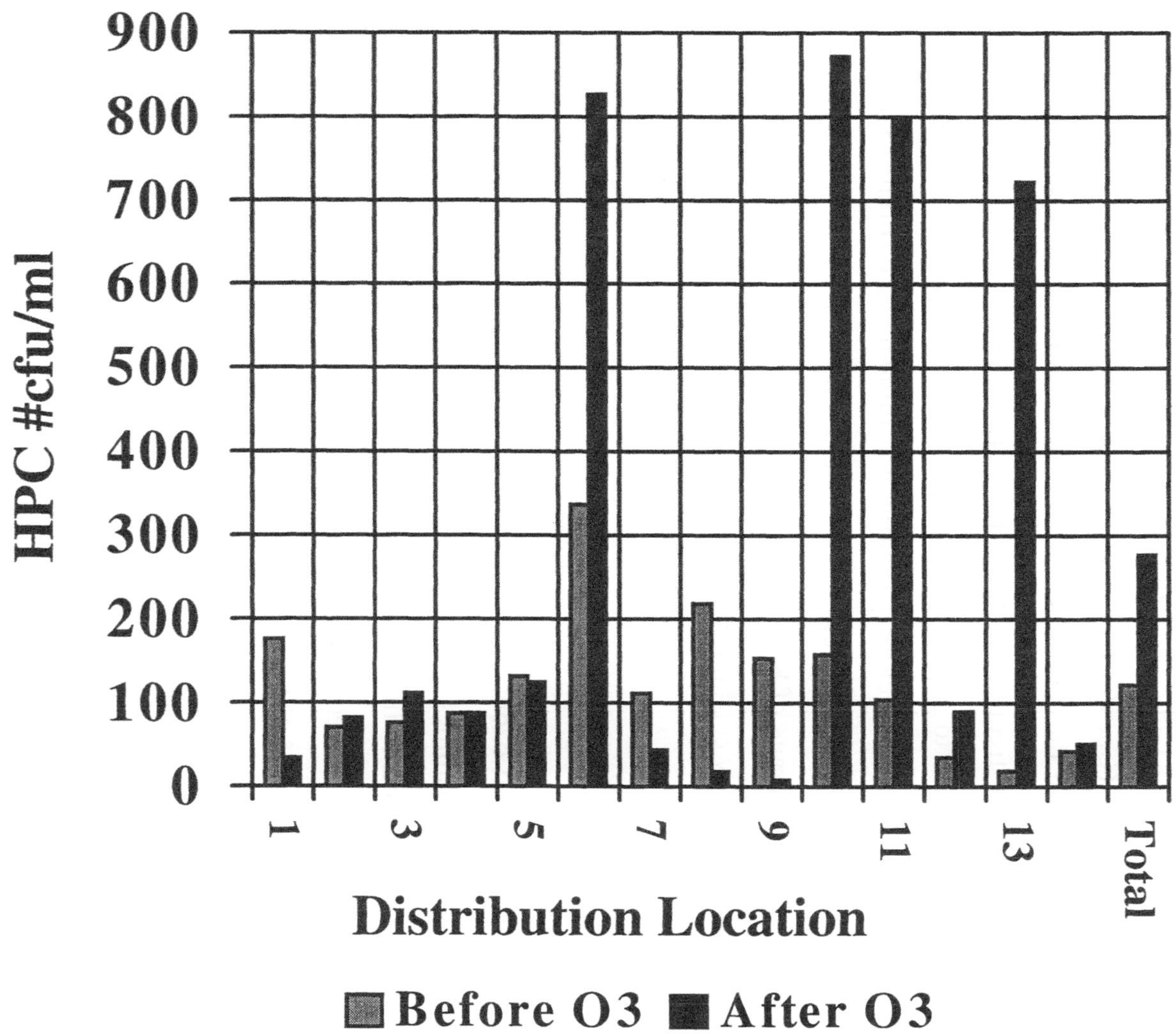

Figure 3.36 Distribution HPC bacteria at Ann Arbor: before and after ozone addition (1995/96 versus 1996/97)

Total coliforms in the influent to the ozonation plant averaged 836/100 mL (range of monthly averages = 203-4120/100 mL). There were zero coliforms observed in the plant effluent; however, the interference rate on the total coliform test for distribution samples increased from an annual average of 1.4 to 6.7 percent (Figure 3.37) (Skadsen, Myers, and Bellamy 1997). The use of ozone has increased AOC levels in the distribution system, which could promote bacterial growth, as reflected in higher HPC numbers at distant locations in the distribution system and interference rates on coliform testing.

Pathogens: Protozoans (*Giardia* and *Cryptosporidium*) and viruses have been detected intermittently at low numbers in the Huron River over the sampling time period (Table 3.16). Viruses appeared to occur more frequently during the winter months when the plant was most vulnerable to *CT* failure. Failure to meet *CT* during the winter months occurred regularly for viruses and intermittently for *Giardia* before ozonation was implemented (Figure 3.31).

Other Parameters: Ozonation resulted in increasing the dissolved oxygen in the distribution system (from an average of 85 percent to 148 percent saturation). Taste and odor of the finished water was reported by consumers to be improved with the use of ozonation.

Summary of System Performance

Primary disinfection with ozone at the Ann Arbor plant proved to be much better than primary disinfection with chloramines at controlling both pathogens and DBPs. It enabled the plant to meet the *CT* requirements in the SWTR and comply with the Stage 1 MCLs with respect to THMs, HAAs, and bromate. Moreover, ozone and chloramines produced DBP concentrations that were less than the placeholder Stage 2 MCLs for THMs and HAAs; although those placeholder MCLs were not promulgated. It provided an unexpected benefit in that THMs and HAAs were lower than predicted. However, current levels of bromate are in a range that could put this plant in jeopardy under some Stage 2 regulatory scenarios. There was no decrease in TOX levels after ozonation was introduced.

Aldehydes and AOC produced by ozone can be removed (e.g., >80 and approximately 40 percent, respectively) by biofiltration in the summer. However biofiltration does not appear to be effective in the winter at this plant.

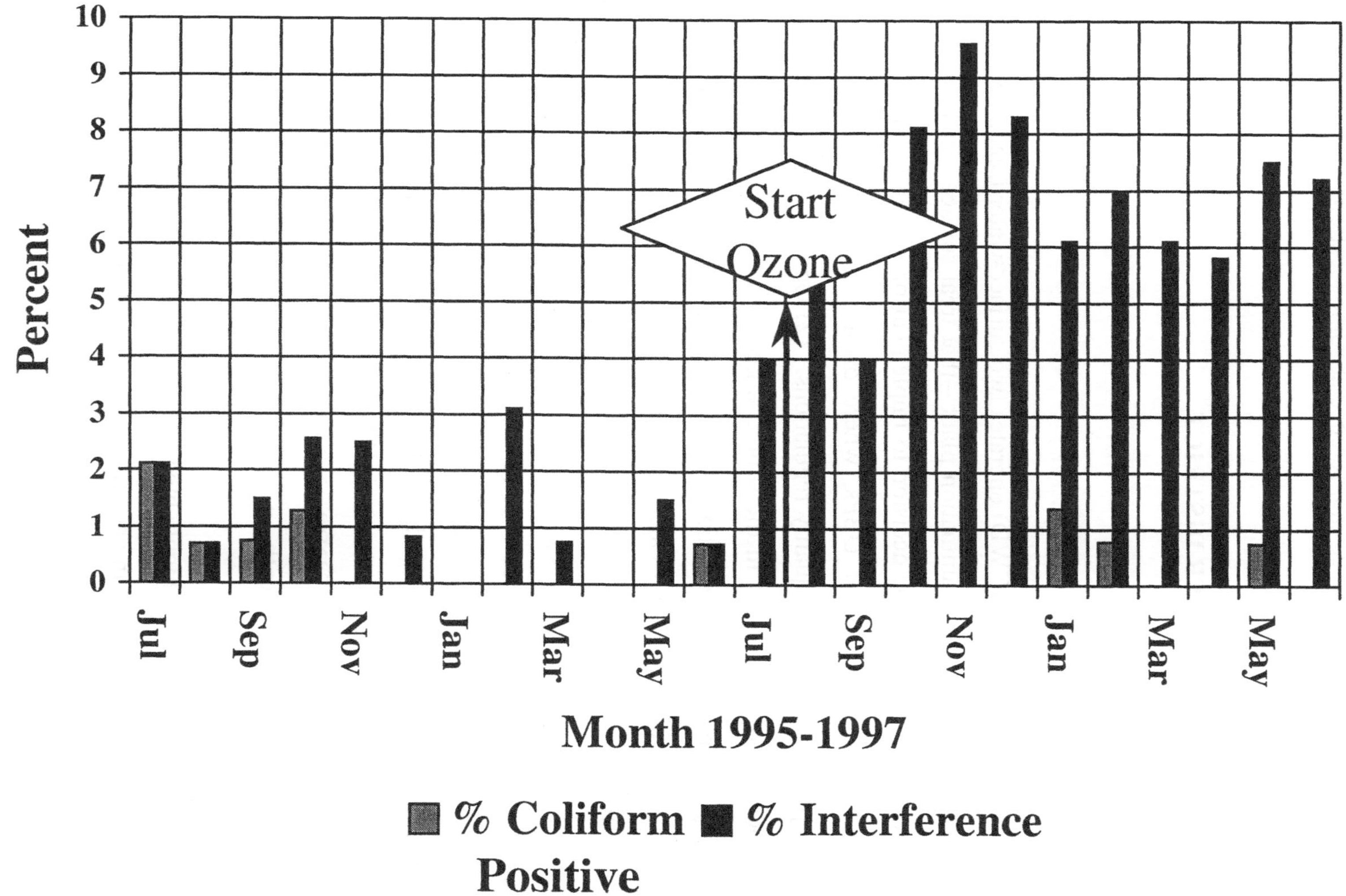

Figure 3.37 Distribution bacteria at Ann Arbor: 1995/96 versus 1996/97

The microbial quality of the water in the distribution system after ozonation showed an increased level of HPC bacteria at distant locations. The closest locations actually have been showing a decrease. There was also increased interference at some locations with total coliform measurements.

OZONE: ALAMEDA COUNTY WATER DISTRICT

Summary

Alameda County Water District (ACWD) operates two surface-water treatment plants—a conventional plant and an ozonation/chloramination plant—that treat surface water periodically high in bromide. The driving forces in the decision to implement ozonation included: (1) The threat of impending federal regulation of DBPs, where ozonation was expected to satisfy anticipated or contemplated future water quality regulations. (2) Ozonation would resolve the persistent taste-and-odor problems present in the source water. (3) Ozonation would enhance coagulation and filter performance. (4) Ozonation would provide an adequate microbial kill. (5) The unique topography of the plant site provided for on-site hydroelectric power, which was a major driver in the decision to go ozone.

The process design was influenced by THMs, turbidity, pathogens, and taste-and-odor issues. Ozonation of raw water was selected over ozonation of settled water as taste-and-odor removal was found to be better with pre-ozonation. In addition, coagulation studies were conducted, and ferric chloride was chosen as the preferred coagulant. Biological filtration and the addition of chlorine to the filtered water and chloramines to the finished water before distribution further addressed the issues of DBPs and pathogens. The cost of the ozone plant (28-mgd design flow) was $40 million, with an annual operating cost of $2.2 million and a maintenance cost of $645,500.

Some of the key facts that were highlighted during this case study are: (1) ACWD selected the ozonation process in 1989-90, in an "era" where the USEPA had only just begun to talk about DBP regulations and not much was known about bromate, HAAs or *Cryptosporidium*. (2) Pilot-plant studies of ozonation were used to optimize process design and proved to be an invaluable tool because of the variability in source-water pH and alkalinity. Rapid shifts in raw-water quality can make the ozone process difficult to control. (3) Ozonation resulted in low levels of THMs and HAAs, improved taste and odor, and more effective coagulation. (4) A less well anticipated factor was the rising concern over the formation of ozonation by-products such as bromate and aldehydes. Bromate is a regulated by-product in the new DBP regulations (USEPA 1998a), and utilities that implement ozonation need to assess the bromide content in their raw water and make design provisions to control bromate formation. (5) The operational experience at the ozone plant has shown that backup equipment is an important aspect of reliability in ozone operations.

Intensive training sessions for the staff were held at the ozone plant when it first went on-line. In addition, every year, the plant is shut down for at least two weeks and all operators attend at least 40 h of training. It took the operational staff a period of one to two years to be "comfortable" with the new technology and operate as an integral unit.

At the conventional treatment plant—which uses prechlorination and post-chloramination—the TTHMs ranged from 43 to 95 μg/L and HAA5 ranged from 23 to 59 μg/L during the case-study period. With ozonation and chloramination, the other plant had a range of TTHMs from 1.3 to 33 μg/L (average = 6.5 μg/L) and a range of HAA5 from 1 to 10 μg/L. The influent bromide levels at the ozone plant ranged from 17 to 200 μg/L, with a mean value of 74 μg/L. The bromate concentrations ranged from <4.5 to 25 μg/L, with the running annual average at 6.9 μg/L for the sampling period studied. Seasonal variations in bromate formation were observed.

The ozone plant achieved an average TOC removal of 47 percent (range = 30 to 60 percent). Some of the TOC removal may be due (in part) to the conversion of TOC by ozone to biodegradable organic carbon, a portion of which can be removed through the biologically active filters. The effluent turbidity at the ozone plant ranged from 0.02 to 0.08 NTU (average = 0.06 NTU). Although the coagulant dose was optimized for turbidity removal, effective TOC removal was also achieved at this plant.

The ACWD case study demonstrated that ozonation made a very difficult-to-treat water easily treatable and was, in this particular case, a cost-effective treatment process. The ozone plant produces a very high quality effluent water with low turbidity, low TOC, low THMs and HAAs, and low taste and odor. However, ozonation introduced bromate as a new concern.

ACWD plans to introduce a carbon dioxide feed system to lower the pH of ozonation, which will enable ACWD to minimize bromate formation.

Background

Case History

ACWD operates two treatment plants, which treat the same surface water (i.e., California SPW augmented by local run-off into Lake Del Valle). ACWD serves over 302,000 people. The 8-mgd Mission San Jose Water Treatment Plant (MSJWTP) is a conventional treatment plant constructed in 1975, which was modified in 1989 to improve post-filtration disinfection contact time and minimize DBP formation. MSJWTP uses prechlorination/postchloramination. WTP No. 2 (WTP2)—a 28-mgd plant—was constructed at a cost of approximately $40 million and went on-line in 1993. WTP2 uses ozone for primary disinfection and chloramines for secondary disinfection. In addition, WTP2 operates with biological filtration.

The problems that the District experienced at the MSJWTP included filter-clogging algae episodes, taste-and-odor problems, episodic changes in turbidity, temperature and pH, and high levels of THMs. SPW is high in bromide—especially during droughts—which results in increased THM formation. A study was performed by Camp Dresser & McKee (CDM) at the MSJWTP pilot plant to evaluate DBP control and other treatment objectives. A portion of the CDM study was part of the AwwaRF project on "Identification and Occurrence of Ozonation By-Products in Drinking Water" (Glaze and Weinberg 1993), which included an extensive study on bromate. One of the recommendations of the CDM and other studies conducted on SPW was to utilize ozone.

Ozonation resolved the problems prevalent at the MSJWTP in that it took care of the taste-and-odor problems and the issue of high THM formation, however this treatment modification introduced bromate as a new issue. Consequently, the MSJWTP has been chosen as the "before treatment modifications" plant and WTP2 has been chosen as the "after treatment modifications" plant for this case study.

Process Selection

A pilot-plant study was conducted by CDM and ACWD in 1989. The study lasted nine months and examined several treatment processes. Subsequently, additional bench- and full-scale studies have been conducted to further optimize the treatment process.

Coagulation studies: The coagulants evaluated were alum and ferric chloride. The coagulant dosage was optimized for turbidity reduction rather than for TOC removal. Ferric chloride had a more robust performance and better process stability, resulting in lower turbidity, particles, TOC, and washwater requirements. The disadvantages of ferric chloride include plant aesthetics and higher overall costs. Ferric chloride was chosen as the preferred coagulant for WTP2

Reduced-pH coagulation was evaluated. Carbon dioxide was used to lower the pH in the raw water and sodium hydroxide was added to the finished water to bring up the pH. However, significant pH drops resulted in poor turbidity removal. Therefore, pH depression before coagulant addition was not considered. Note, the raw water pH (which varies between 7 and 9) can be reduced to a pH of 6.5 to 7.5 with coagulant addition only.

Evaluation of alternative disinfectants: Disinfection studies included pre-ozonation, shifting the point of chlorination to after filtration, and addition of chloramines before distribution. This disinfection scenario was incorporated into the plant design. In the pilot-plant study, pH depression prior to ozonation was evaluated as a method for bromate control (Glaze and Weinberg 1993). This was not implemented in the plant design. Instead a spare chemical system was included as part of the design, which could be used as an acid-feed system if necessary.

Subsequently, ACWD bench-scale tested pH levels of 8, 7.5, 7.0 and 6.5 to determine the acid doses needed to lower the pH to these levels. In addition, ACWD tested the addition of carbon dioxide for pH depression (to pH 7).

Ozone/hydrogen peroxide was evaluated to control taste-and-odor problems. However, ozone alone has been adequate in controlling taste-and-odor problems. ACWD had considered converting the hydrogen peroxide system to an acid-feed system—because this option initially appeared to be less expensive than adding the acid-feed system in the spare-chemical area—to control bromate formation. The plan was if hydrogen peroxide was needed in the future, it would be added in the spare chemical area.

The various treatment studies that dealt with pH-related issues (coagulation process, bromate control) were subsequently used in the decision-making process on treatment-modification selection (see below).

Ozonation of raw water was selected over ozonation of settled water. Although water-quality fluctuations (e.g., pH) would be dampened by the clarification step, other considerations resulted in the selection of pre-ozonation. Taste-and-odor removal was better when ozone was added to the raw water. Also, ACWD was considering direct filtration as an operating mode at that time. In addition, there were concerns about air binding problems with the filters and adverse effects on the coagulated water.

GAC studies: As part of the original construction, GAC was installed in one of WTP2's six filters to test its performance. The anthracite in two other filters has since been replaced with GAC so that all three filters in that train (half of the plant) are the same. This is to allow proper dosing of the filter aid polymer.

Parameters that Influenced Process Design

The parameters that influenced process design were THMs, turbidity, pathogens, and taste and odor.

THMs: Historically, THMs in the conventional plant effluent ranged from 30 to 100 μg/L. A reduction in THM formation was a major goal of implementing ozonation. The process was designed to reduce the THM level to <20 μg/L.

Turbidity: Raw-water turbidity has ranged from 2 to >200 NTU. One of the goals for WTP2 was to produce filtered-water turbidity <0.1 NTU.

Pathogens: The SWTR requires 99.9 percent removal of *Giardia* cysts and 99.99 percent removal of viruses (USEPA 1989a). Ozonation was selected, in part, to meet these regulatory requirements.

Tastes and odors: The treatment process at MSJWTP does not deal well with taste-and-odor problems due to algae or with filter-clogging algae episodes. Ozonation was selected for WTP2 to eliminate the adverse impacts of algae on filtration and on taste and odor.

The Decision-Making Process

Perhaps the most important thing to appreciate about the ACWD decision-making process that led to the selection of an ozonation process was the "era" in which the decision was made. The time frame was roughly 1989-90. This was a time when:

- USEPA had only begun to talk about new DBP regulations; levels as low as 10 µg/L for THMs were being discussed;
- USEPA had not yet begun to work with the drinking-water industry to achieve a balance in the regulatory approach (Means and Krasner 1993);
- most of the industry was not talking about bromate, HAAs or *Cryptosporidium*;
- most of the industry was not talking about membranes, because the technology had not "arrived" yet and was considered to be cost prohibitive;
- and many U.S. engineers were just beginning to seriously consider ozone as a viable, albeit expensive, option for water treatment (Ferguson, Gramith, and McGuire 1991).

These background conditions featured prominently in (and did not deter) ACWD's decision to go to ozone.

Driving Forces

As in many circumstances where compliance decisions are made in advance of regulations, ACWD needed to build new treatment capacity to meet growing demands. ACWD had to decide on treatment for the future before all the returns were in on future standards (Figure 3.38). This was the primary motivation to go forward with the plant.

The largest driving force in ACWD's treatment choice was the threat of impending Federal regulation of DBPs. EPA released a "strawman" discussion paper in 1989 that mentioned possible TTHM standards as low as 25 µg/L (Clark 1989). This trial balloon was distinctly a driver in the decision to go to ozone. The sense of the engineering staff at the time was that going all the way to ozone would provide them with a "belt-and-suspenders" assurance that they would never have to worry about these compliance issues again. Moreover, after

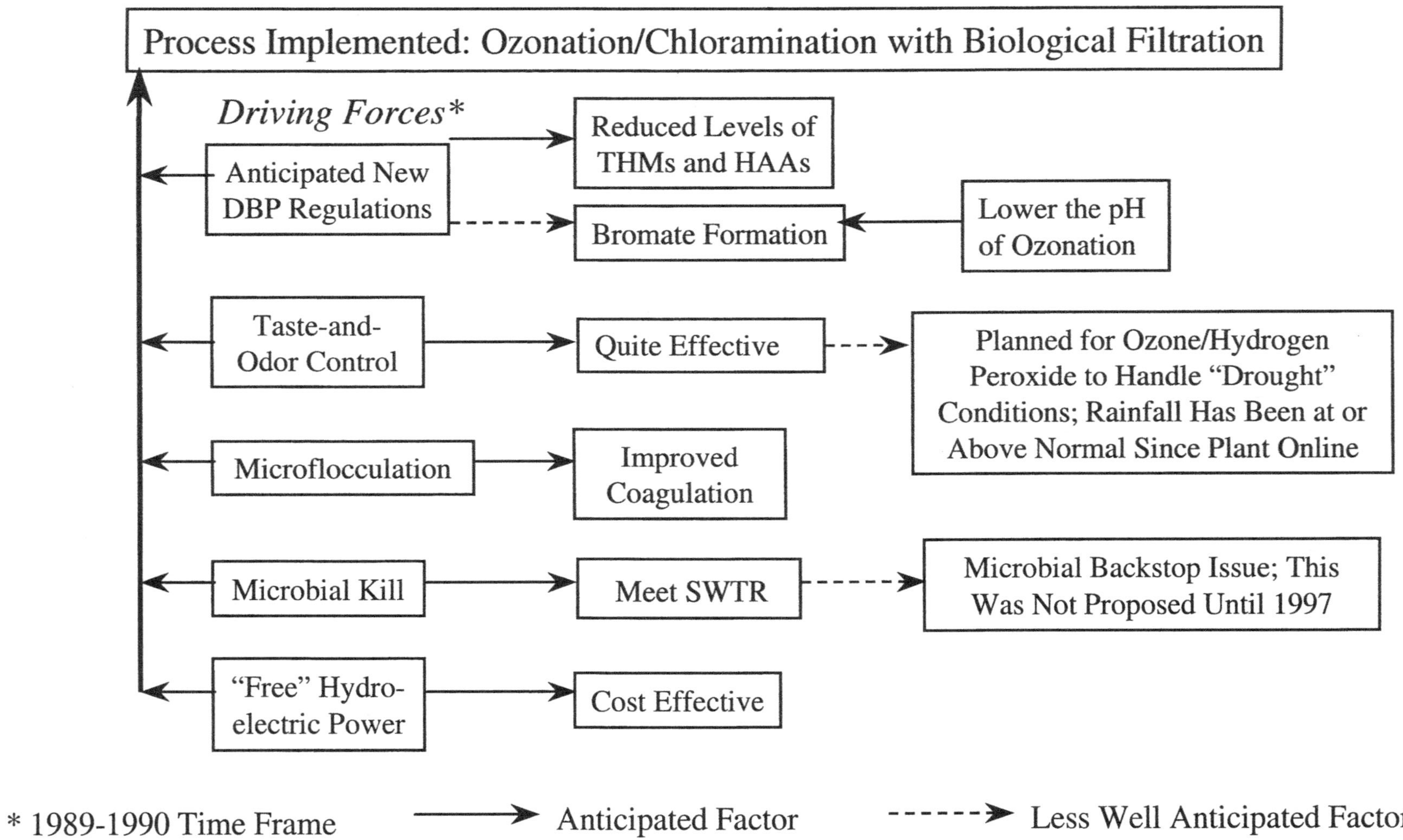

Figure 3.38 The decision-making process at ACWD: implementation of ozonation/chloramination

looking at all other options (chloramines, etc.), there was no other way to get to very low DBP numbers.

A second strong rationale for going to ozone was the fact that the source water was characterized by taste-and-odor problems. The pilot study was done during a six-year drought. The drought created very poor water quality, and the plant was designed to deal with this. Since that time, rainfall has been high, and the raw-water quality has been better. Another ozone/hydrogen peroxide consideration was the desire to save on ozone consumption and cost.

A third benefit of ozone is microflocculation, a process that enhances the effectiveness of particle agglomeration and removal. This flocculation phenomenon is a result of changes in surface chemistry resulting from ozonation; it can result in improved turbidity removal. Microflocculation was an objective of the design.

Savings in coagulant costs were anticipated, but this was not as much of a driver as the other factors mentioned above. Similarly, obtaining adequate microbial kill was a fourth objective of the design and a recognized benefit of ozone, but was not as much of a major driver. Remember, that this was just before the initial SWTR implementation and there was not as much talk about these properties of ozone as there was a few years later after an outbreak in Milwaukee. In addition, *CT* tables had not been produced for ozone at the time.

An unusual driving force existed in this case due to the unique topography of the plant site. There is a 340-ft difference in elevation between the water source (SPW turn-out) and the plant. As a consequence, hydroelectric power generation facilities were incorporated into the design and are capable of providing all of the power needed to run the plant – including ozone generation. At the time, power cost was one of the major obstacles to ozone acceptance. The existence of "free" electricity in this case was a major driver in the decision to go to ozone, almost as important as the potential for further USEPA regulation of DBPs. (The hydroelectric power is not really free, there are maintenance costs.)

Constraints

The major constraints to application of ozone in this "era" would have likely been considerations of cost and doubts over the process performance of a novel technology. The cost hurdle was considerably reduced in this case due to the availability of hydroelectric power.

Doubts about performance were systematically addressed through pilot-plant work. Specifically, there were questions about the benefits of microflocculation that were addressed in pilot testing. In addition, there were concerns about HPC bacteria coming off the biological filter, which were put to rest by demonstrating a feasible technique for reducing the HPC level with a short free-chlorine contact time after the filter prior to ammonia addition.

Another process concern in the design was pH and alkalinity control. The source water exhibits significant variability. In addition, there are many process steps that have significant sensitivity to pH and alkalinity. These variations affect the effectiveness of ozone in microflocculation and inactivation. It affects coagulant performance, biological filtration performance, taste-and-odor control, corrosion control, and DBP formation. This was identified as a major design challenge.

Originally, ACWD planned to convert the hydrogen peroxide storage and feed area into a sulfuric acid storage and feed area. However, due to fire code issues and operator concerns about the safety of sulfuric acid, ACWD was building a raw water carbon dioxide feed system. If concentrated sulfuric acid is stored indoors, the area containing it must be explosion-proof, which is very expensive. It would be cheaper to build a separate structure for a new sulfuric acid feed system. Because the Stage 2 D/DBP Rule negotiations made it clear that the bromate MCL was not going to drop to 5 μg/L for at least another six years (USEPA 2000), ACWD decided that carbon dioxide with a target pH of 7 upstream of ozonation was the preferred option. ACWD was fairly certain that a pH of 7 will get them to a bromate of 10 μg/L or less.

An additional constraint was presented by site conditions in the middle of a developed residential area. This created concerns over the safety of ozone-generating equipment and added costs to the design, but was otherwise overcome as an obstacle. Because of uncertainty about changes in the Uniform Fire Code, a decision was made to house the ozone generators in separate housings. Such precautions were later determined unnecessary and are found nowhere else in more recently constructed ozone plants. Technology pioneers sometimes bear additional costs.

Technology Checklist—Anticipated Factors

The plant designers correctly anticipated that the ozone application would result in very low levels of DBPs (THMs and HAAs typically below 10 μg/L), improved taste and odor, and more effective coagulation. The capability to employ effective process control was confirmed for biological filtration and HPC control.

In fact, the effectiveness of ozone in taste-and-odor control was significantly under-estimated, making the additional step of adding hydrogen peroxide unnecessary to date. It should be noted that the pilot study was done during a six-year drought. Significantly worse taste-and-odor episodes are known to occur during droughts. In the last five years that the plant has been operating, rainfall has been at or above normal levels.

Technology Checklist—Less Well Anticipated Factors

The major unanticipated issue involved in the ACWD case study was the rising concern over additional DBPs that are formed during ozonation, especially when the levels of THMs and HAAs are significantly reduced. Specifically, bromate was a potential compliance problem for ACWD. When source-water bromide levels are higher, ACWD might have trouble meeting a bromate standard of 10 μg/L on a continuous basis. ACWD was adding carbon dioxide feed facilities primarily to enhance process control, but also for bromate control. But compliance may still be a worry. Bromide levels tend to be highest during periods of drought (Krasner, Sclimenti, and Means 1994). The difference between compliance and non-compliance for ACWD might have come to rest as a function of how non-detects are counted in a compliance monitoring requirement for bromate or how bromate levels are averaged across time periods if there are seasonal or multi-year cycles influencing influent-bromide concentrations. However, the California DHS is going to consider non-detects as zeroes for averaging purposes Another compliance opportunity for a plant such as ACWD would be a "risk bubble" approach allowing credit for offsetting greater risk reductions in other DBPs.

Although the process-control issues were anticipated fairly well based on pilot testing, the one area highlighted as needing further enhancement was improvement of capability in on-line monitoring for control of ozone application. This area is now benefiting from improved on-line monitoring technology.

In a much later "era," the deliberations over the IESWTR raised a possible issue of "backsliding" in microbial protection as a potential side effect of DBP compliance (USEPA 1998b). The disinfection scenario (ozone/chlorine/chloramines) at WTP2 resulted in a calculated 2-3 logs of *Giardia* inactivation, and the average disinfection credit at MSJWTP (chlorine/chloramines)—which is primarily achieved through the filters and in the clearwell—is 2.7 logs of *Giardia* inactivation. Roughly half of their total ozone dose is devoted to taste-and-odor control. In view of this, a reduction in ozone dosage when bromide levels are high is a logical strategy for them to comply with a bromate standard. One way to make up the lost log killwould be increased free-chlorine contact at the back of the process – re-introducing some of the THMs and HAAs that they have presently minimized and were planning to never have to worry about again. This might be avoided if a risk bubble approach were applied in evaluating the need for bromate compliance.

Compliance Issues

WTP2 is producing THM and HAA concentrations below the previously proposed placeholder Stage 2 MCLs (USEPA 1994a). It is also meeting the disinfection requirements as set out in the SWTR (USEPA 1989a). However, consistent compliance with the bromate MCL of 10 μg/L is a new issue that will be resolved by ozonating water at a reduced pH (i.e., at pH 7). If and when the bromate standard drops further, ACWD will re-group and make a decision about what to do at that time.

Key Players

The key players involved in the decision-making process were the ACWD staff and consultants from CDM. In addition, Water Quality & Technology Solutions, Inc., worked with ACWD staff on testing the addition of carbon dioxide for pH depression and bromate control.

ACWD serves water to Fremont, Newark and Union City. The median cost of housing (January 1998) in these three cities was $244,000, $227,750, and $192,000, respectively. The median age (1990 census data) in these cities was 32, 30, and 31, respectively. The median household income (in 1989) in these cities was $51,231, $50,471, and $46,988, respectively.

Various efforts were made to inform the public. Neighborhood meetings were conducted in 1988-89 to explain the project to the customers and to receive community support. In addition, a public hearing was held on the Environmental Impact Report. Initially, there was resistance to an "unfunded mandate." The neighborhood was initially against the construction of WTP2 because of appearance. Therefore, the following steps were taken to mitigate their concerns: (1) extra money was put into building a decorative concrete wall with extensive landscaping to shield the construction site from the neighbors' view, (2) mechanical sludge dewatering was added eliminating drying ponds, and (3) the plant's architectural design was improved such that it would present a "lower profile" on the hillside.

Selection Issues and Future Strategies

ACWD has been satisfied with the results of the ozonation process implemented at WTP2. The only thing they would have done differently would have been to install a LOX system to generate ozone rather than the air-feed system that they currently use. Although the air-fed system met their expectation in terms of ozone production, a number of substantial changes had to be made in the system to lower operation and maintenance costs. Perhaps settled-water ozonation would be done now. This was considered but did not appear justified by the pipe work. In addition, the anthracite in two of the filters has been replaced with GAC. Moreover, ACWD is in the process of adding a carbon dioxide feed system in order to control bromate formation and stabilize the coagulation process.

ACWD is currently involved in plans to upgrade MSJWTP. Although the initial studies looked at retrofitting MSJWTP with ozone, adding a third filter, and modifying the clarification process, ACWD is evaluating ultrafiltration as an option for the plant upgrade. Preliminary cost calculations indicate that the cost of ultrafiltration will be similar to adding ozone to the existing plant. However, ACWD still has concerns over DBP formation and taste-and-odor issues, which heavily influenced the process choice at WTP2.

Economic Analysis

WTP2 was constructed for a total cost of $40 million. The capital cost for the ozone process (generators and contactors) was $5.6 million. In fiscal year (FY) 1997-98, the cost of operations at WTP2 (18-mgd typical flow) was $2,193,600, with a maintenance cost of $645,500.

Water is delivered to WTP2 through a 3-ft diameter pipeline. Because of the elevation difference between the raw-water aqueduct and the plant, ACWD installed turbines to generate electricity. The hydroelectric power plant facility at WTP2 produces enough electricity to run all the treatment processes, including ozone generation. At higher plant flow rates (typically 21 mgd), excess power is produced, and this power is exported to MSJWTP, offsetting much of the power costs at that plant.

In FY 1997-98, the cost of operations at MSJWTP (7-mgd typical flow) was $835,100, with a maintenance cost of $262,200.

The chemical costs for dropping the pH to 7 with sulfuric acid is approximately the same as for carbon dioxide (Coffey et al. 1998).

Treatment Facilities

Water Supply

WTP2 has two sources of supply: SPW (water from the Sacramento-San Joaquin Delta) supplies about 79 percent (on average) of the plant-influent blend over the course of the year and the remaining 21 percent (on average) is supplied by Lake Del Valle (local runoff). MSJWTP is supplied with the same raw water. The raw water quality (Table 3.18) is the same for both plants. (The two plants each take water from the same aqueduct 200 ft apart from each other. Any variations in raw-water quality are likely due to different sampling times.)

Table 3.18

ACWD water quality: comparison of conventional treatment (MSJWTP) and ozonation (WTP2)

	MSJWTP*		WTP2†		
Water quality	Influent	Effluent	Influent	Ozonated Water	Effluent
TOC, mg/L					
Minimum	3.0	1.8	3.1	1.3	1.4
Average	4.1	2.6	4.6	4.1	2.5
Maximum	6.4	4.0	6.4	6.6	3.7
UV, cm^{-1}					
Minimum	0.12	0.07			0.06‡
Average	0.19	0.10			0.08‡
Maximum	0.27	0.16			0.13‡
SUVA, L/mg-m					
Minimum	3.0				
Average	4.5				
Maximum	5.7				
Bromide, μg/L					
Minimum	38		17		
Average	63		74		
Maximum	150		200		
pH					
Minimum		7.2§	6.9	6.9	7.5§
Average		8.3§	7.8	7.8	8.4§
Maximum		8.6§	8.6	8.6	8.7§

(continued)

Table 3.18 (Continued)

	MSJWTP		WTP2		
				Ozonated	
Water quality	Influent	Effluent	Influent	Water	Effluent
Alkalinity, mg/L					
Minimum		60[§]	38		35[§]
Average		110[§]	87		80[§]
Maximum		160[§]	170		196[§]
Total hardness, mg/L					
Minimum		60[§]	47		
Average		140[§]	108		
Maximum		190[§]	172		
Turbidity, NTU					
Minimum	1.6	0.04[§]	1.7		0.02
Average	9.8	0.07[§]	16		0.06
Maximum	29	0.20[§]	69		0.08
Temperature, °C					
Minimum	10.4		8.8	8.8	
Average	18.1		16.2	16.2	
Maximum	25.7		25.9	25.9	

[*]Data collected between May 30, 1995 and July 9, 1996 unless noted otherwise.

[†]Data collected between January 3, 1995 and June 25, 1996 unless noted otherwise.

[‡]Data collected between November 1995 and June 1996.

[§]Data collected between January 1996 and December 1996.

MSJWTP

MSJWTP is an 8-mgd conventional plant with upflow clarifiers (Figure 3.39), which uses chlorine and chloramines for primary and secondary disinfection, respectively. It typically operates at a flow of 7 mgd.

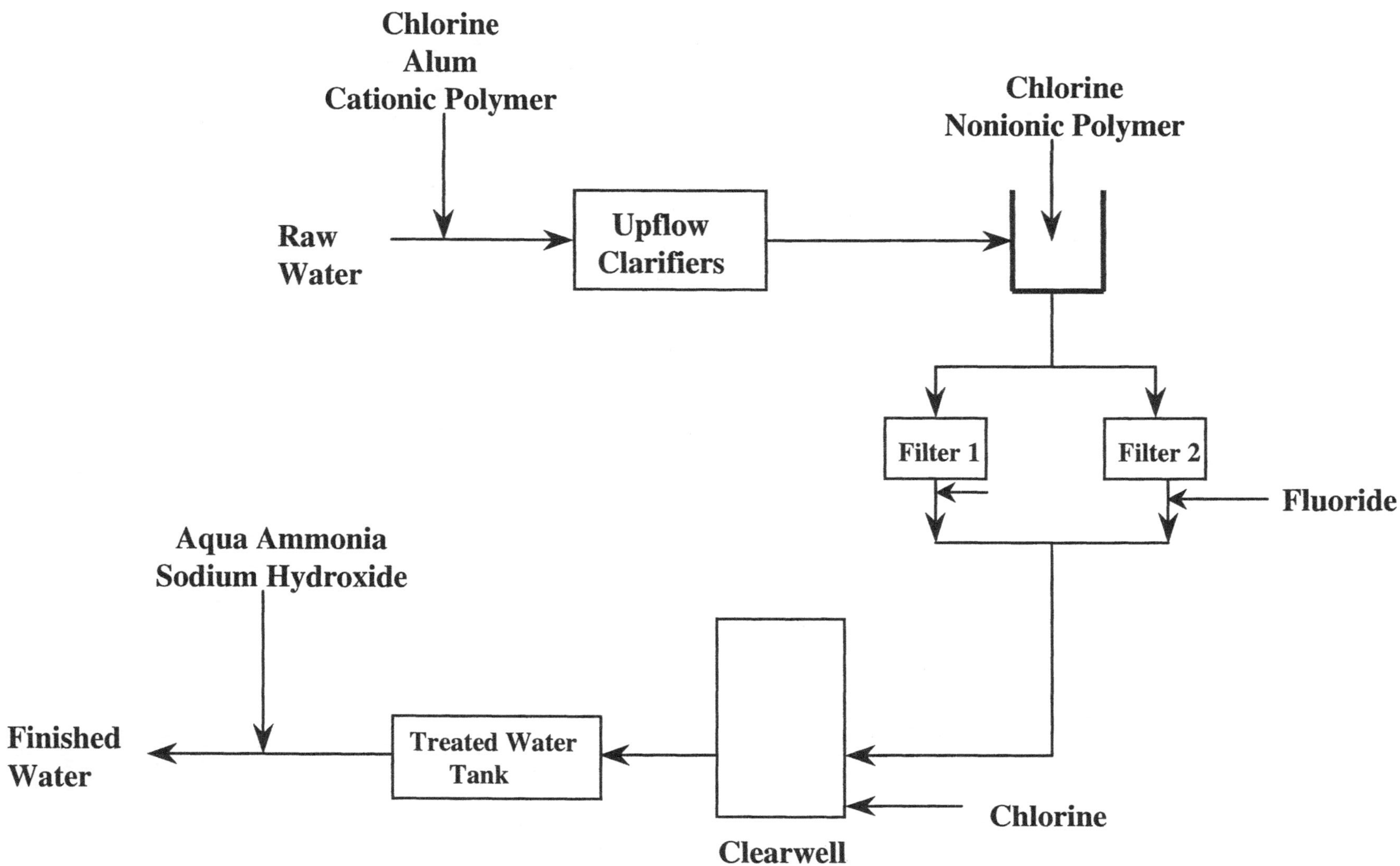

Figure 3.39 MSJWTP plant schematic

The influent to the MSJWTP had moderate levels of TOC (3-6 mg/L) and bromide (38-150 μg/L) (Table 3.18). SUVA—which is an indicator of humic content (Edzwald and Van Benschoten 1990)—was moderate to high (3.0-5.7 L/mg-m).

The raw water (pH varies between 7 and 9) is treated with chlorine mainly for oxidation purposes. The water is coagulated with alum (12-42 mg/L, average = 21 mg/L) and a cationic polymer (Calgon Catfloc C, 1.0-2.1 mg/L, average = 1.4 mg/L) as a coagulant aid. The pH of coagulation ranges between 6.5 to 7.5, depending on dose and alkalinity. A nonionic polymer is added to the clarified water as a filter aid. The water is filtered and disinfected before and after the filters with chlorine. The average disinfection credit—which is primarily achieved through the filters and in the clearwell—is 2.7 logs of *Giardia* inactivation. Before distribution, the water is treated with aqua ammonia to form chloramines, hydrofluosilicic acid for fluoridation, and sodium hydroxide to raise the pH. The finished water is distributed at a pH of 8.3 (on average) and contains a chloramine residual of 1.5 to 2.0 mg/L.

WTP2

WTP2 is an ozonation facility with a design capacity of 28 mgd and a typical flow of 18 mgd. The plant was brought on line in 1993.

The influent to WTP2 had moderate levels of TOC (3-6 mg/L) and bromide (range = 17-200 μg/L) (Table 3.18).

Figure 3.40 shows a schematic of WTP2. The treatment plant consists of two parallel process trains, each with one ozone contactor, two flash mixers (in series), three flocculation basins (in parallel), two sedimentation basins (in parallel), and three filters (in parallel). The total process time to treat water is from two to four hours.

Ozone is used for primary disinfection and for taste-and-odor control. Ozone-rich air is bubbled through the water in a series of ozone-contact chambers. The average ozonation pH was 7.8. However, the pH of ozonation has been as high as 9.2, where ozone demand is typically higher with increasing pH. The temperature of ozonation fluctuated from a minimum of 8.8°C during winter to a maximum of 26°C during summer and by as much as six degrees in a single day. The ozone dosage ranged from 1.7 to 6.5 mg/L, with an average of 3.3 mg/L. This corresponded to a range of ozone-to-TOC ratios of 0.29 to 1.4 mg/mg, with an average of

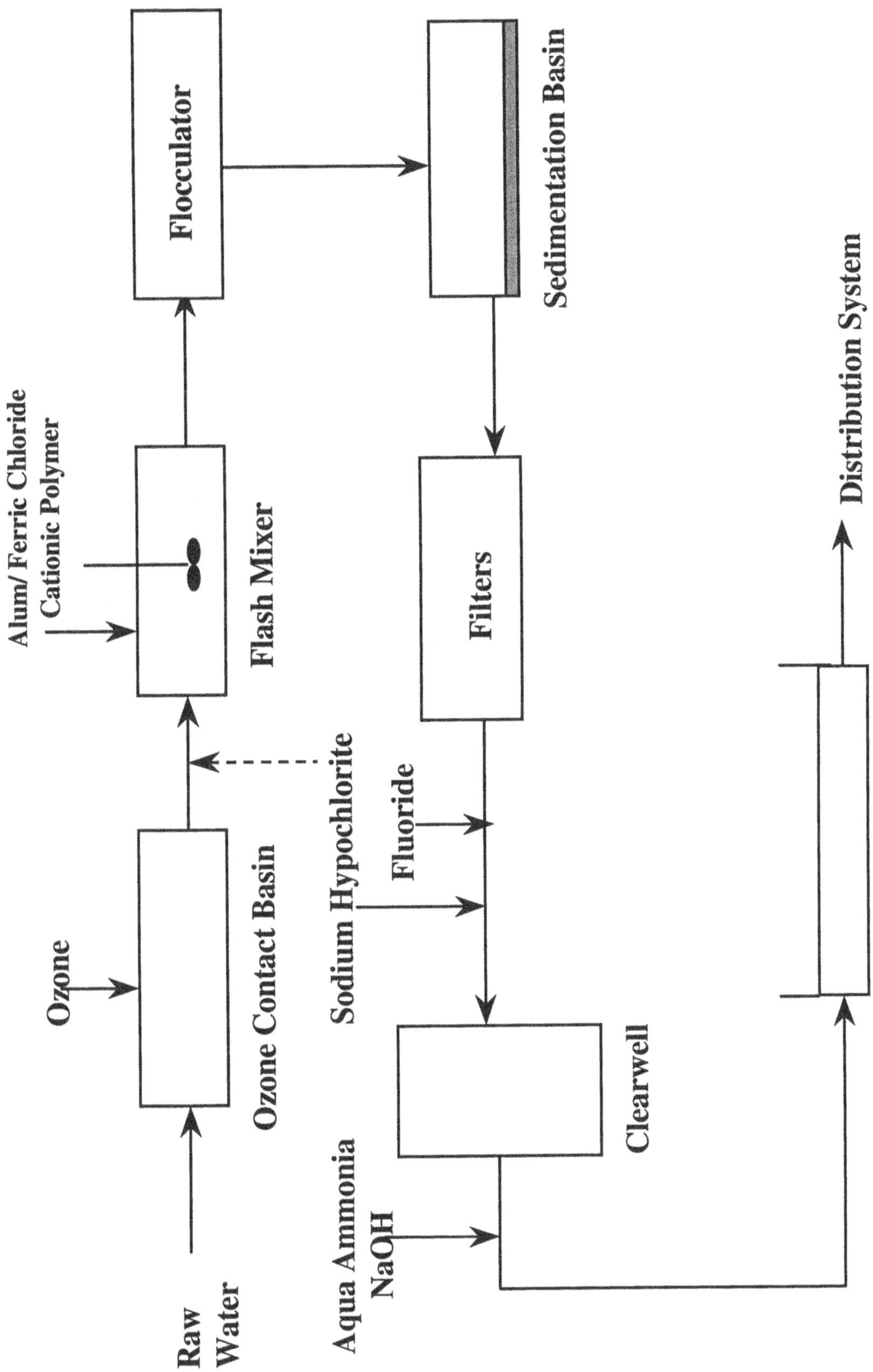

Figure 3.40 WTP2 plant schematic

0.75 mg/mg. The ozone residual in the contactor ranged from 0.3 to 1.2 mg/L, with an average of 0.6 mg/L. (This is typically measured in cell 2; there is generally no detectable ozone in the effluent.)

After ozonation, the water undergoes coagulation, flocculation, and sedimentation. Ferric chloride (8.4-12 mg/L, average = 10 mg/L) and alum (10-26 mg/L, average = 15 mg/L) are the two coagulants used at WTP2. In general, ferric chloride is the main coagulant used. Alum is usually used during periods of low raw-water turbidity or when there are problems with the ferric-feed and/or -storage system. If the coagulant doses are the same, ferric depresses the pH more. However, the ferric dose is generally lower than the alum dose, so the settled-water pH levels are fairly close with each coagulant. In addition, a coagulant aid (Calgon Catfloc C, 1.0-2.9 mg/L, average = 1.5 mg/L) is used.

Following sedimentation, the clarified water is filtered. There are six filters at WTP2, three of them consist of anthracite coal and sand, and three filters consist of GAC and sand. The filters are biologically active and are backwashed with chlorinated water on every third backwash. GAC was used in the filter media to (1) prevent taste-and-odor problems in the finished water, (2) for DBP precursor removal, (3) for the removal of other micropollutants, and (4) for providing a better medium for biological activity. The EBCT of the GAC is 3.2 min at the design flow (28 mgd) and 5.0 min at a typical flow (18 mgd). The original GAC in filter #5 has not been replaced or regenerated since operations began at WTP2; the GAC in filters #4 and #6 was installed in 1998.

The filter effluent is fluoridated and disinfected with free chlorine (sodium hypochlorite) in the clearwell at a dosage of ~1.8 mg/L for ~10 minutes (at 14 mgd) to inactivate the HPC bacteria shed from the filters and to meet the 2-log virus inactivation requirement in the winter. The filter effluent prior to chlorination has had as high as >5500/mL of HPC bacteria. With postchlorination and 5-10 minutes of contact time, typical levels of HPC bacteria in the plant effluent are <1/mL.

After the clearwell, the pH of the water is adjusted for corrosion control (average pH = 8.4) and ammonia is added to form chloramines. The water flows through a 20-MG reservoir prior to the distribution system. The typical detention time in the reservoir is theoretically a day, however it is poorly baffled such that the actual contact time credit is approximately 2-3 hours. The chloramine residual in the finished water is approximately 1.5 mg/L. The disinfection scenario (ozone/chlorine/chloramines) at WTP2 resulted in a calculated 2-3 logs of *Giardia* inactivation and 15-30 logs of virus inactivation.

Operations

Instrumentation and Control

At WTP2, nearly everything can be controlled by the personal computer/programmable logic controller (PC/PLC). The PC is the interface with the PLC, which controls most plant operations.

The WTP2 design included backup equipment. Generally either two pieces of equipment—each at 100 percent capacity—or three pieces—each at 50 percent capacity—were installed. The ozone system has two air compressors, three aftercoolers, two dessicant dryers, three ozone generators, two chillers, and three cooling water pumps. Roughly, half of the cost of the ozone equipment was associated with backup equipment.

Plant Operations

WTP2 is chlorinated (using sodium hypochlorite) on average twice per week for approximately four hours. The chlorine is applied to the effluent of the ozone contactors to help control algae and other growths in the basins. In addition, when the chlorine reaches the filters, it sometimes results in an improvement in filter performance (the rate of headloss buildup slows dramatically).

Challenges

Control of Bromate Formation

The Stage 1 MCL for bromate is 10 μg/L (USEPA 1998a). However, the USEPA may lower the MCL to 5 μg/L in future rule-making efforts (USEPA 1994a; USEPA 2000). Bromate concentrations at WTP2 ranged from "not detected" to 28 μg/L for March 1995 to January 1998.

The lowest measured and reported value was 2.2 µg/L. The average bromate concentration during this time period—when detected—was 10 µg/L. In order to reliably comply with the bromate MCL, ACWD is in the process of modifying the treatment process by installing a carbon dioxide feed system, as lowering the pH of ozonation has been proven to reduce bromate formation (Glaze and Weinberg 1993).

Another solution to the bromate challenge could be source control of the SPW, because during periods of low raw-water bromide (approximately 50-80 µg/L) ACWD has not formed any detectable amount of bromate. If the bromide in the source water were to stay consistently low, WTP2 would not need a carbon dioxide feed system for bromate control and would be able to meet the bromate MCL. (ACWD expects an overall reduction in ozone dose and a general stabilization of the process to be significant additional benefits of a carbon dioxide feed.) ACWD does participate in a California/Federal process to improve SPW. However, if a source-water solution is developed and approved, it would take a long period of time for implementation (most likely decades), so ACWD has decided to implement process changes to minimize bromate formation in the near term.

Aesthetic Concerns

There are aesthetic issues at the plant resulting from the biological process. However, periodic chlorination in the plant helps control algae and other growths in the basins. The application of ferric chloride does stain the basins, impacting the appearance of the plant. However, this is just a plant aesthetic issue.

Implementation and Operational Issues

Operator Training and Start-Up

Training sessions were held at the beginning of WTP2 operations. Both ACWD staff and CDM, the design engineer, did some training. Each equipment supplier provided training on their equipment and Process Applications, Inc., provided specific training for the ozone system. ACWD staff provided training in solids handling and power facility operations and oversaw plant-operations training. CDM developed a fairly general operations and maintenance manual and the ACWD staff developed a binder of standard operating procedures.

Plant operators continuously request additional training in order to search for new and improved ways to do their job. Additional training (post start-up) occurs on both a formal and informal basis. Each year the plant is shut down for at least two weeks and all operators attend at least 40 hours of training. Additional training is done at monthly operators' meetings.

The annual plant shut down is mainly for operator training. Because the operators work 12-hour shifts to provide continuous staffing of the plant seven days a week, it is difficult to get them all to attend training sessions. ACWD tried scheduling multiple sessions, but they had to conduct each session four times, twice during the day and twice at night, which was not practical. When they shut the plant down, they put half of the operators on a Monday through Friday, 8 am to 5 pm schedule and require them to attend a week's worth of training. The remaining operators stay on shift to run the other plant and the distribution system. Then, during the second week, they switch so that all of the operators can attend the sessions. In addition, ACWD gets a lot of maintenance items accomplished during the shutdown. MSJWTP is also shut down each year for at least a week, mainly for maintenance reasons. Both shutdowns are scheduled for the low-demand winter period when system demand is less than half of summer time requirements.

A joint Engineering/Operations start-up team was used at WTP2. The team consisted of five individuals, three from Engineering and two from Operations. They were experts in construction management, electrical engineering/programming, mechanical/structural engineering, process operations, and regulatory compliance. The roles and responsibilities of the team were loosely defined, where the group functioned as a team using their varied skills and backgrounds to get the plant started up. It took the operational staff a period of one to two years to be "comfortable" with the new technology and operate as an integral unit.

Operational Issues

In general, the use of ozone at WTP2 made a very difficult-to-treat water easily treatable. WTP2 produces a very high quality effluent water with low turbidity, low TOC, low THMs and HAAs, and low taste and odor. However, rapid shifts in raw-water quality can make the ozone process difficult to control. The pH can vary by 2 units in 12 hours and the temperature by 6 degrees in the same period. For a constant ozone dose (e.g., 3.1-3.5 mg/L), as the pH rises, the ozone residual drops (Figure 3.41). When the raw-water pH is approximately 7.2, the ozone residual can be approximately 0.7 mg/L, whereas the ozone residual can be approximately 0.2 mg/L for a raw-water pH of approximately 8.5. Also, raw-water turbidity can change rapidly as source water changes are made by the California Department of Water Resources. An ideal solution would be to have a raw-water reservoir to "smooth out" the variations in water quality, but this is not possible. Alternatively, plans have been made to add carbon dioxide to the raw water, which will help with the pH swings. Also, a substantial effort was made to develop an online ozone analyzer that was accurate and reliable. Once the operators could continuously monitor the residual, they could more frequently and accurately adjust the ozone dose.

One of the reasons ACWD selected ozone was to reduce coagulant use. WTP2 uses less coagulant than MSJWTP on average (15 versus 21 mg/L, respectively), which is probably due to microflocculation effects. In general, WTP2 handles the large extreme shifts in raw-water turbidity better than MSJWTP, which is probably due (in part) to the ozone and (in part) to the configuration of the sedimentation basins. The presence of filter-clogging algae in the raw water often resulted in a significant decrease in the flow rates at MSJWTP due to very short filter run times. At WTP2, ozone seems to be able to oxidize these algae such that the filter runs are not that much shorter than normal. It is difficult to clean the filters at MSJWTP with the backwashing system, which uses water only. Air-scour backwashing is used at WTP2, which does a better job of cleaning the filters (fewer mudballs, etc.) as compared to MSJWTP.

The operational experience at WTP2 has shown that backup equipment is an important aspect of reliability in ozone operations. The plant generates ozone nearly continuously in spite of equipment problems. Backup equipment, in addition to maintaining reliability, has allowed the staff to delay maintenance at times. During the initial start-up of the plant, there were a lot of problems with the air compressors, which was mainly due to their large size and the fact that they were cycling so frequently. The problem was solved by adding more air receiver capacity and down-rating one of the units, which made the compressors much more reliable. The water chillers have also posed a problem (in part) because of their large size and the fact that they do not run anywhere near their rated capacity. This problem was addressed by reconfiguring the

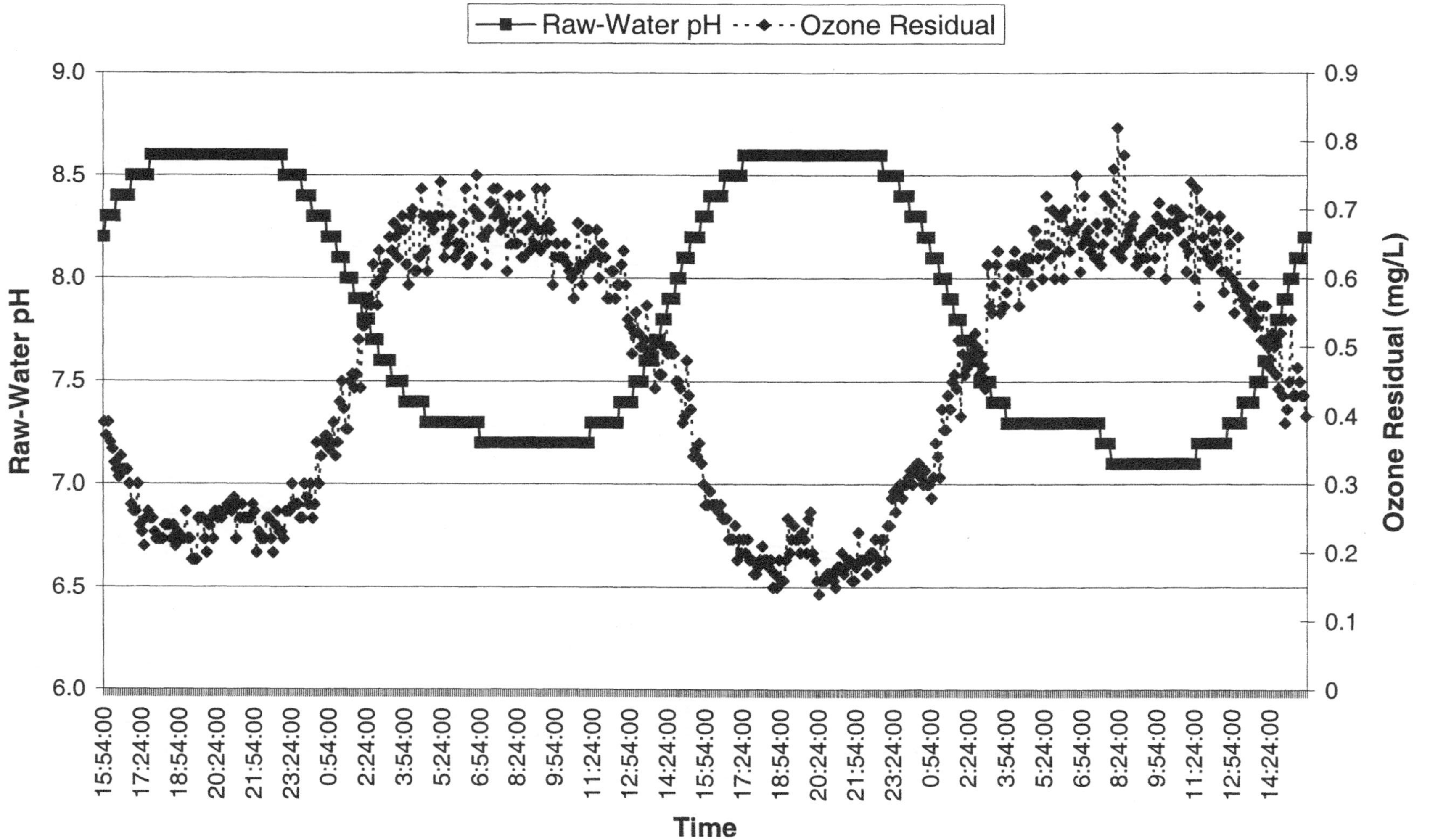

Figure 3.41 Effect of variations in raw-water pH during the day on ozone residual at ACWD on July 13-15, 1997 (applied ozone dose = 3.1 – 3.5 mg/L)

system so that in the winter time, the cold plant utility water is used to cool the ozone equipment and the chillers are turned off.

System Performance

Water Quality

Most of the water-quality data for the case study were from 1995 and 1996 (Table 3.18). The raw-water quality (e.g., TOC, bromide, turbidity, temperature) were similar for both of the plants. Additional TOC, bromide, and DBP data from 1997 were obtained for the case study.

TOC: At MSJWTP, the influent TOC varied from 3.0 to 6.4 mg/L, with an average value of 4.1 mg/L. Influent TOC was lowest in the summer (monthly averages of 3.2 to 4.0 mg/L) and highest in the winter (monthly averages of 4.5 to 5.7 mg/L). In the SPW watershed, many agricultural tracts of land are on peat soil, with agricultural drainage a significant source of TOC (Krasner, Sclimenti, and Means 1994). During the winter, leaching of the soil to remove salt buildup (from summer irrigation) results in agricultural drainage contributing high levels of TOC in the winter. Conventional treatment resulted in plant effluent values ranging from 1.8 to 4.0 mg/L, with an average of 2.6 mg/L. TOC removal typically ranged from approximately 20 to 50 percent at MSJWTP, with an average value of 35 percent.

Box-and-whiskers plots (Figure 3.42) show the variation in TOC concentration at WTP2 for raw, ozonated, and filtered water. At WTP2, the influent TOC ranged from 3.1 mg/L to 6.4 mg/L, with an average of 4.6 mg/L and a median of 4.5 mg/L. Ozonation did not significantly reduce the TOC concentration (range = 1.3 to 6.6 mg/L, average = 4.1 mg/L, median = 3.8 mg/L). At WTP2, the plant-effluent TOC ranged from 1.4 to 3.7 mg/L, with an average of 2.5 mg/L and a median of 2.4 mg/L. At WTP2, TOC removal ranged from approximately 30 to 60 percent (Table 3.19), with an average TOC removal of 47 percent, which was higher than the TOC removal achieved at MSJWTP.

The two plants use different coagulants (ferric versus alum) and have different configurations of the clarification process (horizontal versus upflow). Additional data were obtained for the past two years, which showed (on average) that 28 percent of the TOC was removed through clarification and a total of 36 percent was removed through filtration (with

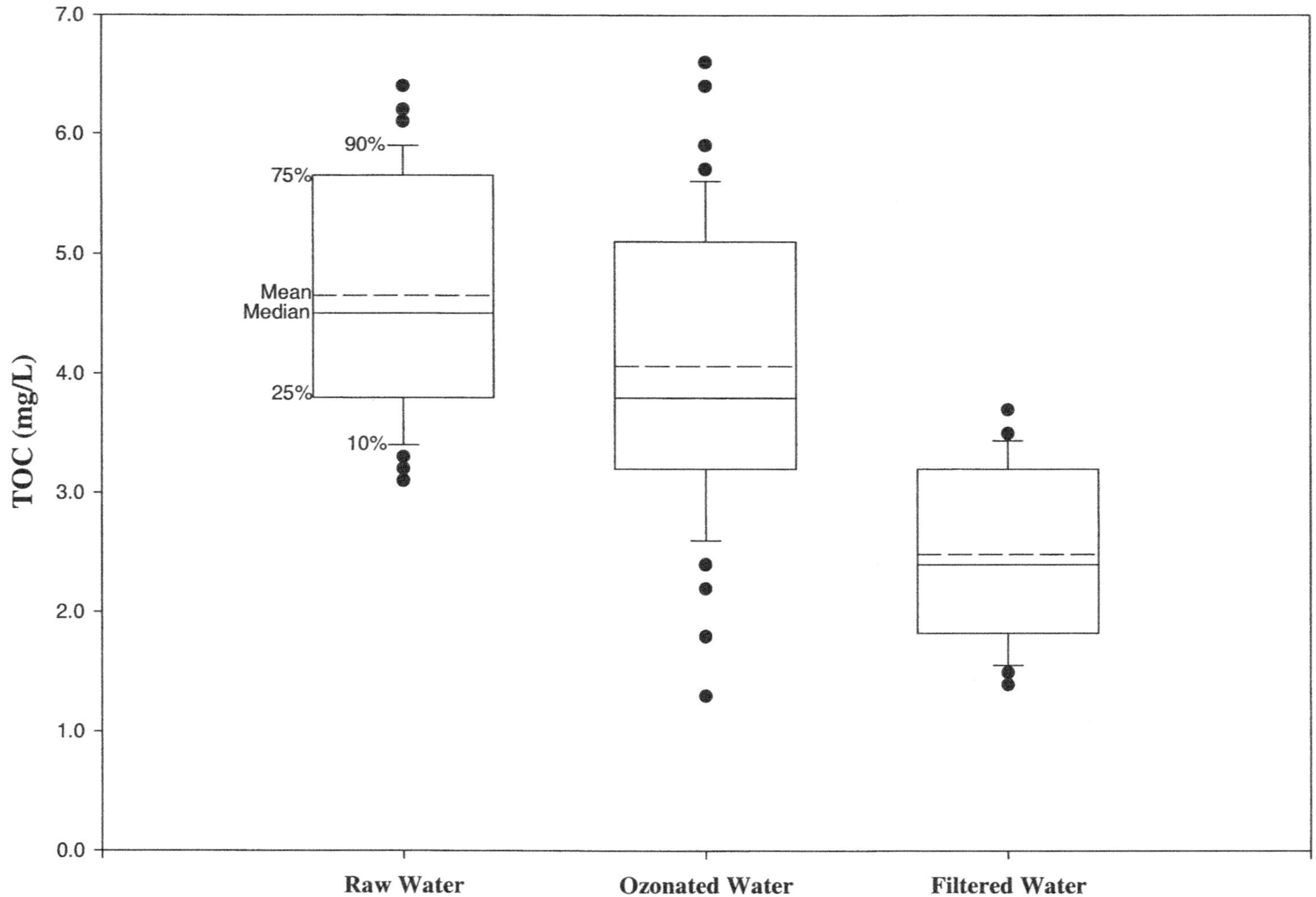

Figure 3.42 Variations in TOC concentration at WTP2

Table 3.19

ACWD WTP2 TOC removal (%) and credit

Date	TOC (mg/L) Raw	TOC (mg/L) Filtered	Actual TOC Removal (%)	Raw-water alkalinity (mg/L as $CaCO_3$)	Required TOC removal (%)	Removal credit: actual removal %/ required removal %	Average monthly credit
January 3, 1995	5.8	3.4	41	82	35	1.18	
January 10, 1995	5.6	2.9	48	74	35	1.38	
January 17, 1995	5.9	3.3	44	139	25	1.76	
January 24, 1995	5.9	3.5	41	140	25	1.63	
January 31, 1995	6.0	3.5	42	170	25	1.67	
January 1995*	5.6-6.0	2.9-3.5	41-48	74-170	25-35	1.38-1.76	1.52
February 1995	5.6-5.8	3.3-3.7	34-43	152-158	25	1.36-1.72	1.57
March 1995	5.8-6.1	3.1-3.7	36-49	102-128	25-35	1.28-1.90	1.51
April 1995	5.6-5.9	3.2-3.7	37-43	90-106	35	1.07-1.22	1.16
May 1995	4.3-5.3	1.9-3.2	40-56	56-112	35-45	1.13-1.60	1.33
June 1995	3.7-4.6	1.6-1.8	51-61	38-48	35-45	1.35-1.68	1.51
July 1995	3.6-4.0	1.5-1.9	52-58	40-45	35	1.50-1.67	1.57
August 1995	3.1-4.6	1.4-1.6	52-65	48-55	35-45	1.45-1.57	1.50
September 1995	3.7-4.5	1.5-2.3	45-60	59-112	35	1.36-1.70	1.47
October 1995	3.9-4.5	2.3-2.6	40-42	89-103	25-35	1.16-1.64	1.30

(continued)

Table 3.19 (Continued)

Date	TOC (mg/L) Raw	TOC (mg/L) Filtered	Actual TOC Removal (%)	Raw-water alkalinity (mg/L as $CaCO_3$)	Required TOC removal (%)	Removal credit: actual removal %/ required removal %	Average monthly credit
November 1995	4.2-4.4	2.5-2.8	36-40	106-114	35	1.04-1.16	1.10
December 1995	3.3-4.6	1.7-3.3	28-48	84-156	25	1.13-1.94	1.40
Sum/12†						January 1995 - December 1995	1.41
January 1996	3.4-6.4	1.9-3.5	38-55	66-110	25-35	1.07-1.76	1.50
February 1996	5.4-5.8	3.4-3.4	37-41	90-147	25-35	1.06-1.66	1.36
March 1996	5.2-5.5	2.6-3.0	46-50	88-96	35	1.30-1.43	1.37
Sum/12						April 1995 - March 1996	1.38
April 1996	3.7-4.0	1.9-2.4	40-49	50-70	25-35	1.20-1.80	1.50
May 1996	3.7-3.8	1.7-1.9	49-55	42-68	25-35	1.39-2.21	1.66
June 1996	3.2-3.8	1.5-1.6	50-58	50-63	25-35	1.55-2.00	1.70
Sum/12						July 1995 - June 1996	1.45

*Range of monthly values (sampled 2-5 times/month; typically sampled weekly).

†Running annual average, updated quarterly and based on monthly average values.

eight percent of the raw-water TOC removed through the filters) at MSJWTP; whereas 33 percent of the TOC was removed through sedimentation and a total of 45 percent was removed through filtration (with 12 percent of the raw-water TOC removed through the filters) at WTP2. The additional TOC removal at WTP2 may be due (in part) to the conversion of TOC by ozone to biodegradable organic carbon, a portion of which can be removed through the biologically active filters.

TOC removal credit: From the data provided (Table 3.19), it was possible to calculate the TOC removal credit—which is required with Stage 1 of the D/DBP Rule (USEPA 1998a)—achieved by WTP2 over the period reported (January 1995 to June 1996). TOC removal credit was calculated (actual TOC removal/required TOC removal) as running annual averages, which were updated quarterly and based on monthly average values. Required TOC removal requirements were based on source-water alkalinity and TOC, as per the D/DBP Rule (Table 3.20). Average monthly TOC removal credit for WTP2 ranged from 1.10 to 1.70, with running annual average values from 1.38 to 1.45. A system is deemed to be in compliance with the DBP precursor control requirement if the normalized running annual averages are at least 1.00. At WTP2, the actual TOC removal was always greater than the required TOC removal.

Table 3.20

Required removal of TOC by enhanced coagulation

for surface-water systems using conventional treatment

Source-water TOC	Source-water alkalinity (mg/L as $CaCO_3$)		
(mg/L)	0-60	>60-120	>120
>2.0–4.0	*(Box 1)* 35.0%	*(Box 2)* 25.0%	*(Box 3)* 15.0%
>4.0–8.0	*(Box 4)* 45.0%	*(Box 5)* 35.0%	*(Box 6)* 25.0%
>8.0	*(Box 7)* 50.0%	*(Box 8)* 40.0%	*(Box 9)* 30.0%

Figure 3.43 shows the distribution of raw-water alkalinity and TOC over time at WTP2. As can be seen, the variations in water quality caused the raw water to fall into five different boxes of the 3x3 matrix for required TOC removal over the time period sampled. Box 5 (intermediate TOC and intermediate alkalinity) predominated, followed by boxes 1 (low alkalinity and low TOC), 6 (high alkalinity and intermediate TOC), 2 (intermediate alkalinity

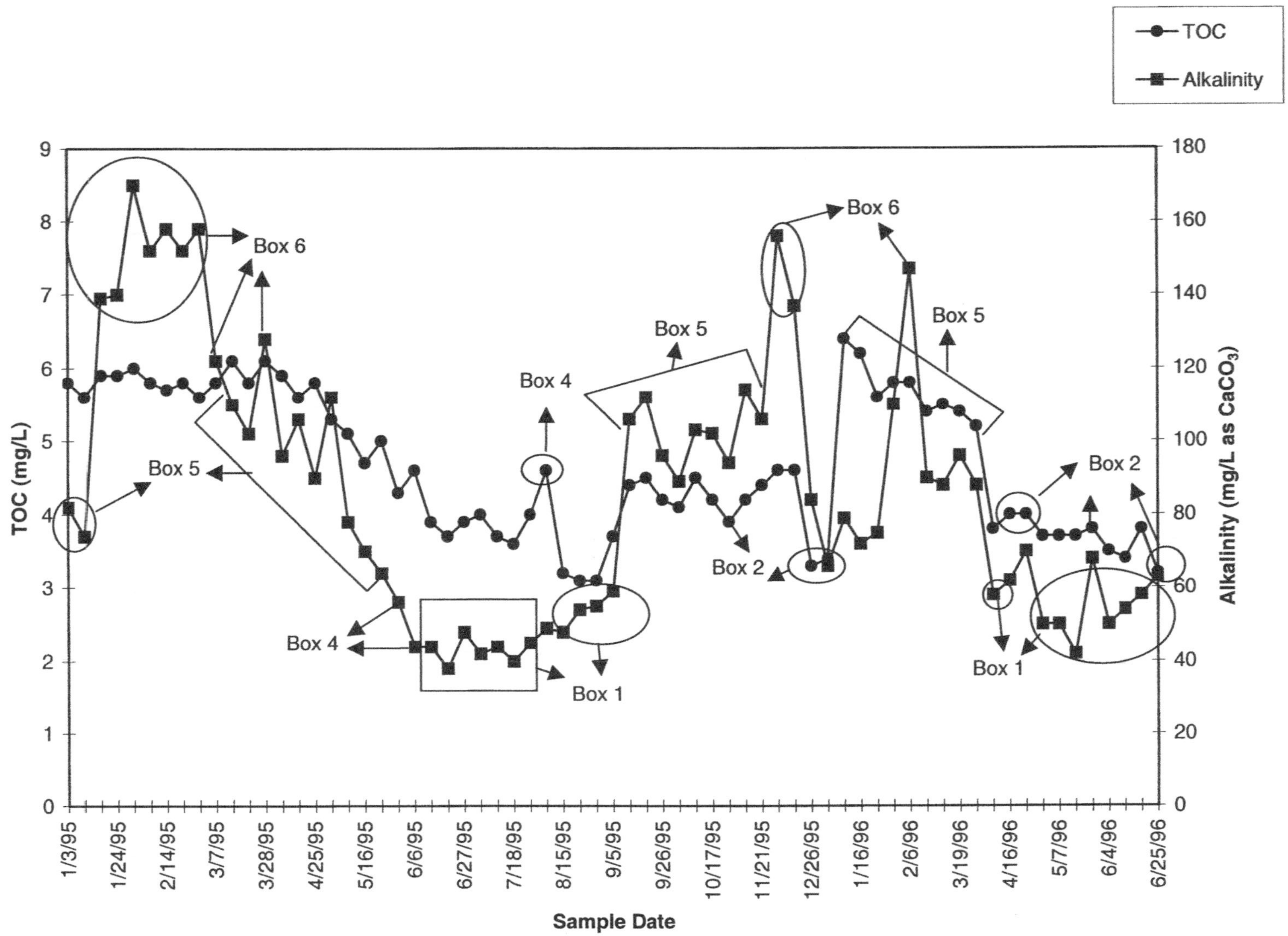

Figure 3.43 ACWD WTP2 raw-water TOC and alkalinity variations over time: boxes correspond to 3X3 TOC/alkalinity matrix (Table 3.20)

and low TOC), and 4 (low alkalinity and intermediate TOC), respectively. Typically, alkalinity and TOC were higher in the winter and lower in the summer. In spite of these wide variations in water quality, WTP2 consistently removed the required amount of TOC.

UV: UV data were available for the raw and finished water at MSJWTP and the finished water at WTP2 (Table 3.18). The MSJWTP influent had an average UV absorbance of 0.19 cm^{-1} (range = 0.12-0.27 cm^{-1}). SUVA—which was calculated using the formula UV x 100/DOC, and it was assumed that the influent TOC was equivalent to the DOC (studies of SPW indicate that ~90-95 percent of the TOC is DOC) (Owen, Amy, and Chowdhury 1993b)—was 4.5 L/mg-m (on average) at MSJWTP (the SUVA values ranged from 3.0 to 5.7 L/mg-m). These values are indicative of moderate-to-high levels of humic substances in the raw water (Edzwald and Van Benschoten 1990). Increasing SUVA and, thereby, increasing amounts of humic material in the water tended to result in an increased level of TOC removal by the treatment process. The maximum TOC removal (54 percent) was achieved at the highest SUVA value (5.7 L/mg-m), whereas the lowest TOC removal (35 percent) was achieved at the lowest SUVA value (3.0 L/mg-m).

At MSJWTP, the raw-water UV was reduced—via physical removal through coagulation and via chemical oxidation from chlorination—to a range of 0.07 to 0.16 cm^{-1} (average = 0.10 cm^{-1}). At WTP2, a similar range of finished-water UV values (0.06-0.13 cm^{-1}; average = 0.08 cm^{-1}) was achieved via ozonation and coagulation.

Turbidity: The MSJWTP influent had a turbidity range of 1.6 to 29 NTU, with an average of 9.8 NTU, and the WTP2 influent had a range of 1.7 to 69 NTU, with an average of 16 NTU (Table 3.18). The MSJWTP effluent had a range of 0.04 to 0.20 NTU, with an average of 0.07 NTU, whereas the WTP2 effluent had a range of 0.02 to 0.08 NTU, with an average of 0.06 NTU. ACWD has been meeting their turbidity goal of <0.1 NTU 99 percent of the time at WTP2. The coagulant dose—which was optimized for turbidity removal and not TOC removal—was resulting in achieving good TOC removals while producing very low-turbidity finished water. It should be noted that turbidity excursions up to 200 NTU have been observed; the time period studied was exceptionally low.

Bromide: MSJWTP had influent bromide ranging from 38 to 150 μg/L, with a mean of 63 μg/L (Table 3.18). At WTP2 (January 1995 - June 1996), the values ranged from 17 to 200 μg/L, with a mean value of 74 μg/L. In U.S. surface waters, the median bromide concentration is 45 μg/L, and the 75th and 90th percentile values are 76 and 160 μg/L, respectively (Amy et al. 1994, USEPA 1994a). SPW—which typically has bromide levels at the high end of the cumulative probability distribution nationwide—has always been a challenge to treat because of the significant impact of bromide on DBP formation. Saltwater intrusion is the major source of bromide in SPW (Krasner, Sclimenti, and Means 1994).

Because WTP2 is an ozonation plant, bromide plays a significant role in compliance strategies due to the oxidation of bromide to bromate, which is a regulated DBP (USEPA 1998a). MSJWTP, being a conventional plant, does not form bromate but has the potential to form significant concentrations of THMs and HAAs—which are also regulated (USEPA 1998a)—as well as more of the brominated species, which are believed to be more toxic than their chlorinated counterparts (Bull 1995, Pegram 1995).

Bromate: During the period July 1995 to June 1996, bromate concentrations at WTP2 ranged from <4.5 to 25 μg/L in the filtered water (Table 3.21). The running annual average—where nondetects (e.g., <5 μg/L) were set at one-half the detection limit (e.g., 2.5 μg/L) when calculating the average—was 6.9 μg/L (Table 3.22). During this period of sampling, the raw-water bromide was 17 to 150 μg/L, with an average of 67 μg/L.

Figure 3.44 shows variations in raw-water bromide and finished-water bromate over a broader period of time (January 1996 to January 1998). During this period, the average and maximum bromide concentrations increased to 103 and 462 μg/L, respectively. In 1997, bromate formation tended to be higher than in the earlier sampling period. When bromate was detected in this study period, the conversion of bromide to bromate (on a molar basis) ranged from 1.3 to 14 percent, with an average of 4.3 percent.

Figure 3.45 depicts the variations in bromate formation over time as a function of selected water quality parameters (alkalinity, bromide, and temperature). During this time period (July 1995 - June 1996), TOC ranged from 3.5 to 5.5 mg/L and the pH was 7.4-8.0. Although bromate formation increased in April and May 1996 (to >10 μg/L) with an increase in influent bromide concentration (to >100 μg/L), a similar increase was not observed during the January 1996 sampling. This may have been due, in part, to temperature, as larger amounts of bromate were formed during spring and summer as opposed to fall and winter. In addition, alkalinity was relatively high (80-126 mg/L) during the fall and winter (low bromate formation seasons) and relatively low (43-93 mg/L) during the spring and summer (relatively high bromate formation

Table 3.21

ACWD DBP data: comparison of conventional treatment (MSJWTP) and ozonation (WTP2)

DBP	MSJWTP	WTP2
TTHM*		
Minimum	43	1.3
Average†	68	6.5
Maximum	95	33
HAA5*		
Minimum	23	1
Average†	45	4
Maximum	59	10
HAA6*		
Minimum	28	2
Average†	52	6
Maximum	68	14
TOX‡		
Minimum	157	<50
Average	267	99§
Maximum	398	257
Bromate*		
Minimum		<4.5
Average†		6.9**
Maximum		25

*Data collected between July 1995 and June 1996 (4 quarters).

†Running annual average on data collected between July 1995 and June 1996.

‡Data collected between December 1996 and April 1998.

§Not detected (<50 μg/L) set at ½ the detection limit (i.e., 25 μg/L) when calculating average.

**Not detected (<5 μg/L) set at ½ the detection limit (2.5 μg/L) when calculating average.

Table 3.22

Compliance with bromate MCL: sensitivity analysis using ACWD WTP2 data

Date	Raw data*	ND = ½ DL†	ND = 0	ND = DL
July 1995	8.0	8.0	8.0	8.0
August 1995	8.8	8.8	8.8	8.8
September 1995	14.3	14.3	14.3	14.3
Summer Quarter	10.4	10.4	10.4	10.4
October 1995	?‡	2.5	0.0	5.0
November 1995	?‡	4.4§	2.5	6.3
December 1995	?‡	3.7§	2.0	5.3
Fall Quarter		3.5	1.5	5.5
January 1996	?‡	4.5§	3.7	5.3
February 1996	Not Sampled			
March 1996	?‡	2.5	0.0	5.0
Winter Quarter		3.5	1.8	5.2
April 1996	12.5	12.5	12.5	12.5
May 1996	?‡	16.4§	15.7	17.2
June 1996	?‡	2.3	0.0	4.5
Spring Quarter		10.4	9.4	11.4
July 1995 - June 1996**		6.9	5.8	8.1

*When bromate was sampled more than once in a month, the average value for that month is reported.

†Detection limits were either 5 or 4.5 μg/L.

‡Indicates that monthly averages for the raw data could not be calculated due to selected data reported as not detected.

§Monthly average of multiple sets of values obtained within the month, some of which were ND and were set to ½ DL and others of which were greater than the DL, resulting in an average value for the month that was greater than ½ DL.

**Running annual averages based on annual average of quarterly averages of monthly samples.

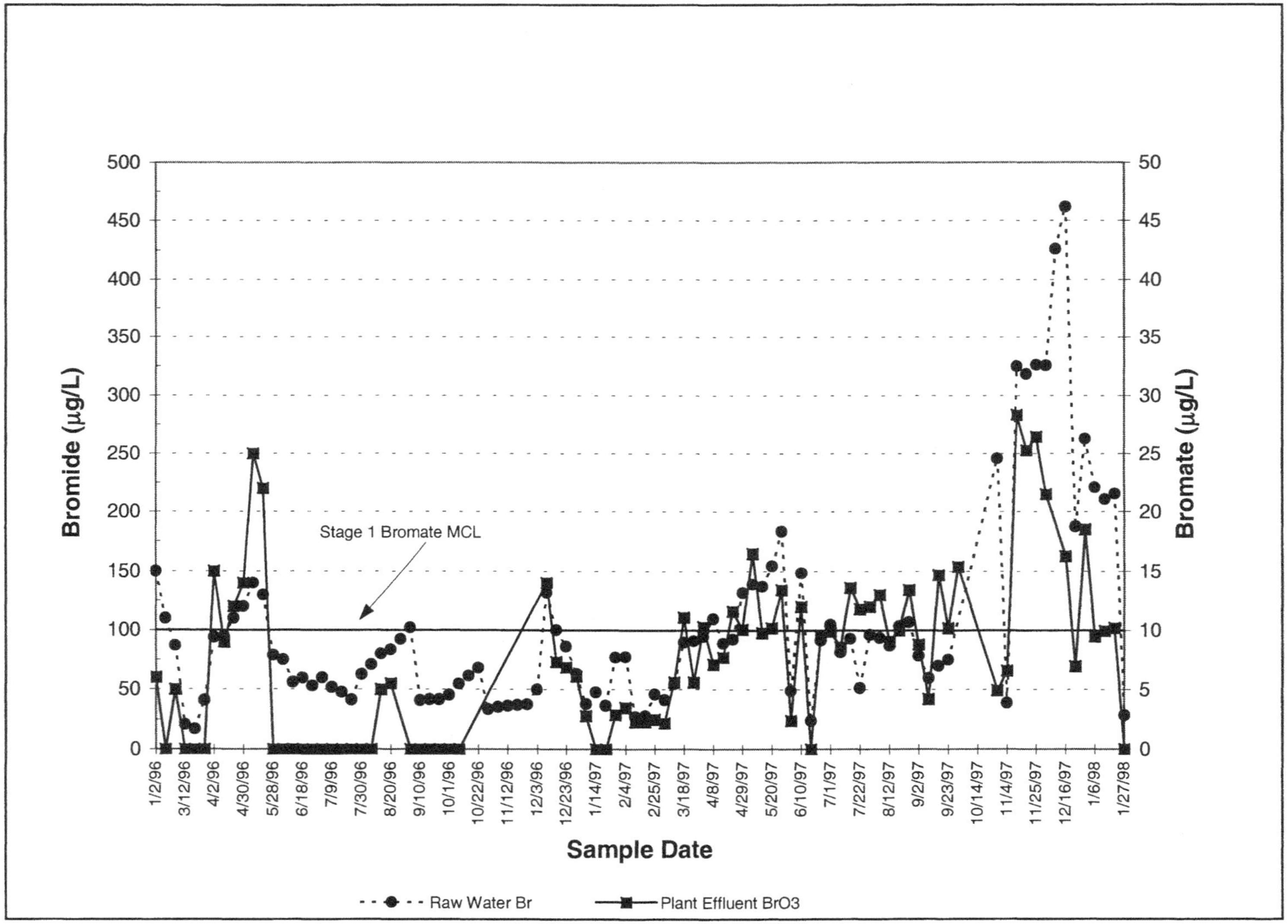

Figure 3.44 ACWD WTP2 raw-water bromide and finished-water bromate over time

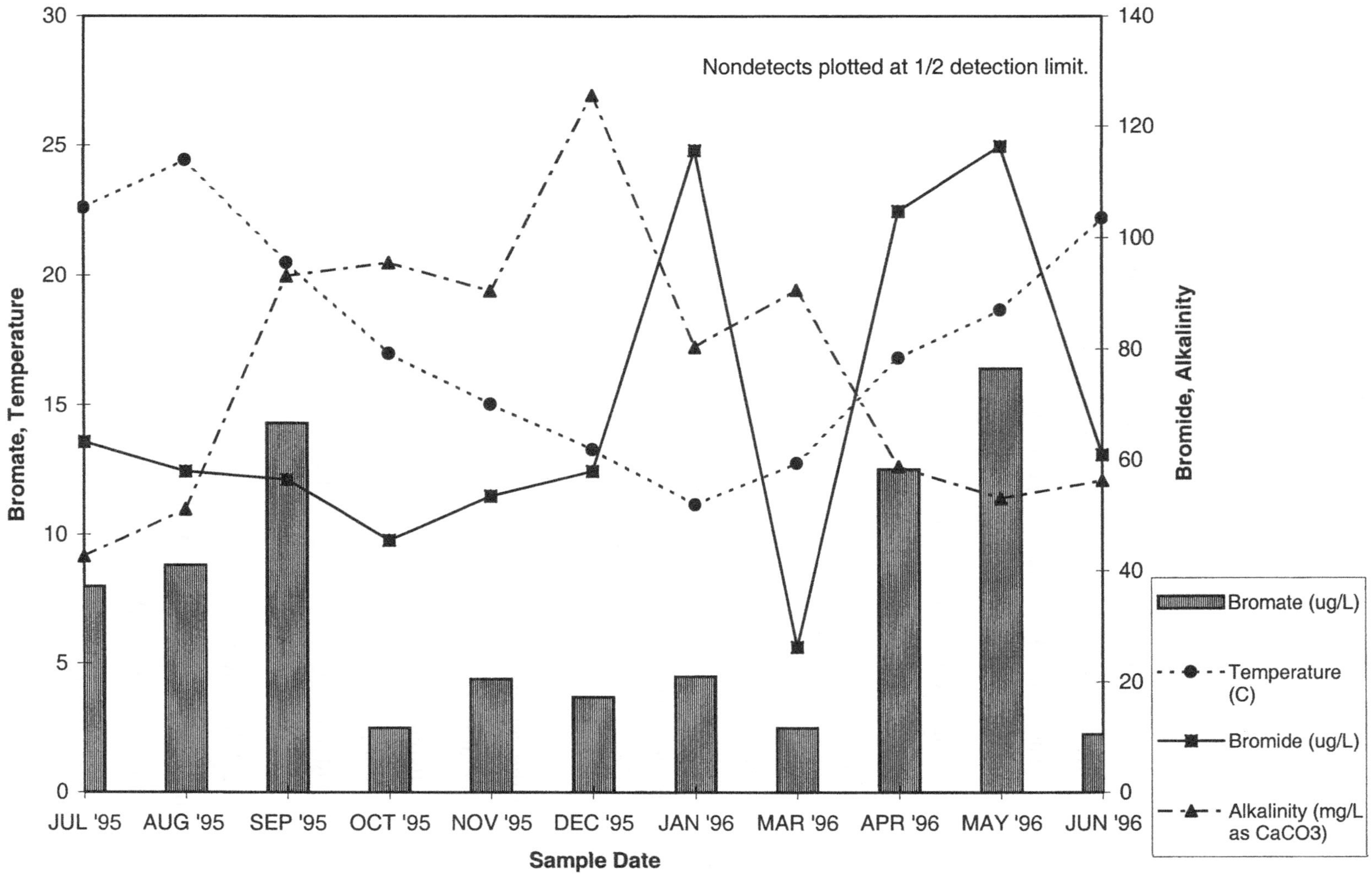

Figure 3.45 ACWD WTP2: variations in bromate formation as a function of selected water quality parameters

seasons). High alkalinity in the influent water could scavenge hydroxyl radicals and, hence, reduce bromate formation. A combination of water quality and operational parameters impact bromate formation (Krasner et al. 1993).

Bromate sensitivity analysis: A significant portion of the bromate data at WTP2 was below the DL. The DL for the analysis was 5.0 or 4.5 μg/L initially. Subsequently, the DL was lowered to 1.5 μg/L. In the ICR, the minimum reporting level for bromate was 5 μg/L (USEPA 1996b). A sensitivity analysis was conducted on the bromate data to examine the effect on bromate compliance with the MCL if values below the DL are set at zero, one-half the DL, and equivalent to the DL (Table 3.22). This is especially important as the MCL for bromate is 10 μg/L (USEPA 1998a). Moreover, during the negotiated regulation for the D/DBP rule-making, it was assumed that a utility would need to design for treatment processes that meet the MCL with a 20-percent safety factor (i.e., <80 percent of the MCL or <8.0 μg/L for bromate) (Krasner, Owen, and Cromwell 1996).

When the DL was 1.5 μg/L, bromate was almost always detected at WTP2. Based on results that have been reported at less than 5 μg/L, setting the NDs for a DL of 4.5-5.0 μg/L at zero—at least at WTP2—is likely an underestimation of possible formation. However, setting NDs at the DL can be an overestimation. Setting the NDs at one-half of the DL may be an appropriate estimate of occurrence, at least when averaged over time (e.g., a one-year compliance period).

The results of the sensitivity analyses (Table 3.22) indicate that if the NDs are set to the most conservative value at the DL, the running annual average for bromate formation (8.1 μg/L) is not reliably within the safety margin of the Stage 1 bromate MCL. Assuming NDs equal to zero yields a value of 5.8 μg/L. Alternatively, choosing an ND value equivalent to half the DL results in a running annual average of 6.9 μg/L. For the purpose of this case study, any bromate value reported below the DL was assumed to be equivalent to half of the DL.

TTHMs: Figure 3.46 shows TTHM concentrations (January 1996 to January 1998) in the effluents of WTP2 and MSJWTP. At MSJWTP, the 10th to 90th percentile TTHMs were 54 to 90 μg/L, with a median of 74 μg/L. At WTP2, the 10th to 90th percentile TTHMs were 2.3 to 13 μg/L, with a median of 3.4 μg/L. Periodic spikes (up to 129 μg/L) in TTHM formation at WTP2 correspond to samples collected when the basins are chlorinated.

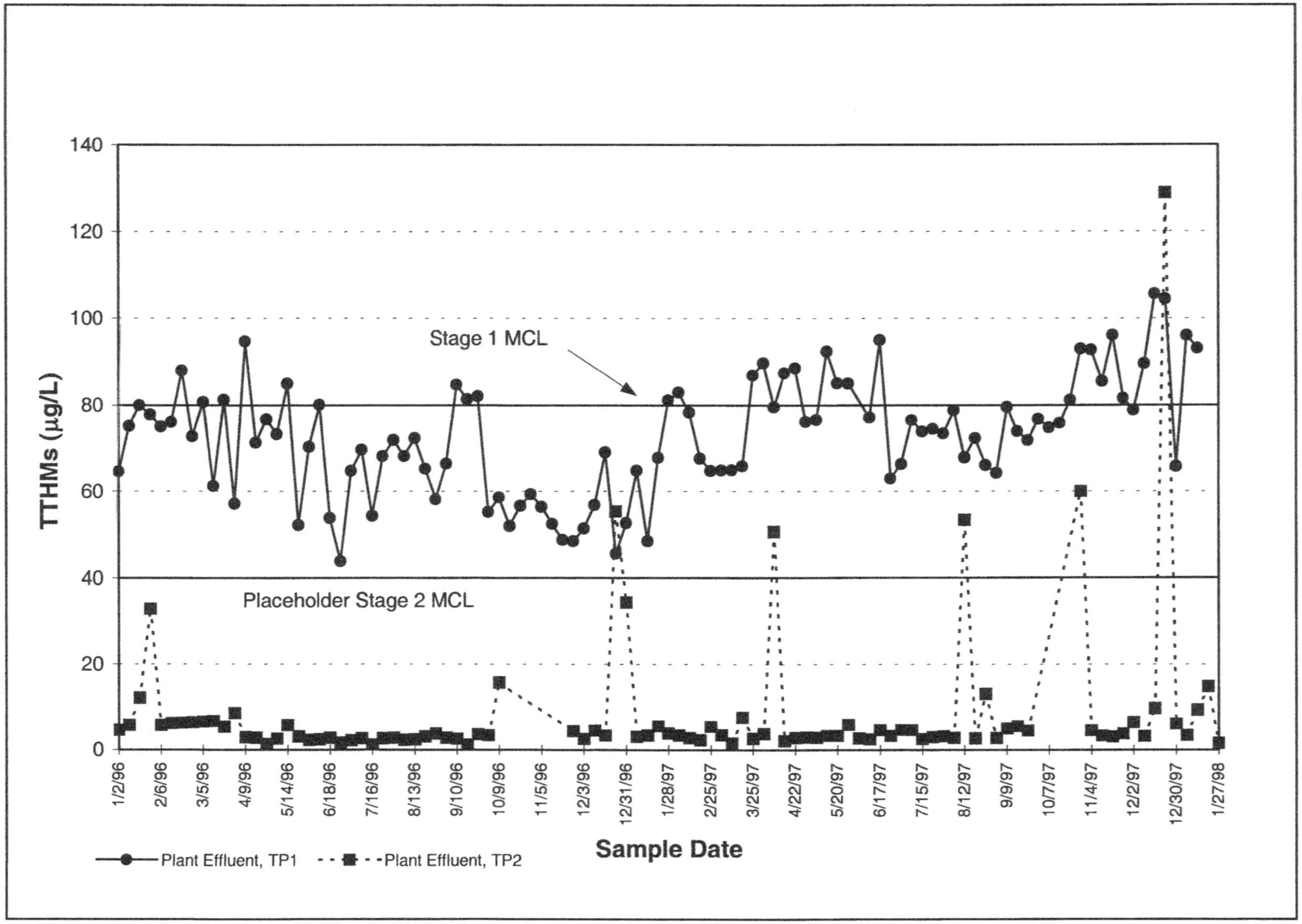

Figure 3.46 ACWD: TTHM concentrations at MSJWTP (TP1) and WTP2 (TP2)

For the time period of July 1995 to June 1996, the running annual average at WTP2 was 6.5 μg/L, whereas at MSJWTP the running annual average was 68 μg/L. At its current level of TTHM formation, it would be hard for MSJWTP to meet the Stage 1 TTHM MCL of 80 μg/L with a 20-percent safety factor (i.e., <64 μg/L) and it would not be able to meet the previously proposed placeholder Stage 2 TTHM MCL of 40 μg/L with the current treatment.

HAAs: Figure 3.47 shows the concentrations of HAA5 (January 1996 to January 1998) in the effluents of WTP2 and MSJWTP. At MSJWTP, the 10th to 90th percentile HAA5 concentrations were 28 to 56 μg/L, with a median of 36 μg/L. At WTP2, the 10th to 90th percentile HAA5 concentrations were 2.1 to 12 μg/L, with a median of 4.0 μg/L.

At MSJWTP (July 1995 to June 1996), the HAA5 running annual average was 45 μg/L and the HAA6 running annual average was 52 μg/L. At WTP2, the HAA5 running annual average was 4 μg/L and the HAA6 running annual average was 6 μg/L. The Stage 1 and previously proposed placeholder HAA MCLs are currently based on HAA5 values: Stage 1 and 2 MCLs of 60 and 30 μg/L, respectively (USEPA 1994a, 1998a). However, future HAA MCLs may be based on HAA6 values. With current treatment, the MSJWTP meets the Stage 1 HAA5 MCL with a safety factor (i.e., <48 μg/L)—although exceeds this level when based on HAA6—and does not meet the previously proposed placeholder Stage 2 MCL.

Aldehydes: The following seven aldehydes are analyzed for at WTP2: acetaldehyde, butanal, formaldehyde, glyoxal, methyl-glyoxal, pentanal, and propanal. After ozonation, the sum of these aldehydes was typically 45-65 μg/L. After biological filtration, most of the results are NDs. When high aldehyde concentrations are seen in the filter effluent, it is generally due to formaldehyde and glyoxal.

In the winter and spring of 1995, aldehyde concentrations in the anthracite filters were quite high (up to approximately 40 μg/L), whereas the levels remained low in the GAC filter effluent (typically <10 μg/L). This was believed to be temperature-related, that the microbial activity in the anthracite filters slowed down significantly during the colder weather. The microbes in the GAC filter, however, seemed to remain more active—as measured by aldehyde uptake—during the winter. During the winters of 1996 and 1997, ACWD has not experienced a difference in performance between the filter media types (Figure 3.48). It may be that the microbes in the anthracite filters took longer to get established, and now they can handle cold weather better.

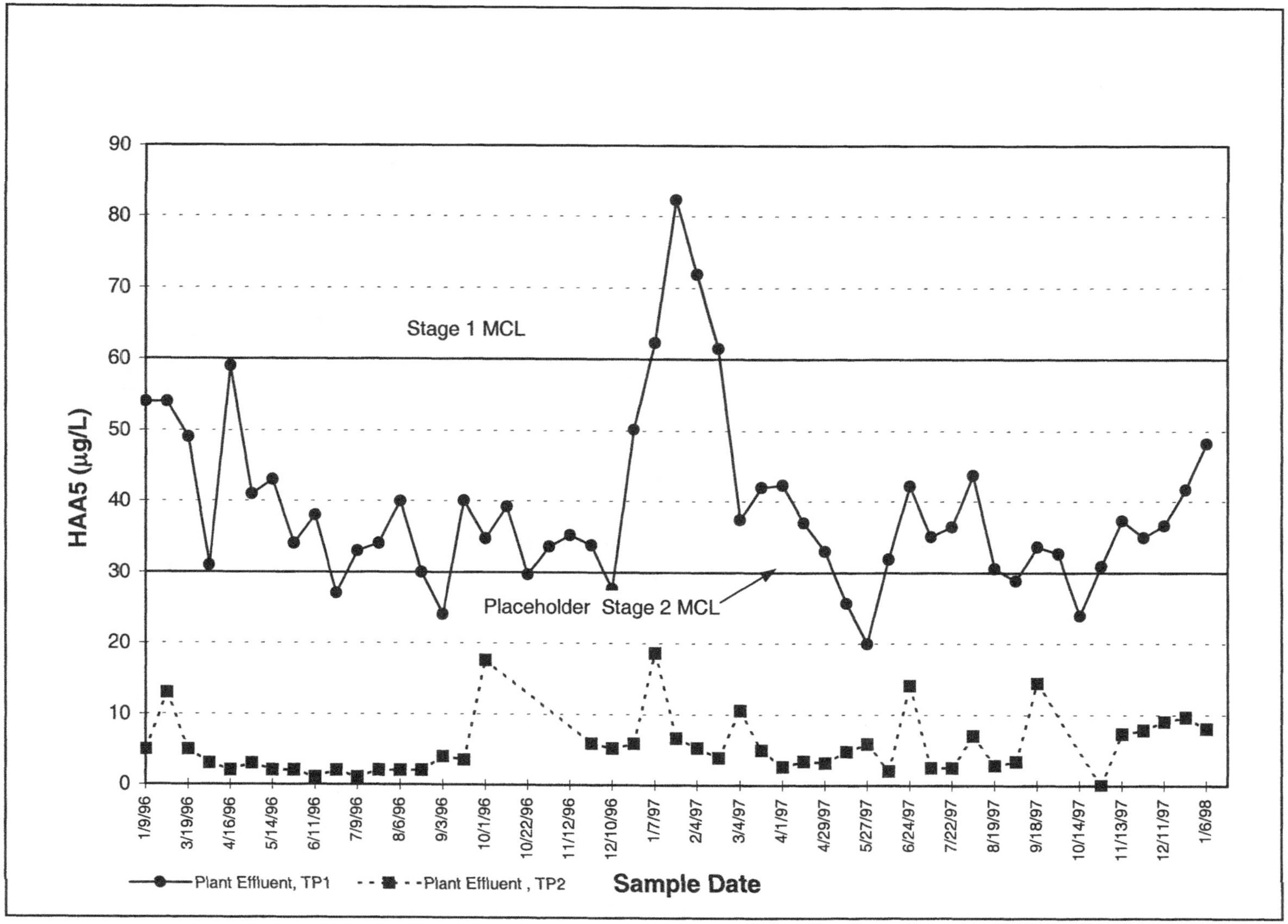

Figure 3.47 ACWD: HAA5 concentrations at MSJWTP (TP1) and WTP2 (TP2)

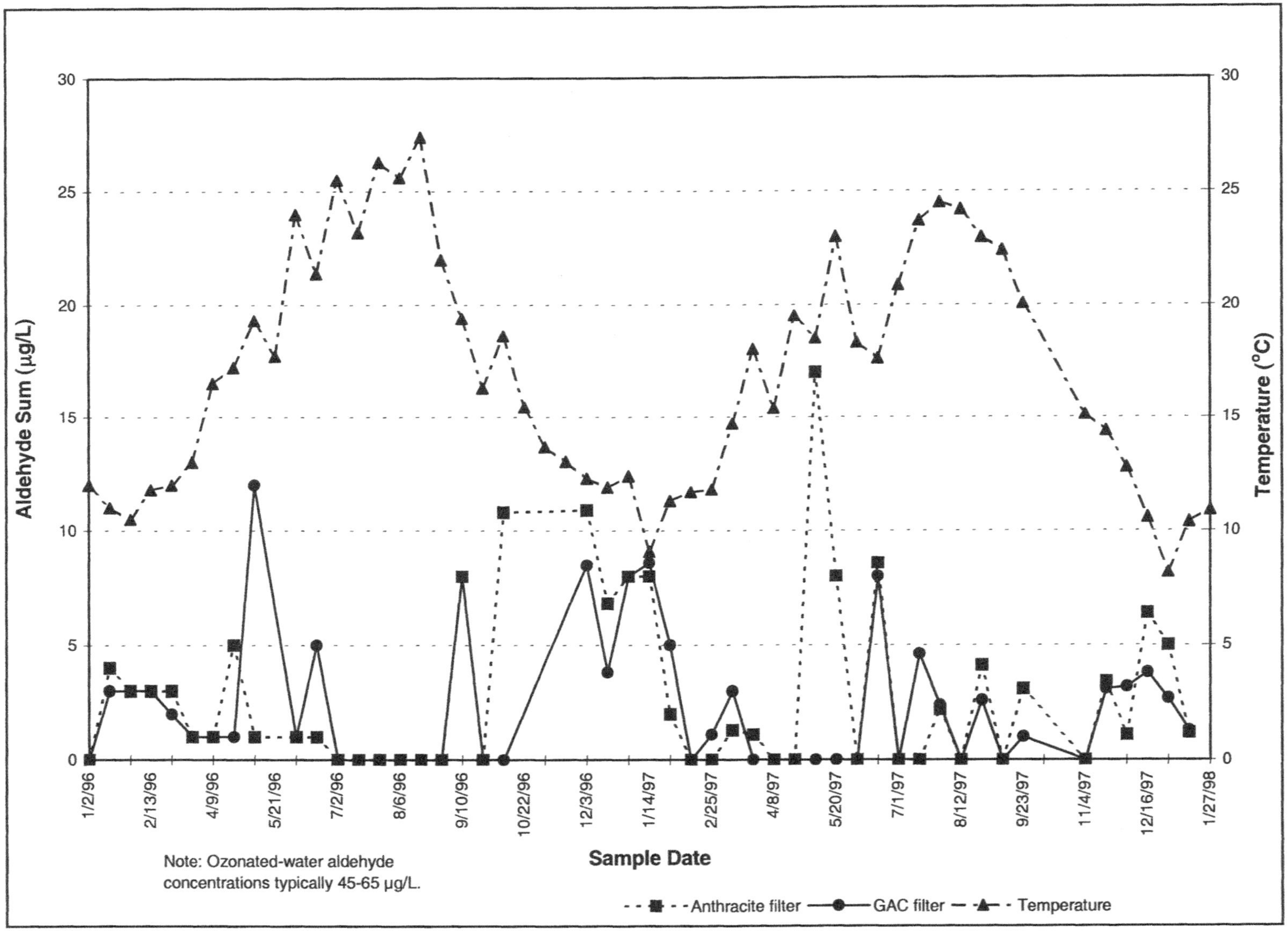

Figure 3.48 ACWD WTP2 filter-effluent aldehyde concentrations (all data reported as sum of seven aldehydes analyzed)

During the periodic chlorination of the basins, there is typically 0.2-0.6 mg/L of chlorine in the filter influents. Periodically, chlorine is added to the backwash as well. In general, it does not appear as if periodic exposure to low chlorine residuals has a significant impact on the biological filtration process at ACWD WP2.

CNCl: CNCl—a by-product of chloramination (Krasner et al. 1989)—was detected at 1-2 μg/L in the finished water of both plants (October 1997 - April 1998 monitoring). Although ozonation may result in increased CNCl formation—as formaldehyde can react with chloramines to form CNCl (Pedersen et al. 1999)—the use of biological filtration at WTP2 removed most of the formaldehyde produced by ozone.

Chlorate: On average, there is 162 μg/L of chlorate in the finished water at WTP2. Chlorate is a decomposition product found in hypochlorite solutions (Gordon, Adam, and Bubnis 1995). This decomposition is accelerated when hypochlorite solutions are stored at warm temperatures. At WTP2, the tank of hypochlorite is located indoors. Although the chemical building is not temperature-controlled, its temperature stays fairly constant because the building is partly below grade and the tanks are not exposed to sunlight. Chlorate formation is also minimized by getting fresh solutions frequently. ACWD gets a new delivery approximately every three weeks. In California, the Department of Health Services has an action level of 200 μg/L for chlorate. ACWD tries to stay below this action level.

TOX: At MSJWTP (December 1996 - April 1998), the TOX ranged from 157 to 398 μg/L (average = 267 μg/L), whereas the TOX at WTP2 ranged from <50 to 257 μg/L (average = 99 μg/L) (Table 3.20). Ozone/chloramines can reduce the formation of unknown halogenated organic by-products in addition to lowering THM and HAA concentrations.

Summary of System Performance

Ozonation/chloramination and ferric chloride instead of alum at WTP2 is much better than the conventional treatment process at MSJWTP at controlling THMs and HAAs—as well as TOX—but ozonation introduced bromate as a contaminant (trade-off issue). WTP2 is capable of complying with the Stage 1 and previously proposed placeholder Stage 2 MCLs with respect to THMs and HAAs—as well as the DBP precursor control requirement—but will barely comply with the Stage 1 bromate MCL of 10 μg/L with current treatment. A carbon dioxide feed system should enable ACWD to meet the bromate MCL. In addition, biological filtration has allowed ACWD to control other ozone by-products such as aldehydes.

CHAPTER 4
CASE STUDIES FOR IMPROVING PRECURSOR/DBP REMOVAL

INTRODUCTION

Technologies to improve precursor removal for the control of DBPs include enhanced coagulation, GAC, and membranes. Two case studies for enhanced coagulation and one case study for membranes were conducted (Table 2.1). Although utilities have installed GAC for taste-and-odor and/or SOC control, utilities are not installing GAC for DBP-precursor removal *per se*. Therefore, one case study was conducted on a utility that installed GAC for non-DBP reasons, but has been able to utilize biological activity on the GAC for the removal of DBPs (Table 2.1).

ENHANCED COAGULATION: CITY OF PHOENIX

Summary

The City of Phoenix owns and operates a drinking-water supply and distribution system that serves more than 1.1 million customers. Surface-water supplies for the City are treated using conventional treatment facilities. In recent years, the City implemented significant modifications to its treatment facilities in order to meet the increasing demand and to improve water quality. Many of the water quality improvements were aimed at achieving greater compliance with various drinking water regulations such as SWTR, LCR, and D/DBP Rule. This case study summarizes the City of Phoenix's experience in implementing treatment process modifications to achieve greater compliance with respect to the D/DBP Rule. The process studies conducted to identify necessary process modifications are summarized here along with a description of actions taken to implement process study recommendations. Costs involved at various steps of the process modifications and the evolution of the decision-making process are also discussed.

The deliberative process leading to the selection of a DBP-control-treatment strategy for the five treatment plants occurred within the context of the development of a water quality master plan (Bontrager, Chowdhury, and Hayes 1992). The planning process was carried out by a consultant under the direction of key members of the senior management team, involving the utility director and the directors of engineering and operations.

The planning process was intended to provide a comprehensive analysis of all technological options for dealing with DBPs, encompassing the comparative advantages and disadvantages. Part of the impetus for launching such a study in 1988 was the belief that enough was known about alternative technology choices to provide a foundation for a fully informed decision. Today, the Phoenix decision makers feel this study serves as a beginning, but intend to conduct more studies in the near future to evaluate new technology and revisit the previous study's recommendations.

Background

Case History

The City of Phoenix's water supply is from the Salt River Project (SRP), the Central Arizona Project (CAP), and water withdrawn from the Phoenix Valley aquifer. Water from the Salt and Verde Rivers is managed through a series of dams and canals by the SRP, whereas the CAP supplies CRW from Lake Havasu. More than 90 percent of the City's water supply is provided by surface water. The City's allotment of SRP water is treated at four conventional treatment plants (Verde, Val Vista, Squaw Peak, and Deer Valley), all featuring coagulation, flocculation, sedimentation, filtration, and chlorination. The city's allotment of CAP water is treated through direct filtration at one treatment plant (Union Hills). Major unit processes at this plant include coagulation, flocculation, filtration, and chlorination. Table 4.1 summarizes some key design features of the treatment plants. Current treatment plant capacity and the 1988 capacity are listed in Table 4.1. Schematic diagrams of processes at each treatment plant are provided in Figures 4.1 through 4.5 for the Verde, Val Vista, Squaw Peak, Deer Valley, and Union Hills WTPs. The majority of the process study took place in 1988. Some of the recommended changes from this study were subsequently implemented (1989 - 1995) along with the expansion of treatment capacity at several treatment facilities.

Table 4.1 also indicates the disinfection and DBP-precursor-removal processes in use when the process studies were undertaken. Free chlorine was used as primary and residual disinfectant

Table 4.1

Key design features of City of Phoenix's WTPs

WTP name	Source water	Disinfection process	DBP-precursor-removal process	Capacity (mgd) 1988	Capacity (mgd) 1998
Verde	Salt and Verde Rivers	Chlorine	Conventional coagulation/filtration	30	50
Val Vista	SRP	Chlorine	Conventional coagulation/filtration	140	140
Deer Valley	SRP	Chlorine	Conventional coagulation/filtration	120	150
Squaw Peak	SRP	Chlorine	Conventional coagulation/filtration	110	140
Union Hills	CAP	Chlorine	Coagulation/direct filtration	80	160
Total capacity				480	640

throughout the water supply system. Primary mechanism employed for DBP-precursor removal was coagulation using aluminum salts. Well-head chlorination was practiced for groundwater before blending with surface waters within the distribution system.

Driving Forces for Process Modifications

Plant expansions at the City of Phoenix were primarily initiated by the increase in water demand; however, process changes were primarily driven by the need for improving water quality to comply with increasingly more stringent water quality standards. The City of Phoenix developed a Water Quality Master Plan (WQMP) to set the strategy for future process improvements in the late 1980s when the regulatory direction with respect to DBPs was relatively uncertain.

The major motivating force in conducting a master plan study of DBP-control options was anticipating Safe Drinking Water Act (SDWA) compliance needs. The 1989 "Strawman Rule" paper produced by the USEPA (Clark 1989) discussing potential TTHM MCLs in the 20 μg/L range was a factor in shaping expectations. At that time, capital financing in Phoenix required a public referendum on bond issues. It was therefore necessary to begin the multi-year process of planning and obtaining approval well in advance of compliance deadlines.

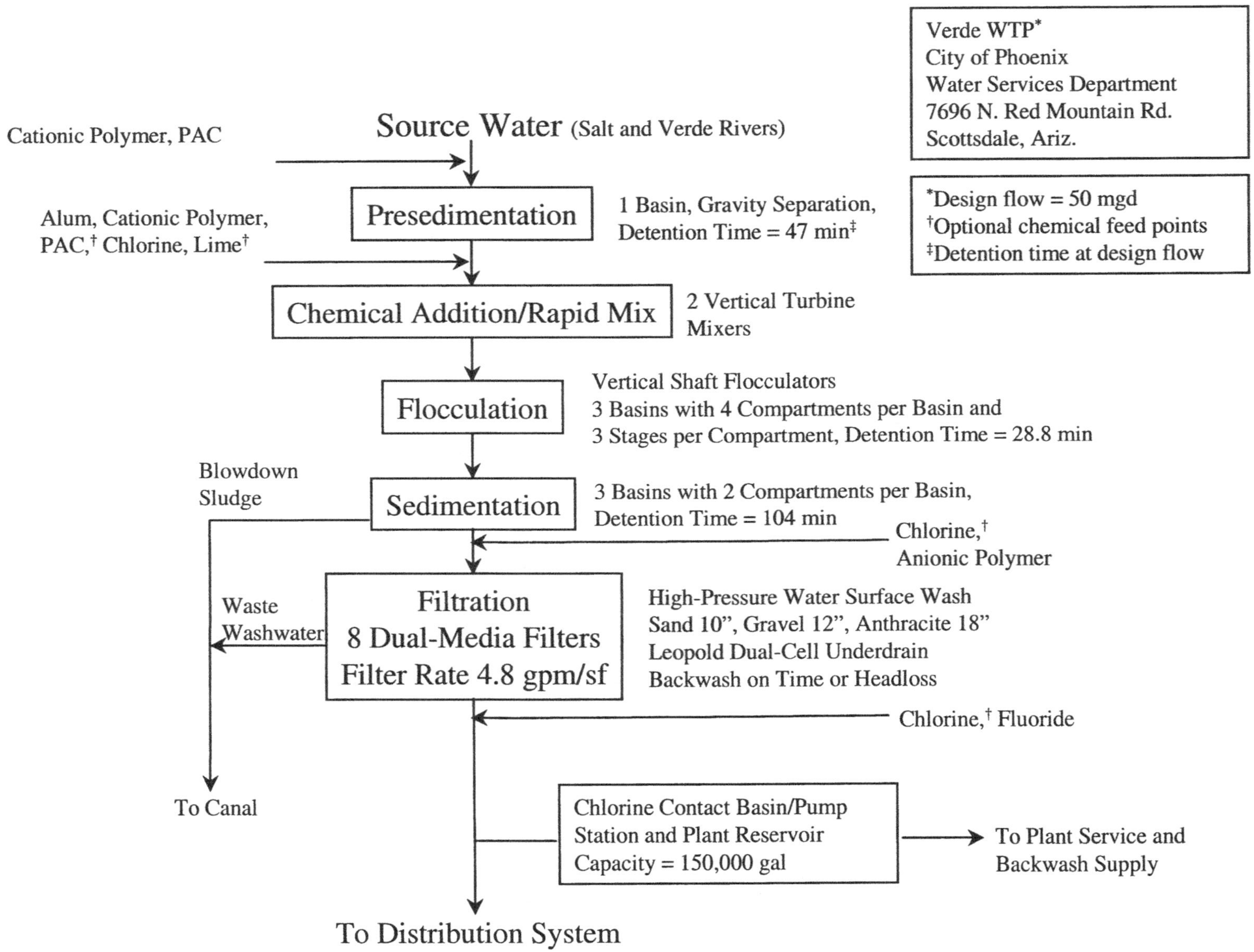

Figure 4.1 Verde WTP process description

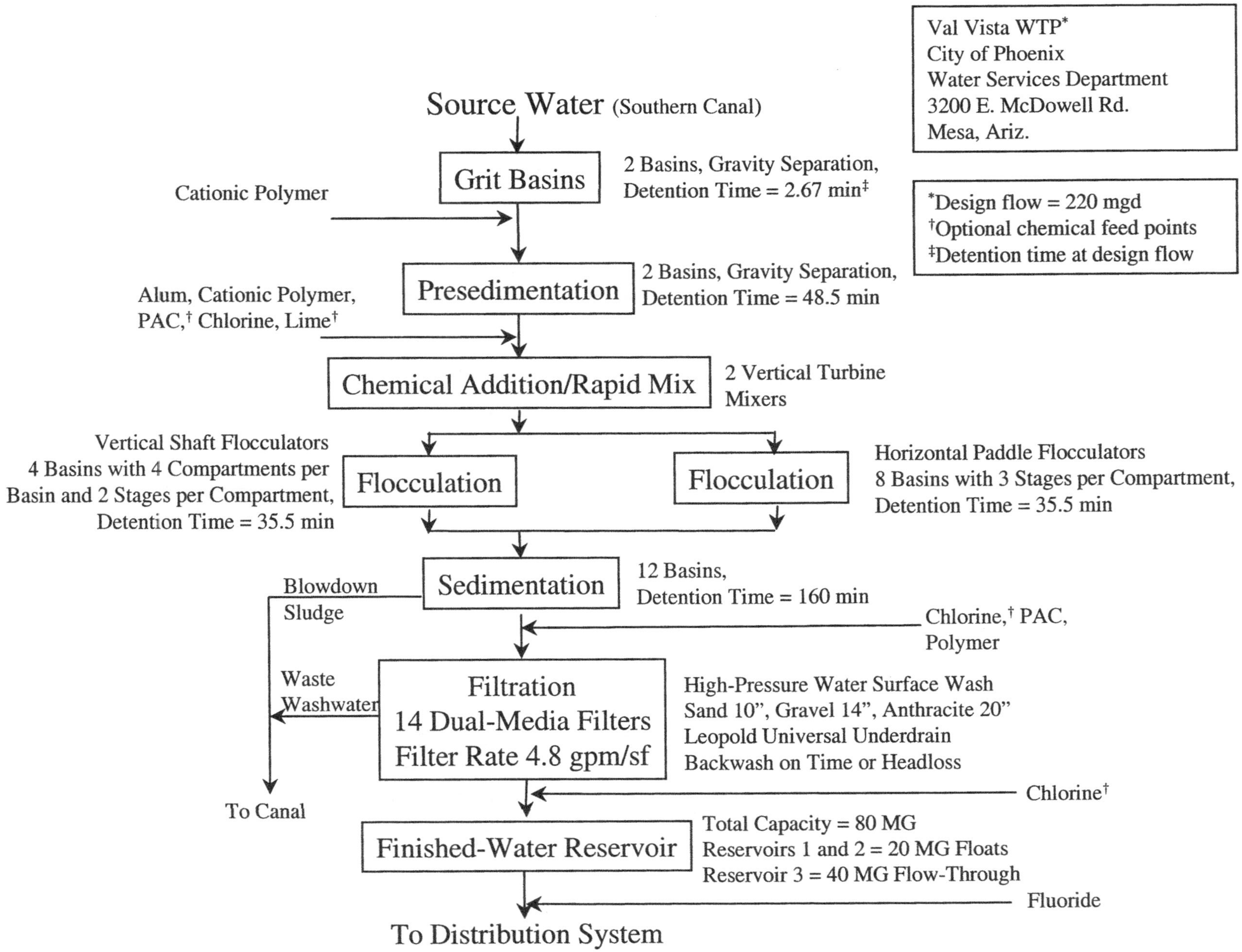

Figure 4.2 Val Vista WTP process description

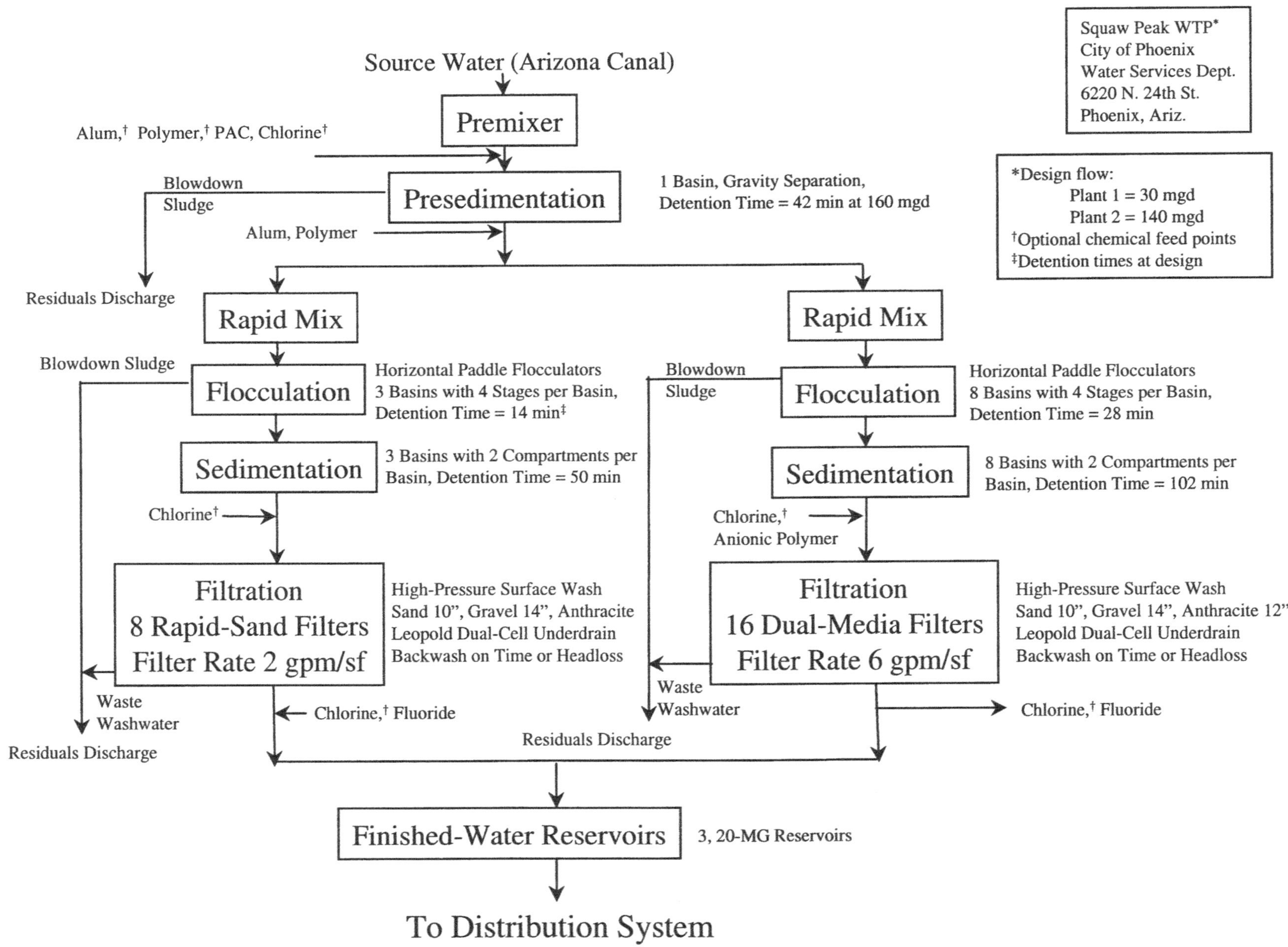

Figure 4.3 Squaw Peak WTP process description

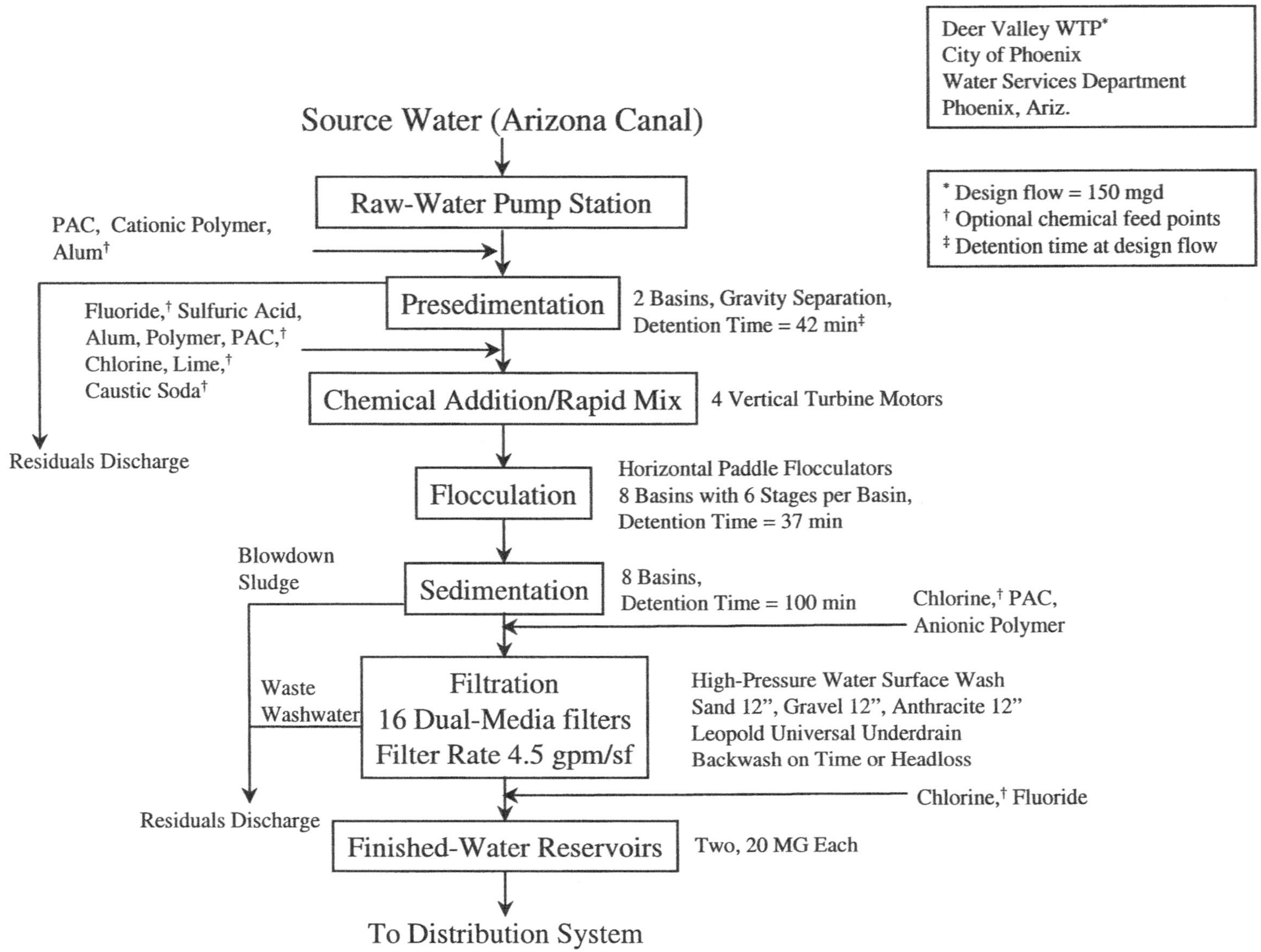

Figure 4.4 Deer Valley WTP process description

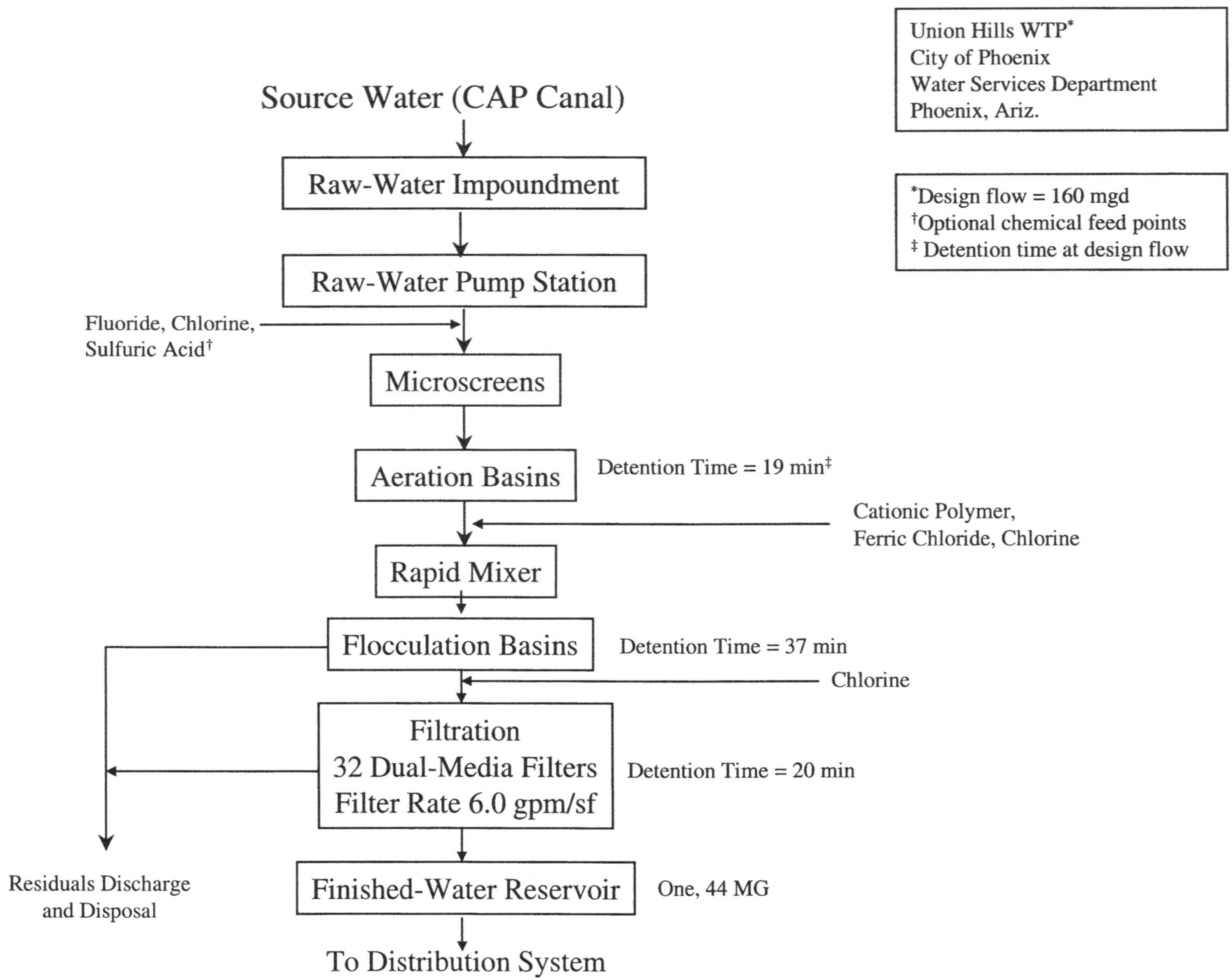

Figure 4.5 Union Hills WTP process description

An equally potent motivation concerned the fact that Phoenix was growing and needed to upgrade key components of plants as well as to add significant new capacity. These investment requirements stimulated a desire to determine a compliance strategy without waiting for final regulations. A compliance strategy was needed to provide some assurance that upgrade and expansion decisions made before final regulations would be in agreement with the compliance approach. It was particularly important to have a glimpse of the future in order to provide flexibility in the areas of instrumentation and space planning at the five plant sites.

An additional motivation for developing a master plan was derived from trends in running annual averages for TTHMs measured in the distribution system. Although the average range was comfortably below the TTHM standard of 100 μg/L and Stage 1 limits (80 μg/L for TTHMs), seasonal variability was sufficient to nudge even the running annual average up over 100 μg/L in an extreme year. It was not clear if these data indicated a multi-year cycle in TOC variability or if an upward trend in TOC was being observed. Concern over TOC variability was reinforced by complaints from a bottled-water manufacturer in the system who was experiencing increasingly noticeable production problems at his membrane plant.

TTHMs were the main drivers in the analysis. Although HAAs and bromate were discussed as part of the master plan analysis, they were not the primary consideration. Moreover, the focus of analysis was on controlling TTHMs; individual THMs were not evaluated in depth.

Summary of Master Plan Strategies

The City of Phoenix historically relied on prechlorination at the presedimentation basins and at the rapid mix processes to provide disinfection and to maintain the aesthetic appearance of the plant. Monitoring results for THMs, combined with testing to develop the master plan, suggested that continuing with pre-chlorination would result in violation of the currently applicable THM Rule. Consequently, it was suggested that pre-chlorination be discontinued as long as disinfection requirements of the SWTR could be met with post-chlorination. The master plan also included future implementation of improved precursor removal processes to comply with lower MCLs. The reason for such a process improvement was the desire to provide the best quality water to customers within financial constraints. In addition, the City, because of its geographical location in a high-temperature climate, wanted to continue free chlorination as primary and residual disinfectant. This desire to use free chlorine disinfection also biased the decision to select treatment technologies geared toward greater removal of NOM.

The major constraint affecting consideration of technology options was concern over taste-and-odor issues associated with chloramines (Krasner and Barrett 1985), that the use of a number of blending wells throughout the distribution system would create a significant operational problem maintaining the correct ratio of chlorine to ammonia. The City was also concerned about the likelihood of encountering nitrification problems in the hot climate of Phoenix (Wolfe et al. 1990). These concerns weighed heavily against the choice of either a chlorine/chloramines option or an ozone/chloramines option.

The need to provide extended contact time is often an issue when considering switching from pre-chlorination to post-chlorination. This can be a major constraint because space may not be available to build additional water-storage facilities. In this case, however, there was ample finished-water contact time that could be further increased through baffling modifications.

Treatment Practices Before Process Modifications

Treatment processes employed before process modifications were primarily designed to remove suspended particles from water by coagulation, sedimentation, and filtration (except Union Hills, which is a direct-filtration plant). Plants used alum as the primary coagulant with polymeric flocculation and filter aid. Average alum dosages before process modifications ranged from 15 to 25 mg/L. Variations in alum dose were primarily caused by changes in source water turbidity caused by storm events or SRP canal maintenance activities. Disinfection was provided by chlorination through the treatment processes.

To develop an understanding of the chlorination practices for the City's WTPs, the City's consultant interviewed WTP superintendents on May 23, 1988. The following interview summaries represent chlorination practices before any process modifications. Figures 4.1 through 4.5 describe treatment processes at each WTP.

Chlorination Practices at Verde WTP

Chlorine can be added at up to three locations in the plant. Most was added immediately downstream of the alum feed point, the Parshall flume. Chlorine dosage is based on maintaining a 0.4 mg/L free chlorine residual in the settled water. Chlorine dosages during the day were substantially higher than at night because of the exposure to sunlight in the sedimentation basins, varying from 1.5 to 2.0 mg/L at night to 2.3 to 2.8 mg/L during the day. The rate of chlorine feed was closely monitored and typically adjusted three to five times per 8-h shift. Presumably due to the limited penetration of sunlight in the flocculation basins, little reduction in free chlorine residual occurs.

The motivation for pre-chlorination was to conform to traditional practices for good water treatment. Microbiological growth control in the process basins was not given as a principal reason for pre-chlorination.

Chlorine also was added to settled water to maintain a free chlorine residual of 0.7 mg/L in the finished water. Chlorine can also be added to filtered water, but this was not usually practiced. Previously, the finished water goal for free chlorine residual was 1.4 mg/L, but this was reduced to minimize THM formation.

Chlorination Practices at Val Vista WTP

Chlorine can be added at almost any stage of treatment. The usual practice was to add enough chlorine before pre-sedimentation to produce a free residual across the basin, usually less than 0.1 mg/L. Chlorine was then added at the static mixers with the goal of producing a free residual of 0.3 mg/L in the settled water. Chlorine was then applied to settled water to produce a free residual of 0.6 to 0.7 mg/L in the filtered water. Algal growth control was the principal reason for practicing pre-chlorination.

After filtration, water flows to reservoirs in an unusual configuration that makes additional chlorine necessary. A portion of the flow passes through a 40-MG reservoir and was normally chlorinated with an additional 0.1 mg/L after storage. However, before entering this reservoir, water passes by two other 20-MG reservoirs that are allowed to fill during periods of low demand. Once water has entered these reservoirs, it may remain for a long detention time because of reserved hydraulics, with a decrease in free chlorine residual. Only during periods of high demand do the 20-MG reservoirs empty. When this water with low chlorine residual and possible elevated concentrations of THMs is mixed with freshly treated water, the total chlorine residual decreases. These chlorine dosages have been found to maintain the desired residual in the system.

The total chlorine dosage at Val Vista averaged approximately 2.8 mg/L, with more chlorine used during the summer, and the treatment goal was to produce a finished water with a free residual of 0.8 mg/L.

Chlorination Practices at Squaw Peak WTP

Approximately 2 to 3 mg/L chlorine was typically added upstream of the pre-mixers to maintain a free residual of 0.3 to 0.6 mg/L in the settled water. A second dosage (approximately 1.1 mg/L) was applied to settled water to maintain a free residual of 0.8 mg/L in the filtered water. Although the plant can add chlorine to filtered water, it was not normally necessary. As noted at the other plants, less chlorine was used during winter.

Chlorination Practices at Deer Valley WTP

Average total chlorine dosage was 2.6 mg/L, with a significant portion (1.5 to 2 mg/L) applied before flocculation. Pre-flocculation chlorine dosage was adjusted approximately twice a day, but remained fairly constant. As a result, the settled water residual varied from 0.0 to 1 mg/L). After filtration, a second dosage of chlorine was applied (0.7 to 1 mg/L), and this dosage was adjusted more frequently, even hourly. The treatment goal was to produce a free chlorine residual of 0.8 to 0.9 mg/L in the finished water. Once a month, high dosages of chlorine were applied to the grit basins to control algal growth. The plant also has the ability to add chlorine after filtration.

Chlorination Practices at Union Hills WTP

Chlorination practices at Union Hills WTP included pre-chlorination before the source water impoundment, which has a long residence time. The normal operation included aeration at the source-water conduit leading from the impoundment to the treatment plant. Aeration was provided primarily for air stripping odor-causing compounds. The operations staff, however, found significant reductions of THMs during aeration. Chlorine was added at the rapid mix to maintain a residual in the filtered water. Additional chlorine was added to the finished water to maintain the target chlorine residual. A portion of the finished water is delivered directly to the distribution system, whereas the remaining water passes through a finished-water reservoir. The directly delivered portion can experience significant detention time in the pipeline from the plant.

Quality of Treated Water Before Process Modifications

The City of Phoenix routinely monitors for TTHM concentrations at the effluents of the WTPs and within the distribution system. The finished-water THMs at the effluent of the WTPs ranged from approximately 25 to 60 μg/L. Table 4.2 summarizes the average distribution-system THMs measured in Phoenix. These values, along with the running annual average concentrations of THMs, are shown in Figure 4.6. For reference, Table 4.3 summarizes several source water quality parameters for each Phoenix plant, including TOC, UV, and bromide. Although the average detention time for the surface water in the Phoenix distribution system is approximately 48 h, increases in TTHM concentrations after the treated water left the plants appeared to be small.

As shown in Table 4.2 and Figure 4.6, the distribution-system THM level follows a cycle of increasing and decreasing values. It is believed that this cycle is caused by long-term variations of organic content of the water. When source water quality is poorer, the disinfection practices before process modifications resulted in annual average distribution-system THM values that were close to the applicable MCL.

Table 4.2

Average THM concentrations in the Phoenix distribution system*

Year	First (January to March)	Second (April to June)	Third (July to September)	Fourth (October to December)
	Sampling quarter			
1985	66	69	66	65
1986	64	62	57	53
1987	52	45	52	55
1988	54	60	57	NA†
1989	NA	NA	NA	NA
1990	36	52	NA	62
1991	66	103	120	53
1992	56	69	80	73
1993	52	55	64	36
1994	30	40	72	38
1995	55	56	88	56
1996	50	65	70	53
1997	41	67	71	47
1998	43	72	74	51

*Running annual average TTHM data are shown in Figure 4.6.

†Data not available in City records.

Summary of Process Studies

Both bench- and pilot-scale studies were conducted for the City's SRP WTPs. Process studies included evaluating enhanced coagulation, PAC adsorption, GAC adsorption, and nanofiltration for improving DBP precursor removal. Alternative primary disinfectant (ozone) and alternative residual disinfectant (chloramines) were also experimentally evaluated. The following

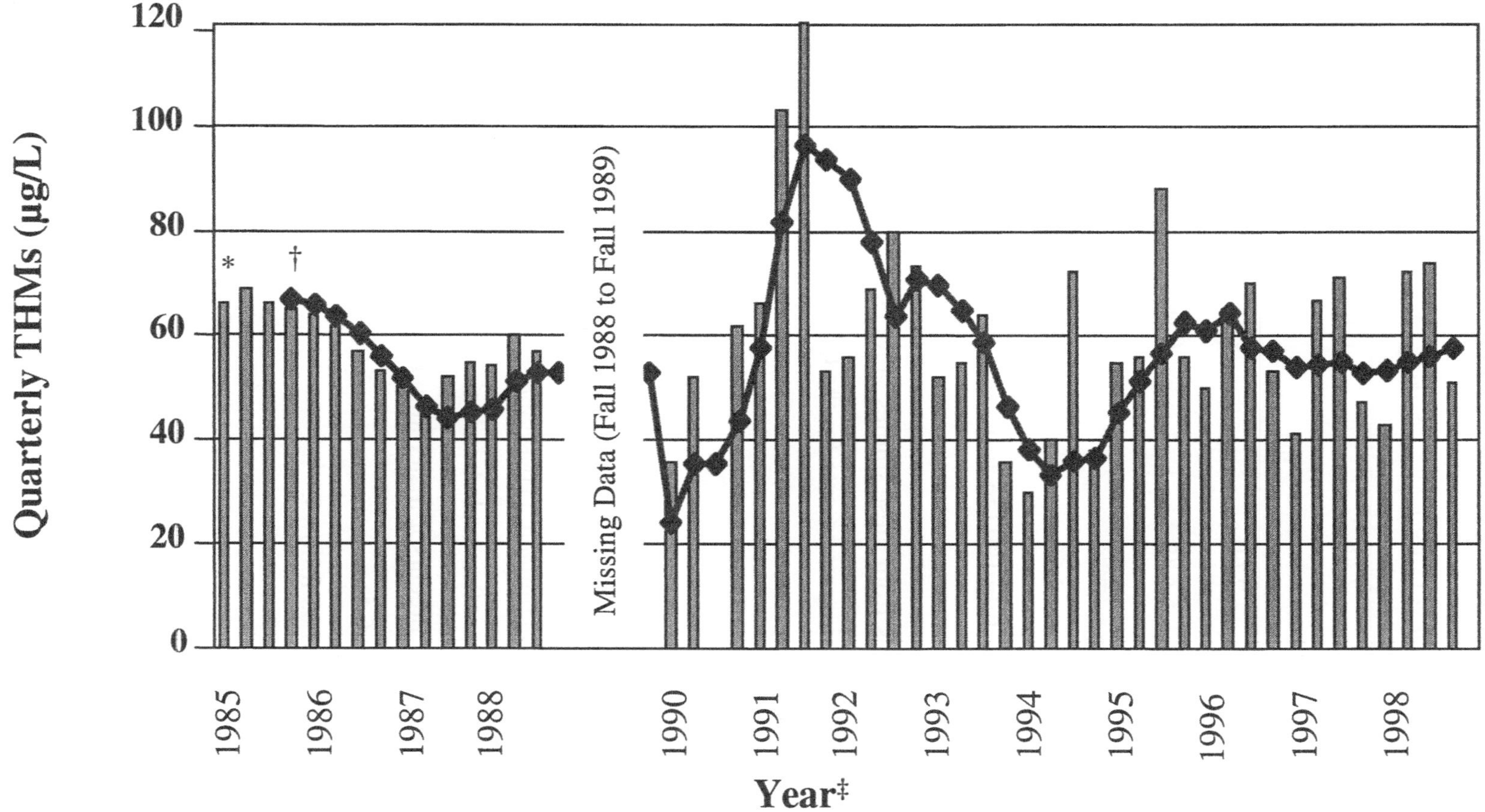

*Bars represent quarterly averages of all distribution-system samples.
†The line represents running annual average THM concentrations.
‡For each year, the numbers are in the sequence of winter, spring, summer, and fall.

Figure 4.6 Distribution-system TTHMs for City of Phoenix

Table 4.3

Average source water quality for Phoenix WTPs*

WTP name	Source water	Average TOC level (mg/L)	Average UV (cm^{-1})	Average bromide level (μg/L)
Verde	Salt and Verde Rivers	2.46	0.065	63
Val Vista	SRP	2.60	0.058	63
Deer Valley	SRP	2.76	0.067	81
Squaw Peak	SRP	2.67	0.059	67
Union Hills	CAP	2.60	0.041	82

*Data represent 12-month period from July 1997 to July 1998.

summarizes the results and conclusions from these studies. Process-study results formed the basis to develop a strategy for water quality improvements for the City of Phoenix's treatment system.

Bench-Scale Study Conclusions

- Without pH adjustments (keeping the pH in the natural range):
  - Alum coagulation can remove 13 to 15 percent of the DOC from the water.
  - Pre-chlorination increases the estimated THM concentration in the distribution system by approximately 10 μg/L (average of several observations). Presumably this is due mainly to a higher concentration of DBP precursor material in the water at the time of chlorination.
- Reducing the pH of coagulation to 6.5:
  - Improved the removal of NOM as follows:
    - ♦ DOC: 20 percent
    - ♦ TTHMs: 40 percent
    - ♦ UV: 40 percent
  - There were no significant differences in the removal of DOC between coagulation with or without pre-chlorination.

- Applying ozone or ozone/hydrogen peroxide before or after coagulation:
  - Had no significant impact on the removal of NOM.
  - Increased the THM concentration that might form in the distribution system by up to 15 μg/L (average of several observations).
- Applying PAC at the rapid mix did not improve NOM removal; the dosages of PAC evaluated were up to 50 mg/L.
- GAC was effective for removing DOC and THM precursors when conventional treatment was followed by GAC adsorption:
  - Reduced the DOC concentration of the finished water to less than 1 mg/L.
  - Depending on the GAC regeneration frequency, THM concentrations of 20 to 50 μg/L are possible in the distribution system. The carbon usage rate for a finished-water THM concentration of 25 μg/L was found to be 1200 lb/MG for Verde River and 400 lb/MG for Salt River water. The usage rates were based on an EBCT of 10 min.
- The use of membrane filtration:
  - Was found to be effective in removing NOM.
  - The product water generated using the Filmtech NF 170 membrane had DOC values less than 1 mg/L, THMFP values in the range of 15 to 30 μg/L, and non-detectable levels of UV-absorbing substances.
- The use of chloramines as a secondary disinfectant:
  - Was found to be effective in controlling THM formation.
  - The addition of ammonia after one hour of free chlorine contact time stopped the THM-formation reaction almost completely, although in some instances, the THM values were found to increase by up to 10 μg/L after the addition of ammonia (average of several observations).
  - The application of chloramines as the residual disinfectant in the distribution system produced average system THM concentrations of 50 μg/L or less for all the alternatives evaluated and below 25 μg/L for some alternatives.

Pilot-Scale Study Conclusions

- Reducing the pH of coagulation:
  - Was an effective alternative for removing DOC, UV, and THM precursors.
  - Reduced THM levels in the distribution system by 20 to 30 μg/L if pre-chlorination was also eliminated. Seasonal pre-chlorination was not evaluated to a large extent.
  - May result in shorter filter runs than the existing treatment strategy.
  - The application of PAC did not improve removals of DOC or THM precursors.
  - A finished-water pH of 7.5 did not reduce the amount of THMs that formed in the finished water as compared to a finished water pH of 8.0.
- Ozonation of settled water:
  - Produced premature turbidity breakthrough in the filters.
  - Increased finished-water THM levels if free chlorine was used as the secondary disinfectant, possibly because of an increase in active NOM sites during ozonation.
  - The use of ozone as the primary disinfectant, followed by the application of preformed chloramines as the secondary disinfectant, produced THMs in the distribution system of less than 10 μg/L (average of several observations).

Water Quality Strategy

The recommended water quality strategy shown in Figure 4.7 contains five basic treatment options to improve water quality: four disinfection options and five key decision points. Estimated water quality and costs associated with the various options are found in Table 4.4. The range of estimated quality reflects changes in source water for SRP supplies. The time frame for process improvements envisioned at the time of master plan development is included in Table 4.5.

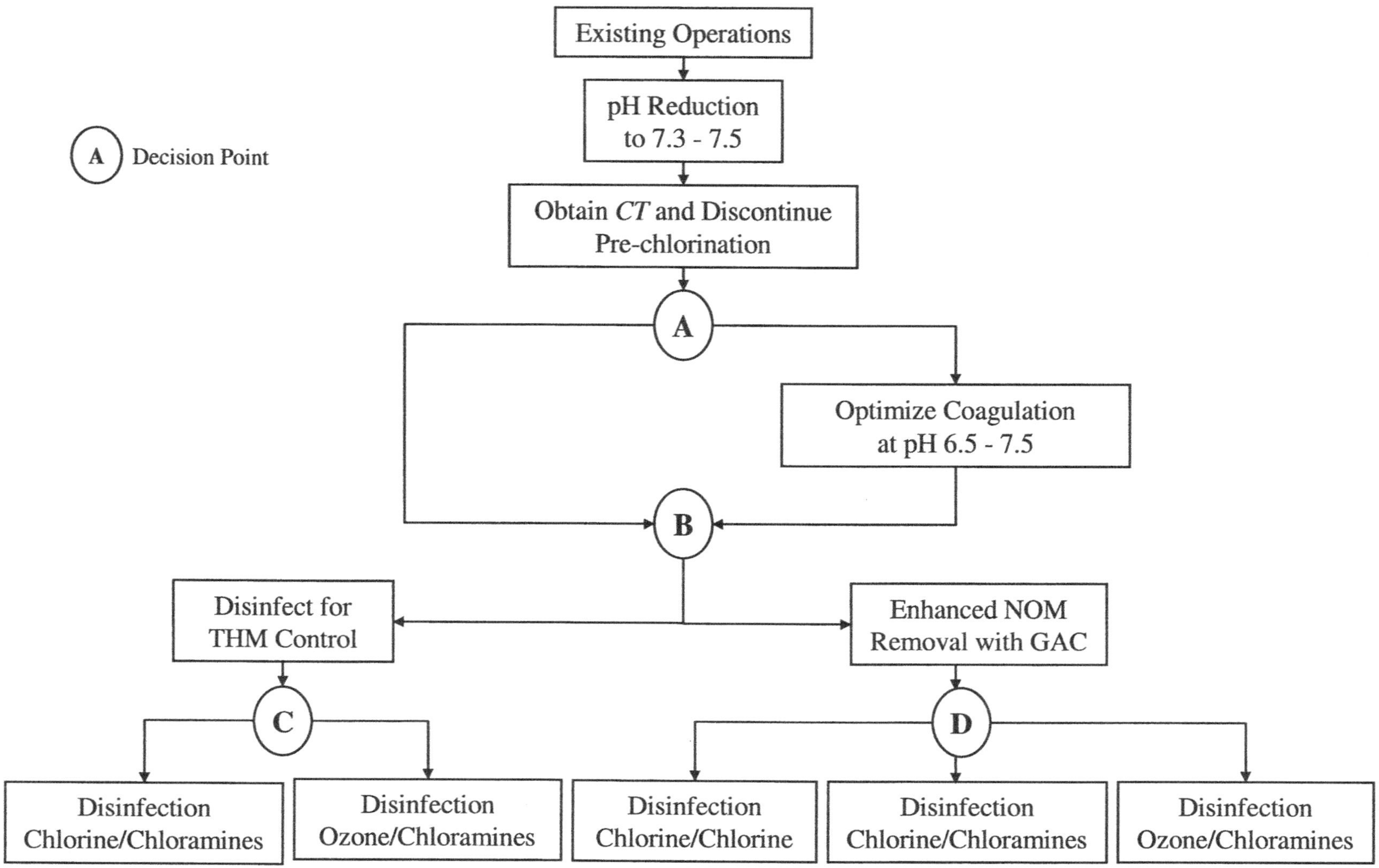

Figure 4.7 Water quality strategy at City of Phoenix

Table 4.4

Estimated water quality and costs based on water quality strategy at City of Phoenix

Water quality strategy	Optimize coagulation	Disinfectant		Treated-water DOC (mg/L)		Distribution THMs (μg/L)		Estimated cost*	
		Primary	Secondary	Low	High	Low	High	$/MG	$/yr
Elimate pre-chlorination†									
No	No	Chlorine	Chlorine	2.2	2.7	70	105	0	0
Yes	No	Chlorine	Chlorine	2.2	2.7	60	95	4	1
Yes	Yes	Chlorine	Chlorine	2.0	2.4	40	60	150	54
Disinfect for THM control									
Yes	Yes	Chlorine	Chloramines	2.0	2.4	20	35	160	58
Yes	No	Chlorine	Chloramines	2.2	2.7	20	40	10	4
Yes	Yes	Ozone	Chloramines	1.9	2.3	5	15	210	76
Yes	No	Ozone	Chloramines	2.2	2.7	5	15	60	22
Enhanced NOM removal with GAC									
Yes	Yes	Chlorine	Chlorine	1.0	1.2	20	50	420	150
Yes	No	Chlorine	Chlorine	1.0	1.2	20	50	270	98
Yes	Yes	Chlorine	Chloramines	1.0	1.2	15	30	420	150
Yes	No	Chlorine	Chloramines	1.0	1.2	15	30	280	100
Yes	Yes	Ozone	Chlorine	0.8	1.0	25	35	455	160
Yes	No	Ozone	Chlorine	0.8	1.0	25	35	310	110
Yes	Yes	Ozone	Chloramines	0.8	1.0	5	15	470	170
Yes	No	Ozone	Chloramines	0.8	1.0	5	15	330	120

*Costs are in addition to present practices. Costs in $/MG include capital and operations and maintenance costs. Capital costs are amortized at 8.5 percent over 20 years. Costs in $/yr are an annual cost for a typical household of four using 250 gpcd.

†Includes pH reduction to 7.3-7.5.

Table 4.5

Steps in following implementation plan at City of Phoenix

Step	Action	Time frame*
1	Continue existing operation until acid and caustic feed facilities are in place. Continue the use of chlorine as both primary and secondary disinfectant.	Until mid 1991.
2	Accomplish pH reduction to 7.3 to 7.5 when acid and caustic feed facilities are available. Continue the use of chlorine as both primary and secondary disinfectant.	Will vary at each plant. All plants by mid 1991.
3	Discontinue pre-chlorination when adequate *CT* after filtration is available. Continue the use of chlorine as both primary and secondary disinfectant.	Will vary at each plant. All plants before 1992.
4	At decision points A and B, select between the options: • Optimize coagulation (pH 6.5 to 7.5). • Disinfect for THM control. • Enhanced NOM Removal with GAC. The treatment option to optimize coagulation may be used alone (using chlorine as both primary and secondary disinfectant) or in conjunction with disinfection for THM control or enhanced NOM removal with GAC.	Make decision during second half of 1992.
5	If disinfection for THM control route is chosen, at decision point C select between ozone and chlorine as primary disinfectant.	Place in service before mid 1995.
6	If enhanced NOM removal with GAC route is chosen, at decision points D select primary/secondary disinfection combination.	Place in service before the end of 1995.

*Time frame envisioned at the time of the master plan development.

The five options to improve water quality were:

1. Reduce pH of existing operations (pH of 7.8 to 8.0) to between 7.3 and 7.5 before coagulation.
2. Discontinue pre-chlorination before coagulation. Associated with this option was the need to ensure that adequate contact time was provided for disinfection. To accomplish this, reservoir modifications would likely be required at all treatment plants except Verde WTP.
3. Optimize coagulation by significantly lowering pH to between 6.5 and 7.3 before coagulation.
4. Disinfect for THM control by selecting a disinfection scheme that produces lower TTHMs.
5. Enhanced NOM removal by use of GAC adsorption.

The four factors influencing decisions on water quality strategy were:

1. The choice of primary and secondary disinfectant based on microbiological quality and reasons other than reducing DBPs.
2. Whether to meet probable future TTHM regulations, which were anticipated to be more stringent, by changing disinfectants.
3. Whether to significantly increase DOC removal to produce lower levels of TTHMs and other DBPs and reduce DOC in the distribution system.
4. The cost of producing higher-quality water.

Microbiological quality, the most important water-quality consideration, is provided by primary and secondary disinfection. The two choices for primary disinfection were chlorine and ozone. Both disinfectants could fully satisfy—based on the time frame of the development of the WQMP—all present and anticipated future microbiological quality requirements. The choice of the primary disinfectant for the City of Phoenix was based on its impact on the production of DBPs and cost considerations.

Studies conducted during this project on SRP surface-water supplies indicated that the use of ozone as a primary disinfectant followed by chlorine as the secondary disinfectant did not reduce the formation of THMs. Similar results were obtained by others on studies of CAP water. In addition, the use of ozone as a primary disinfectant is more expensive than chlorine. Only when ozone is used as the primary disinfectant and chloramines as the secondary disinfectant are significantly less THMs formed.

The two choices for secondary disinfection are chlorine and chloramines. From a microbiological quality perspective, chlorine is preferred based on the following:

- Chlorine is a stronger disinfectant than chloramines and therefore provides an added measure of safety.
- Phoenix has had a long and satisfactory experience with the use of chlorine.
- A change to chloramines at the surface-water plants would likely necessitate a change at all groundwater production facilities to preclude taste-and-odor problems where waters containing the two disinfectants mix.
- The potential exists for growing nitrifying bacteria in the distribution system with a chloramine residual, causing a depletion of chloramine residual and possibly requiring annual conversion to free chlorine for certain periods of time.

Summary of Implemented Process Changes

Process changes implemented at the City of Phoenix's WTPs included the installation of acid- and caustic-feed facilities at all plants, and the installation of baffles and other modifications to the finished-water reservoirs to ensure adequate primary disinfection under post-chlorination conditions. Table 4.6 summarizes the process improvements along with the amount of capital funds. As shown in this table, the City of Phoenix has spent a total of approximately $9 million for process improvements primarily related to improvements in DBPs.

Technology Checklist—Anticipated Factors

The effectiveness of the combination of eliminating pre-chlorination and optimizing coagulation was correctly anticipated. The running annual average of TTHMs has stayed consistently beneath the TTHM standard of 100 μg/L and is also within the Stage 1 limit of 80 μg/L. Given the complexity of the Phoenix systems—with five plants and multiple blending wells

Table 4.6

Summary of process modifications and capital costs at City of Phoenix

Plant name	Process improvement	Implementation time	Capital cost
Verde WTP	Acid/caustic feed	1993	\$3,425,000
Val Vista WTP	Acid/caustic feed	1993	\$1,730,000
	Reservoir modifications	1998	\$500,000
Squaw Peak WTP	Acid/caustic feed	1992	\$2,838,000
Deer Valley WTP	Acid/caustic feed	1990	\$455,000
	Reservoir modifications	1996	\$500,000
Union Hills WTP	Reservoir modifications	1999	\$500,000

throughout the distribution system—it was not a simple matter to predict TTHM levels in the distribution system. Some limited distribution-system "modeling" of DBP formation after treatment was performed to confirm this analysis.

Eliminating pre-chlorination was a concern to operators because it had been a standard practice. Pre-chlorination was considered important for algae-and-bacteria control, especially in the hot climate. Alternative procedures for cleaning were developed, however, and operational problems have not been material.

GAC10 was estimated to be about twice as costly and twice as effective in DOC removal as optimized coagulation. The Technology Workgroup that provided input to the Stage 1 regulation negotiation committee estimated the costs and effectiveness of enhanced coagulation and GAC10 to be in about the same proportions (Krasner, Owen, and Cromwell 1996).

Technology Checklist—Less Well Anticipated Factors

Handling and maintaining of the sulfuric acid feed system proved to be somewhat problematic and involved compared to what was expected. For example, some incidences with leaking tanks resulted in some waste-disposal issues. It is interesting to note that Phoenix committed to optimized coagulation long before the Stage 1 Rule provisions concerning high-alkalinity sources were envisioned.

History of Use of Modified Processes and Operator Perception

After modifications, acid- and caustic-feed facilities were available at all of the City of Phoenix's WTPs; however, these facilities are used sparingly because of the additional operational cost of chemicals. When these facilities were used, operators found additional chemicals did not pose any operational problems. Coagulation optimization testing at the Deer Valley plant showed that lowering the pH to 6.8 did not increase sludge production, but doubling the alum dose (from 15 to 30 mg/L) increased sludge production by about 15 percent.

Process improvements related to increased detention times in the finished water reservoirs have made positive changes in the operations of the treatment facilities and resulted in improvements in finished-water quality. In most cases, pre-chlorination is no longer used for disinfection. Disinfection is generally achieved in the finished-water reservoir with some disinfection being still carried out through the filters. Pre-chlorination is practiced occasionally for basin-maintenance purposes. Operators do not report any difficulties due to the changes in disinfection strategy.

Water Quality Impacts of Process Modifications

Process modifications at the City of Phoenix's WTPs have resulted in lower THMs and HAAs in the distribution system. Figure 4.6 summarizes the quarterly THM numbers for the Phoenix distribution system showing a decrease in overall THM levels in the distribution system as a result of process improvements.

ENHANCED COAGULATION: CITY OF TAMPA

Summary

The City of Tampa operates a 100-mgd conventional treatment plant, which switched to ferric sulfate as the coagulant and optimized the coagulation process in 1991. The influent surface water is high in color and TOC. The source water is subject to extreme variability in its raw water characteristics due to seasonal swings.

The driving forces in the decision to switch coagulants and optimize the coagulation process were: (1) Coagulation with ferric sulfate instead of alum would solve residuals disposal problems as it produced a ferric humate residual with a beneficial reuse. (2) Optimized coagulation was expected to satisfy future anticipated requirements with the DBP Rule. (The term "enhanced coagulation" was defined and conditions for implementation were developed during the proposal process for the Stage 1 D/DBP Rule in 1993 [USEPA 1994a]. Tampa investigated the feasibility of optimizing the coagulation process beyond what was required for turbidity removal as a method for reducing DBP precursors before any regulatory requirement for enhanced coagulation was proposed.) (3) The ferric coagulant manufacturer worked with Tampa to devise a product that would perform well in the treatment plant and as a reusable product for agricultural use.

The process design was influenced by residuals disposal, TOC, DBPs, turbidity, and color. The capital cost of optimizing the coagulation process (i.e., converting to "enhanced coagulation" with ferric sulfate from standard coagulation with alum) cost the City of Tampa less than $100,000. The annual cost of operation increased by approximately 18 percent.

Some of the key facts that were highlighted during this case study are: (1) Step-wise approach to investigating and meeting different anticipated regulatory scenarios. The decision-making process included the investigation of future modifications—such as alternative disinfectants, GAC, and membranes—which looked beyond the immediate implementation of enhanced coagulation. (2) Implementation of enhanced coagulation with ferric sulfate solved their concern over residuals disposal and at the same time addressed issues such as TOC removal and lower DBP formation. (3) The City of Tampa undertook studies with the ferric coagulant manufacturer to investigate various applications for the ferric humate residual to allow for beneficial reuses.

The treatment strategy implemented did not significantly change the operational needs of the plant and no additional staff was added. Provisions for training the operating staff were provided by the ferric sulfate vendor. It took a period of one year for the treatment plant to operate like an "integral" unit.

TOC removal before modifications ranged from 21 to 77 percent, with an average value of 50 percent. With enhanced coagulation (ferric sulfate), TOC removal ranged from 70 to 88 percent, with an average TOC removal of 81 percent. Enhanced coagulation with ferric sulfate was better than sweep coagulation with alum at removing TOC (DBP precursors).

Before treatment modifications—for the year 1990—the TTHMs ranged from 27 to 111 μg/L and averaged 59 μg/L (no data were available for the HAAs). (In 1990, Tampa did have a minimum daily finished-water TTHM level at 27 μg/L, however this value was considerably lower than normal for any year that they were utilizing alum.) After treatment modifications—for the time period July 1997 to December 1998—the TTHM levels ranged from 47 to 67 μg/L (average = 60 μg/L) and HAA5 ranged from 32 to 66 μg/L. (The minimum TTHM level with ferric-sulfate treatment of 47 μg/L was a typical lower level found in Tampa's water under either treatment scenario.) The new treatment scenario reduced the maximum THM formation from 111 to 66 μg/L.

Care must be taken in comparing the year 1990 with the years 1997 and 1998, as Tampa finds not only seasonal variations in raw-water quality but also yearly cycles of variation. To determine the effect of THM formation under full-scale treatment conditions with the same source-water quality, a half-plant comparative study was conducted (Gianatasio, Hook, and Hjersted 1996). Enhanced coagulation with ferric sulfate and acid feed was evaluated in one basin, whereas standard alum (sweep) coagulation without the use of acid feed for pH control was utilized in the second basin. A 1-hour THMFP test was performed on the settled-water effluents. The 1-hour THMFP for iron-treated settled water averaged 34 μg/L, whereas alum-treated settled water produced an average 1-hour THMFP of 81 μg/L.

The City of Tampa case study demonstrates how enhanced coagulation with ferric sulfate can be used to meet issues as diverse as beneficial residuals reuse, improved TOC removal, and a reduction in DBP formation. Enhanced coagulation with ferric sulfate produced a better quality and a more consistent quality finished water than conventional coagulation with alum and allows for compliance with the TTHM and HAA5 MCLs in Stage 1 of the D/DBP Rule. The City of Tampa is currently investigating future modifications (i.e., ozonation/BAF) that will allow its treatment plant to meet the anticipated LT2ESWTR and Stage 2 of the D/DBP Rule.

Background

Case History

The City of Tampa operates the Hillsborough River WTP (HRWTP), which serves approximately 450,000 people. Built in 1924, the HRWTP—a conventional treatment plant—has a capacity of 100 mgd and utilizes the Hillsborough River as the source water. The Hillsborough River is a highly colored water, with raw-water color as high as 400 pcu and TOC as high as 38 mg/L during the peak summer rains.

From 1924 to 1991, the HRWTP had relied upon alum for coagulation and color removal. Alum residuals, historically, had been disposed in landfill sites throughout the City. In the late 1980s, landfill sites were being exhausted and closed, and permits for new landfill sites were becoming difficult to obtain. In 1988, the City of Tampa Water Department undertook an extensive three-year search to find an alternative coagulant. Ferric sulfate—utilized in an enhanced coagulation mode—was chosen and brought on-line in 1991, with 100-percent reuse of the ferric humate residual material.

In addition, enhanced coagulation increased TOC removal and will allow the HRWTP to meet the Stage 1 D/DBP Rule MCLs. Further studies with ozone, GAC, and membranes are in the process of being conducted to enable HRWTP to meet potential new requirements for the Stage 2 D/DBP Rule and ESWTR.

Routine monitoring for water quality and microbial parameters and DBPs was conducted before and after treatment modifications were introduced. The calendar year 1990 was chosen for the "before treatment modifications" data analysis and the calendar year 1997—with additional supporting data from the Information Collection Rule (ICR) (July 1997 to December 1998)—was chosen for the "after treatment modifications" data analysis. Care must be taken in comparing the year 1990 with the years 1997 and 1998, as Tampa finds not only seasonal variations in raw-water quality but also yearly cycles of variation.

Process Selection

The 1986 and 1996 SDWA amendments included provisions for additional rulemaking that have resulted or will result in new drinking water standards, treatment standards, and testing and monitoring requirements. Some of the provisions that apply to HRWTP include: SWTR (USEPA 1989a); IESWTR (USEPA 1998b) and LT2ESWTR (USEPA 2000); D/DBP Rule (USEPA 1998a); ICR (USEPA 1996a); and the LCR (USEPA 1991). Compliance with the D/DBP Rule will be especially challenging at the HRWTP due to the need to balance the new requirements with the disinfection and treatment requirements of the current SWTR, the IESWTR, and the anticipated LT2ESWTR.

In order to achieve compliance with the various regulations, the Tampa Water Department created a Water Quality 2000 (WQ2000) program in July 1997. WQ2000 consists of a series of water treatment improvement projects that will chart the course of Tampa's future water treatment. The program includes the development of a Water Quality Master Plan, the operation of several innovative pilot-plant facilities that will allow Tampa to compare alternative water treatments, and the selection and construction of an improved treatment process to be completed by 2003.

The regulatory process over time has affected process selection at HRWTP, which has been completed (or scheduled) in several phases:

Phase 1 (1924-1981). HRWTP was a conventional treatment plant utilizing alum for coagulation and color removal and free chlorine as the primary and secondary disinfectant.

Phase 2 (1982-1991). Chloramines were introduced for secondary disinfection to achieve compliance with the THM Rule.

Phase 3 (1991-present). Enhanced coagulation with ferric sulfate was implemented, which reduced the TOC level in the finished water and will allow the HRWTP to meet Stage 1 D/DBP Rule MCLs and meet the IESWTR requirements.

Phase 4 (after 2002). Future modifications—GAC, ozonation with biologically active filtration (BAF), membranes—are being assessed, which will allow HRWTP to meet the Stage 2 D/DBP Rule MCLs and LT2ESWTR requirements to be promulgated in May 2002.

A number of studies (bench-, pilot-, and full-scale) have been conducted to evaluate the various treatment processes considered. The evaluations are ongoing and the cost of conducting the current evaluations (1998 – ongoing) has been estimated at $350,000. The current studies are being conducted by Black & Veatch, CH2M Hill, and the University of Central Florida. The plant staff are active participants in the studies and are involved in the operation of the pilot plant and data collection. Pilot-plant studies of the three technologies being evaluated for Phase 4 modifications (to be completed in 2002) are in the process of being conducted. Preliminary results and comparisons of the tests conducted are listed below.

Coagulation studies (1988-1990). DBP precursor (TOC) removal experiments were conducted, which evaluated different coagulants (polyaluminium chloride, ferric chloride, and ferric sulfate). Ferric sulfate was chosen as a viable alternative coagulant to alum. An extensive testing program was conducted consisting of jar testing to determine enhanced coagulation treatment parameters, a 10-mgd pilot study to access operational parameters, and a half-plant pilot study to compare enhanced ferric sulfate coagulation with the existing conventional alum treatment (Gianatasio, Hook, and Hjersted 1996).

The results of the jar testing indicated that enhanced coagulation (via charge neutralization) with ferric sulfate occurred within the pH range of 3.5 to 4.8, whereas sweep-floc (sweep) coagulation could be performed within the pH range of 5.8 to 7.0. (Aluminum or iron coagulation can remove TOC according to two general mechanisms [Hall and Packham 1965, Amirtharajah and Mills 1982, Edwards and Amirtharajah 1985]: [1] adsorption onto aluminum or ferric hydroxide floc and [2] formation of insoluble complexes [aluminum or iron humates or fulvates] in a manner analogous to charge neutralization. For the first mechanism, the humic or fulvic acids must first complex aluminum and iron cationic species under most pH conditions before precipitation of aluminum or ferric hydroxide occurs. The second mechanism is influenced by the acidity of the NOM representing ligand sites for metal complexation. Generally, the first mechanism is more dominant at higher coagulant dosages and higher pH conditions, whereas the second mechanism is relatively more dominant at lower dosages and pH conditions.) A considerable reduction (50 to 70 percent) in coagulant dose was realized when enhanced coagulation was utilized over sweep coagulation. Further, the production of an iron humate solid—with a beneficial reuse—during charge neutralization resulted in the choice of enhanced coagulation over sweep coagulation.

Sulfuric acid was used for pH reduction, whereas carbon dioxide failed to supply the needed pH reduction due to the relatively high natural alkalinity of the water. Final pH adjustment used a combination of caustic soda and soda ash.

Pilot-plant results indicated that the general treated water quality was equivalent to or better with ferric sulfate than alum. Ferric sulfate was operationally versatile and TOC removal was consistently better with ferric sulfate (72 percent on average) than with alum (59 percent on average).

Primary disinfection occurred using free chlorine at the effluent of the settled-water basins before final pH adjustment. Final pH adjustment occurred after filtration in a baffled blending chamber. Final post chlorination and ammoniation (to produce a monochloramine residual in the distribution system) also occurred within the blending chamber. The contact time for free chlorination from the end of the settling basins through the filters usually averaged 30 to 60 min. It is during this period that the majority of THM formation occurred.

To determine the effect of THM formation under full-scale treatment conditions, a 1-h THMFP test was performed on the settled-water effluents. These were performed at ambient pH (averaging 4.5 for ferric sulfate and 6.5 for alum treatment), as would be done full scale. The 1-h THMFP for iron-treated settled water averaged 34 μg/L, whereas alum-treated settled water produced an average 1-h THMFP of 81 μg/L. The THMFP for ferric-sulfate-treated settled water was significantly lower than for alum-treated water, due to a combination of higher TOC removal and chlorination at a lower ambient pH.

Further, the switch to ferric sulfate was favored by the development of a beneficial reuse of the iron humate residual material produced. Application studies were performed utilizing ferric humate on turf grasses, citrus, miscellaneous fruits and vegetables, and animal feed supplements (Gianatasio, Hook, and Hjersted 1996).

Alternative disinfectant studies (1989, 1994, 1998). Ozonation with BAF has been evaluated in a series of pilot-plant studies. Pilot-plant testing evaluated ozone demand and decay. In addition, dual-media filters of anthracite and sand and GAC over sand have been evaluated. The projected capital cost of an 80-mgd ozonation system was estimated at \$11.2 million (\$6.6 million for ozone generator building, \$1.2 million for low-lift pump station, \$0.7 million for chillers, \$2.8 million for ozone-contact chamber) plus \$6.3 million for filter modifications. (This was based on a 15-min ozone-contact time and two 2000-lb/day ozone generators [3 mg/L dose, 80 mgd, one redundant generator].) The projected additional annual cost of operations (no labor) was estimated at $0.43 million.

The advantages of the ozonation/BAF system are that it (1) could provide substantially greater disinfection capability (*Cryptosporidium* inactivation) (Finch et al. 1994), (2) remove taste-and-odor compounds, (3) provide some additional TOC removal (approximately 0.5 mg/L), (4) result in minimal halogenated DBP formation, and (5) has the lowest capital and annual operating costs of the three advanced treatment technologies being evaluated. The main disadvantages are it (1) provides a lower degree of cyst removal/inactivation than membranes, (2) requires closer monitoring, and (3) is not as absolutely effective in removing taste- and odor-causing compounds as GAC and membranes.

GAC studies (1998). A bench-scale GAC test was conducted. With an EBCT of 30 min, a regeneration frequency of 169 days was required. It was estimated that retrofitting HRWTP (80 mgd) with GAC would cost $37 million in capital costs and an additional $10 million in annual operating costs. The advantages of GAC are that it will substantially remove both TOC and DBP precursors (McGuire et al. 1989) and taste- and odor-causing compounds. The main disadvantage of this process—in addition to its high capital and operating costs—is that it does not provide any pathogen-reduction benefit.

Membrane studies (1998 – ongoing). The membrane information is confidential at this time (the data are not yet available for public use). The capital cost of installing membranes (80 mgd) was estimated between $74 and $85 million, with operating costs between $9 to $10 million per year. (The membrane process would be an integrated process with nanofiltration, where nanofiltration would be preceded by coagulation, sedimentation, and granular-media filtration or by microfiltation. In order to deliver 80 mgd to Tampa's distribution system, the system would have to be sized to treat between 99 and 125 mgd, depending on which integrated membrane system was selected and the percent recovery of the different membranes within the process. Moreover, between 19 and 45 mgd of reject water would need to be discharged every day. Finally, capital and operating costs associated with the conveyance and disposal of membrane reject water were not included in these costs.)

The membrane system will produce the highest quality finished water of the three technologies evaluated due to the high degree of cyst removal and the substantial removal of TOC, DBP precursors, and taste- and odor-causing compounds. However, the disadvantages are its high cost, high percentage (15 percent) of reject water requiring disposal, and the need to expand the capacity of the pretreatment processes.

Parameters that Influenced Process Design

The parameters that influenced process design were residuals disposal, TOC, DBPs, turbidity, and color.

Residuals disposal. Enhanced coagulation with ferric sulfate was selected as it provided the HRWTP with the means for a 100-percent reuse of the residuals. The beneficial reuse of iron solids is possible, in part, because of Florida's iron-poor soils. The City of Tampa entered into a contract with a private company (Kemiron), which hauls the residuals to a private facility for further processing into an agricultural and livestock feed material. Water from the de-watering process is used to either irrigate a golf course or is recycled to the HRWTP for treatment.

TOC removal. The D/DBP Rule requires surface-water systems with conventional treatment to utilize enhanced coagulation to achieve designated TOC removal levels (USEPA 1998a). HRWTP typically has influent TOC levels ranging from 4 to 26 mg/L and alkalinity levels ranging from 42 to 143 mg/L (Table 4.7). Under a worst-case scenario (high TOC, low alkalinity), the plant may be required to achieve a 50-percent removal of TOC. Conventional treatment with alum did not consistently meet this requirement, whereas enhanced coagulation with ferric sulfate substantially outperformed this requirement (Table 4.7).

DBP control. Enhanced coagulation was also selected for DBP control as it substantially improved DBP precursor removal.

Turbidity. Enhanced coagulation was selected as it would lower the turbidity of the finished water and enable optimization of the plant.

Color. HRWTP is primarily a color-removal facility. Enhanced coagulation with ferric sulfate was as good as alum for color removal.

Table 4.7

Comparison of water quality before and after enhanced coagulation treatment at City of Tampa

Water quality parameters	Influent*	Before modifications Finished water†	After modifications Finished water‡
TOC, mg/L			
Minimum	4.3	1.8	1.6
Average	13	6.2	2.9
Maximum	26	8.9	5.1
TOC removal, %			
Minimum		20	70
Average		50	81
Maximum		77	88
UV, cm^{-1}			
Minimum	0.149		
Average	0.688		
Maximum	1.270		
SUVA, L/mg-m			
Minimum	2.94		
Average	4.03		
Maximum	5.06		
Bromide, mg/L			
Minimum	<0.02		
Average	0.057		
Maximum	0.113		
Temperature, °C			
Minimum	15	13	19
Average	25	25	25
Maximum	29	31	30

(continued)

Table 4.7 (Continued)

Water quality parameters	Influent	Before modifications Finished water	After modifications Finished water
pH			
Minimum	6.8	7.1	7.2
Average	7.6	7.6	7.6
Maximum	8.5	8.2	7.7
Alkalinity, mg/L as $CaCO_3$			
Minimum	42	80	48
Average	93	122	92
Maximum	143	187	125
Total hardness, mg/L as $CaCO_3$			
Minimum	58	118	102
Average	150	166	207
Maximum	216	244	266
Ammonia, mg/L as NH_3-N			
Minimum	<0.1		
Average	<0.1		
Maximum	0.31		
Turbidity, NTU			
Minimum	1.2	0.04	0.04
Average	2.1	0.32	0.11
Maximum	40	1.13	0.43
Total coliforms, # /100 mL			
Minimum	<10		
Average	1100		
Maximum	6000		<1

(continued)

Table 4.7 (Continued)

		Before modifications	After modifications
Water quality parameters	Influent	Finished water	Finished water
Fecal coliforms, # /100 mL			
Minimum	<1		
Average	152		
Maximum	1600		
Giardia, # cysts/100 L			
Minimum	<8		
Average	<8		
Maximum	25		<0.1
Cryptosporidium, # oocysts/100 L			
Minimum	<8		
Average	<8		
Maximum	25		<0.1
Viruses, pfu/100 L			
Minimum	<1.02		
Average	<1.02		
Maximum	1.03		<1

[*]Data from ICR sample collection from July 1997 - December 1998.

[†]Data collected for calendar year 1990.

[‡]Data collected for calendar year 1997; ICR data from July 1997 - December 1998 was used for organic analyses.

The Decision-Making Process

The City of Tampa implemented treatment modifications at HRWTP in response to increasingly stringent regulations and other needs related to raw water quality characteristics and residuals disposal dilemmas. The implementation strategy was designed to incrementally meet changes in regulations one at a time. Chloramines were introduced in 1982 for secondary disinfection to meet the THM Rule (USEPA 1979). Enhanced coagulation was implemented in 1991 for non-regulatory reasons; however, it is helping in meeting the Stage 1 D/DBP Rule. Further improvements were made as part of the "Partnership for Safe Water" program (1995), which helped in meeting the IESWTR. WQ2000 program was implemented in 1998 in order to evaluate three advanced treatment technologies that may be required to meet the Stage 2 D/DBP Rule and/or LT2ESWTR. Plant expansion with ozonation and BAF may be implemented in 2002 to allow the plant to meet future anticipated regulations.

Several studies were undertaken in the decision-making process, such as a benchmarking study (staff reductions), automation/instrumentation upgrade study, WQ2000 Master Plan evaluation, and future plant modification and expansion study. In addition, studies were undertaken on chlorine dioxide, ozonation, BAF, membranes, enhanced coagulation, and GAC to help with the decision-making process.

Driving Forces

The major reason for choosing enhanced coagulation with ferric sulfate in 1991 was the issue of disposal of the alum residuals produced by the plant (Figure 4.8). Historically, alum residuals were disposed in landfill sites, but due to the closure of several landfill sites and the reluctance of regulatory agencies to issue permits for new sites, the City of Tampa was forced to look at alternative technologies for residuals disposal and/or water treatment. The Tampa Water Department investigated alum recovery and reuse options and the option of alternate coagulants. Extensive testing and research over a three-year period determined that enhanced coagulation with ferric sulfate would meet anticipated requirements with the DBP Rule and produce a ferric humate residual with a beneficial reuse.

In addition to the closing of landfill sites, the utility had to make some replacement and expansion decisions regarding the plant facilities. The need to make these decisions helped them towards taking action before the Stage 1 D/DBP Rule.

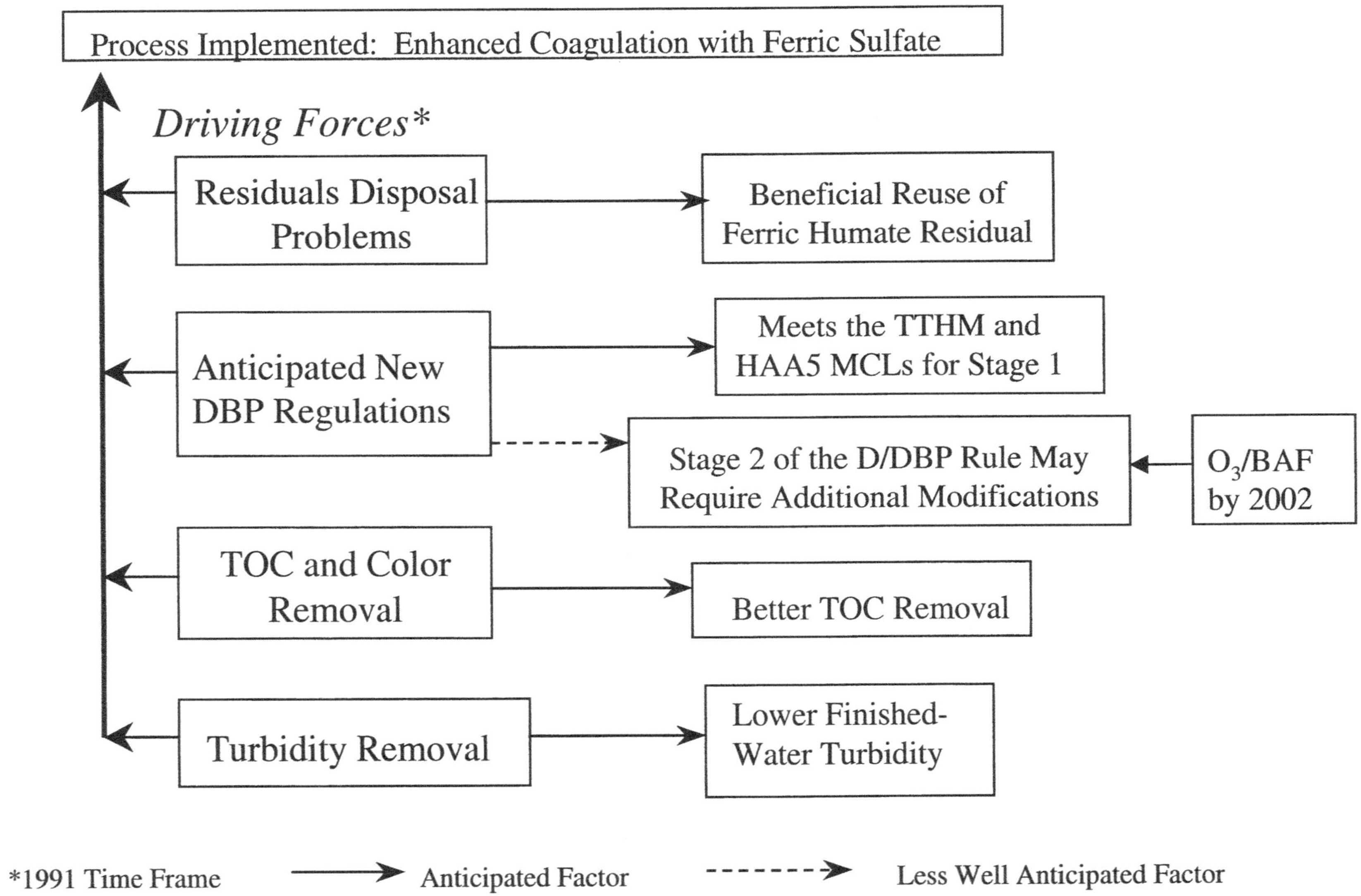

Figure 4.8 The decision-making process at Tampa: switch to ferric sulfate as the coagulant and implementation of enhanced coagulation

An additional driving force in the decision-making process was that the ferric coagulant manufacturer worked diligently alongside Tampa staff to devise a product that would perform well both in the water treatment plant and as a reusable product for agricultural use.

Constraints

The major constraint to enhanced coagulation was disposal of residuals. Beneficial reuses of residuals were needed rather than the conventional landfill disposal route. Prior to the work performed by Tampa and the ferric coagulant manufacturer, the potential to produce a reusable by-product had not been demonstrated. The research was needed to demonstrate the concept. Lack of proof of concept was therefore an initial constraint.

Another constraint was staffing limitations being imposed at the plant due to competitiveness benchmarking study results. The research work performed with the ferric coagulant manufacturer proved useful in coping with this constraint. The current procurement practice for ferric sulfate requires the supplier to remove and dispose of the residuals as part of the supply contract, saving the utility precious employee time at the plant.

Technology Checklist—Anticipated Factors

They correctly anticipated that enhanced coagulation with ferric sulfate would allow for better TOC removal than standard coagulation with alum. In earlier times, ferric coagulant was not considered for use at Tampa due, in part, to impurities in its manufacture. Tampa worked with the ferric coagulant manufacturer to develop a new product. New production processes using virgin iron ore solved much of this problem.

They correctly anticipated that enhanced coagulation with its efficient removal of DBP precursors would allow HRWTP to meet the Stage 1 TTHM MCL. However, they were not paying any attention to HAAs.

They correctly anticipated that enhanced coagulation would lower the effluent turbidity and allow for plant optimization to meet the "Partnership for Safe Water" goals and IESWTR requirements.

Technology Checklist—Less Well Anticipated Factors

Tampa staff is well aware of the treatment challenge posed by the extreme variability in their raw water. However, in switching to ferric sulfate they had no base of knowledge built from experience. As a result, they had to develop this base of knowledge as they went along. The source water goes through seasonal swings that present different challenges. In the spring and early summer rainy seasons, there is high color, high TOC, and low alkalinity. During the rest of the year the surface-water flows are influenced by groundwater recharge that provides water with less color and TOC, but higher alkalinity. The daily swings in influent quality can also be significant. Even though the Tampa staff was well aware of these characteristics, the change over to ferric sulfate was very challenging due to the important differences in its impact on pH compared to alum.

They did not anticipate the fact that the low pH levels used during enhanced coagulation would allow manganese to pass through the filters and affect the control of manganese at HRWTP. There was also an issue arising from the use of some product water in an aquifer storage and recovery project, where pH had to be re-adjusted to prevent leaching of natural arsenic. With the promulgation of an HAA MCL, another pH balancing concern has been introduced. They have had to adjust their seasonal balancing of pH control and coagulant dose in a manner that is impacted by multiple interacting constraints. In some operating modes, they are not achieving as much with ferric sulfate as might be theoretically feasible in terms of microbial/DBP objectives (*Cryptosporidium* removal and TOC removal) because of the other battery of constraints on the range of *pH* that is appropriate to different circumstances. In some modes, they are not able to optimize the economics of their use of ferric sulfate in terms of minimizing the dose and chemical costs because of many of the same constraints.

Tampa staff did not anticipate any problems downstream in their distribution system as a result of the changeover in coagulants. It was not a major factor in their changeover. The change from alum to ferric sulfate does imply placing a very different water into pipes that have been treated very differently up to that point. There are questions and concerns here (e.g., interference with corrosion control, dissolution of aluminium from scale, changing the iron chemistry of the pipe ecosystem) that are being considered by utilities in the Tampa region as they look at DBP-control technologies, but they were not a factor in Tampa's switch to ferric sulfate. (Tampa did not experience any corrosion problems with the changeover.) Part of the reason for this may be the fact that Tampa has an aggressive flushing program that rotates through the entire system on a five-week cycle. This may dampen the effect of such changes. It is worth noting that the prior changeover to chloramines went without difficulty (e.g., nitrification) as well. The aggressive flushing program may be the thing that keeps these problems at bay.

Compliance Issues

With enhanced coagulation, HRWTP is in compliance with the Stage 1 D/DBP MCLs and is also meeting the TOC removal requirements in the Rule. Lower effluent turbidities should enable compliance with the IESWTR. Although the level of THMs has been greatly reduced, HAAs have not been as significantly reduced and remain much closer to the Stage 1 MCL. Because the coagulation pH was lower with enhanced coagulation and primary disinfection occurred before final pH adjustment, chlorination at a lower pH would also have some effect on lowering THM formation and potentially raising HAA formation. This is an important feature of TOC removal as a DBP-control strategy, it does not guarantee the same level of control for all DBPs when primary disinfection occurs before final pH adjustment. Nonetheless, implementation of enhanced coagulation allowed the City of Tampa to achieve compliance with existing regulations while investigating future treatment modifications (WQ2000) that will allow for compliance with the anticipated Stage 2 D/DBP Rule MCLs and the LT2ESWTR.

Key Players

The decision-making process involved a variety of WTP staff and consultants. Stakeholder panels were formed and kept informed all along the way. Stakeholder meetings were held to explain new technologies, with taste testing of the water produced in the pilot plant.

To introduce the flow of public opinion into the WQ2000 process, a Public Advisory Committee was established in November 1997. Individuals represent a broad range of community interests, including neighborhoods, public health, plumbing, education, agriculture, business, environment, industry, and commerce. The 13-member committee meets at critical times during the

WQ2000 planning and implementation schedule to review issues and develop recommendations regarding costs, rate impacts, public health and safety, environmental issues, and aesthetic qualities.

Demographic information from 1990 were obtained for the City of Tampa. The median household income was $22,772. The median family income was $27,662. The median cost of housing was $58,600. The median age of the population was approximately 33 years.

Selection Issues and Future Strategies

The Tampa Water Department has been satisfied with the results of the enhanced coagulation process and the decisions taken to implement future modifications at HRWTP.

Future strategies include staff reductions to meet benchmarking, automation/ instrumentation upgrade to meet WQ2000 objectives, and future plant expansion in 2002 to include ozonation/BAF.

Economic Analysis

The capital cost of converting to enhanced coagulation from standard coagulation with alum cost the City of Tampa less than $100,000. The economic impact was mostly contained to providing an acid-feed system and a caustic-soda-feed system. The acid-feed system was less of a cost impact because the City was able to utilize a rubber-lined steel storage tank that had been used for liquid chlorine storage. The feed pumps used were fairly inexpensive (only around $1000 each) and they last approximately three years. Tampa has two pumps in the system with one pump capable of feeding the maximum dose. The second pump is for backup. Tampa keeps a third pump in stock for quick change out. The caustic-soda-feed system was more expensive in that the City had to purchase two 20,000-gal fiberglass tanks. The feed system was less costly in that Tampa used air diaphragm pumps for caustic feed. These pumps are only $400 each and are changed out as needed, with spare pumps kept in stock. They also seem to last approximately two to three years.

The additional annual cost of operations increased by approximately 18 percent. The 18-percent increase in operation costs reflects the purchase and use of sulfuric acid and caustic soda. As indicated above, maintenance costs of these systems are relatively low. This cost was greatly offset by the reduction in residuals handling (see below). The cost for training operators was not considered as the City is constantly training the WTP personnel on all systems, with annual updates by chemical vendors as part of their chemical specifications.

An economic advantage realized by switching to ferric sulfate from alum was the production of an iron-rich residuals. The City of Tampa entered into an agreement with a private company (Kemiron), which treats, processes and sells the residual product for soil application and as livestock feed material. Initially, the Tampa water department shared in the profits realized from the sale of the dewatered residuals. With the City's current contract, Kemiron is responsible for all aspects of residuals removal after Tampa's belt filter presses. In the City's new contract, Tampa is no longer being paid for the residuals by Kemiron. Instead, Kemiron is now supplying the personnel and front end loader equipment to turn residuals piles to promote additional drying, and load and haul off the residuals (iron humate). The Tampa water department is in a down-sizing mode and this saves Tampa an additional two people. With the reduction of two positions, no truck hauling charges, and no landfill tipping and monitoring charges, Tampa estimates their annual saving to be \$350,000 to \$500,000.

The WQ2000 program has been estimated to cost from \$50 to \$80 million. The final cost will be based on actual capacity and treatment process improvements selected. The projected capital cost of the ozonation system is \$9.7 million. Funding for WQ2000 will come from the sale of municipal bonds. It is anticipated that the added cost of WQ2000 will be about \$2 per month for the average Tampa residential water customer. Currently, the City has maintained water rates that are among the lowest within Florida: monthly water bills (for a residential meter using 6,730 gal/month) inside and outside the City are \$9.36 and \$11.70, respectively, and are from \$10.84 to \$45.87 in other parts of Florida.

Treatment Facilities

Water Supply

The City of Tampa's water supply is obtained from a 12-mile long reservoir that impounds 1.7 billion gallons of water and is supplied by the Hillsborough River. Withdrawals from the reservoir are regulated under a water use permit, which limits annual average day withdrawals from the Hillsborough River Reservoir to 82 mgd. The quality of water withdrawn from the reservoir varies seasonally. Typically, the water is high in color during periods of normal to above normal rainfall and low in color following periods of low rainfall. The reservoir is also subject to algal blooms under certain conditions. Copper sulfate is applied periodically to control algae growth in the reservoir.

Typical water quality parameters for the plant influent are presented in Table 4.7. The influent to the plant is highly colored (typical range = 30-300 pcu), with a moderate-to-high TOC (4.3-26 mg/L, average = 13 mg/L), an average total hardness of 150 mg/L (range = 58-216 mg/L), and an average alkalinity of 93 mg/L (range = 42-143 mg/L). The SUVA—which is an indicator of humic content (Edzwald and Van Benschoten 1990)—was moderately high (2.9-5.1 L/mg-m).

HRWTP: Before Treatment Modifications

The HRWTP is a 100-mgd conventional treatment plant (Figure 4.9), which uses chlorine for primary disinfection and chloramines for secondary disinfection. It typically operates at a flow of 80 mgd to voluntarily meet lower turbidity goals. Water treatment processing systems included: rapid mix basins, flocculation basins, setting basins, two accelerator basins (only used in emergencies), filtration facilities, and finished water disinfection systems. The water went through a six-step treatment process before it became finished water and was pumped through 2,100 miles of distribution pipes.

Raw water (pH varies between 6.8 and 8.5) was treated with potassium permanganate for taste-and-odor control. Then sweep coagulation took place for color removal with alum (range = 50-200 mg/L, average = 120 mg/L) at a pH of 5.7 (range = 4.9-6.6). (NSF Standard 60 lists the maximum use of alum at 150 mg/L. This is based on an evaluation of health effects data, as the level of aluminum in the finished drinking water should not exceed 2 mg/L. However the pH of minimum solubility for aluminum is ~6 at 25°C [Snoeyink and Jenkins 1980], which was the average coagulation pH for Tampa.) An organic polymer was added to enhance the flocculation process. The settled water was treated with lime for partial pH adjustment.

Primary disinfection was attained by applying chlorine to the settled water to produce a free residual of 1.0 mg/L on top of the filters, with 0.1 mg/L on the bottom. The water passed through

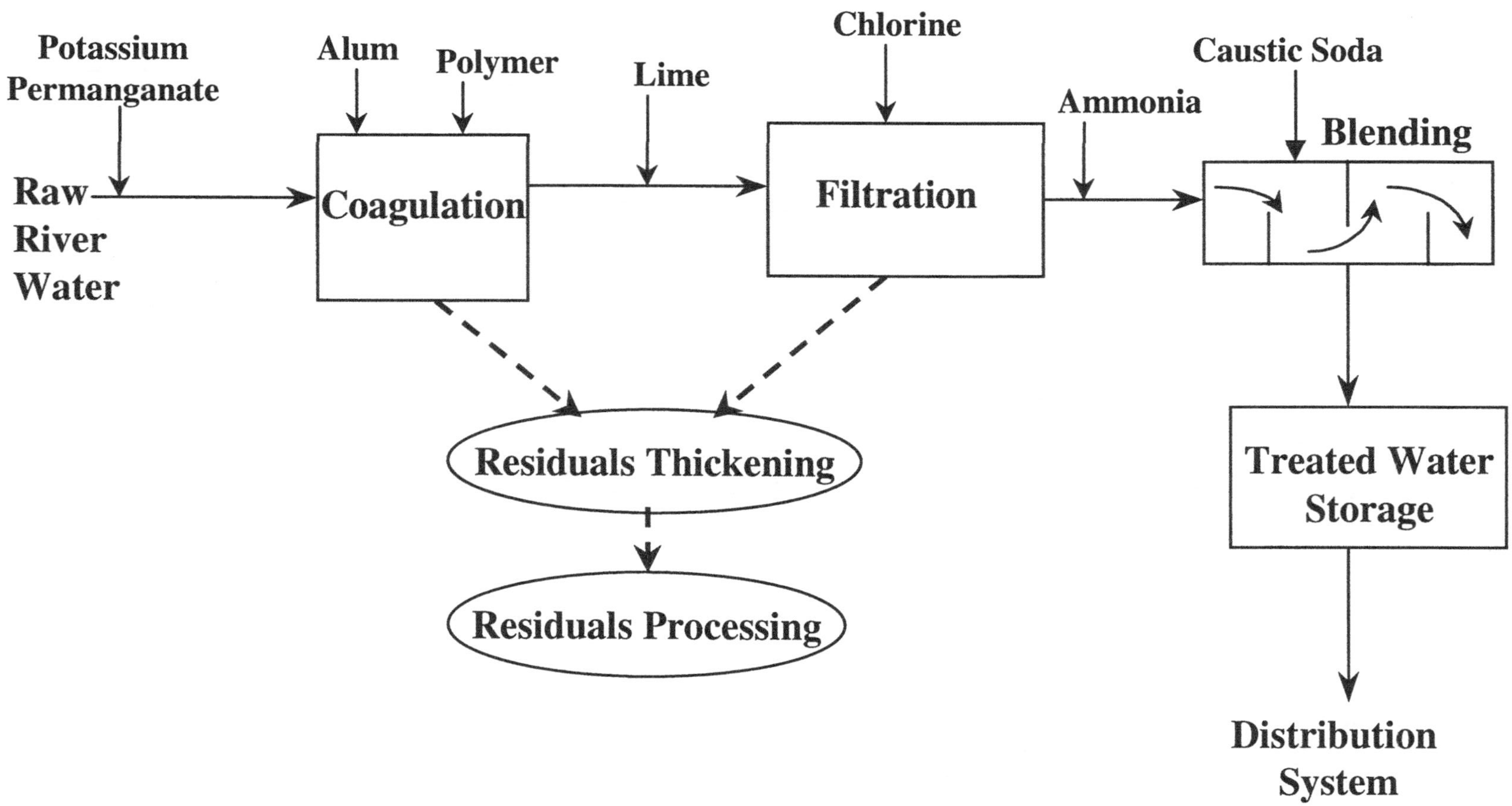

Figure 4.9 City of Tampa: sweep coagulation (1982-1990)

filters containing layers of sand, gravel and anthracite coal to remove any remaining particles. The contact time for free chlorination from the end of the settling basins through the filters usually averaged 30 to 60 min. After filtration, more chlorine was added as necessary and ammonia was added to form chloramines. The pH of the finished water was adjusted to 7.6 (range = 7.1-8.2) with caustic soda in the baffled blending chamber to produce a Langelier Index of +/-0.2. The finished water was then put in treated-water storage before distribution.

HRWTP: After Treatment Modifications

Figure 4.10 shows a schematic of the treatment plant after modifications. Raw water is treated with potassium permanganate. Enhanced coagulation with ferric sulfate (range = 40-300 mg/L, average = 140 mg/L) is practiced at a pH of 4.0 (range = 3.5-4.8). Sulfuric acid is added to attain the coagulation pH. The settled water is treated with lime for partial pH adjustment. The residuals are thickened and then pumped to a residuals-processing facility for further dewatering, processing, and disposal.

Primary disinfection is with free chlorine to produce a 1-2 mg/L residual on top of the filters. After filtration, chloramines are used for secondary disinfection. The finished water disinfectant residual ranges between 3 to 4.5 mg/L of monochloramine. The required *CT* is achieved with credit for the treatment plant (2.5 log) and additional *CT* credit for chloramine contact time and treated-water storage. The final pH (range = 7.2-7.7, average = 7.6) of the finished water is adjusted with caustic soda and soda ash in the blending chamber to meet the Langelier Index of +/-0.2.

Operations

Instrumentation and Control

The operations center at the HRWTP functions as the control center for the plant and for the water transmission and distribution system. The staff utilizes a SCADA system to control and monitor plant operating equipment. SCADA is also used to monitor water quality parameters at

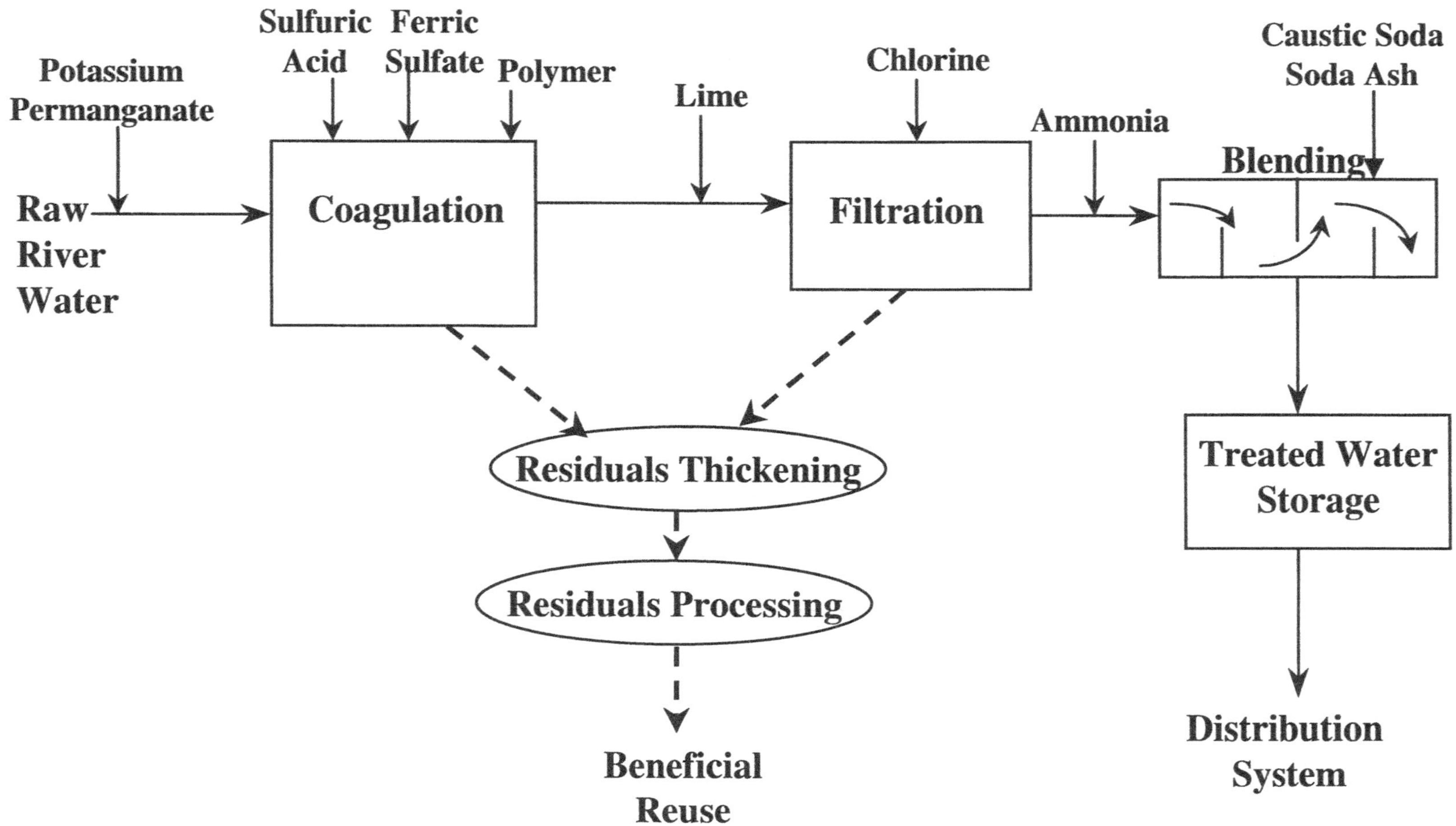

Figure 4.10 City of Tampa: enhanced coagulation (1991-present)

HRWTP, as well as selected components and operating conditions at remote pump stations and storage facilities.

A computer model is used to determine operating needs at HRWTP to achieve compliance with *CT* requirements. The model allows plant personnel to change operating parameters and to evaluate whether adjustments made will produce the *CT* values required.

Plant Operations

CT. The clearwell facilities and disinfection practices at HRWTP were evaluated to determine compliance with the *CT* requirements in the SWTR and the Florida Administrative Code. A conventional plant receives 2.5-log credit for removal of *Giardia* and 2.0-long credit for removal of viruses. Therefore, under the SWTR, the plant must provide a 0.5-log inactivation of *Giardia* and a 2.0-log inactivation of viruses through disinfection. The HRWTP uses chloramines for disinfection of the filtered water. For purposes of the *CT* evaluation, only a 2.0-log inactivation of viruses was reviewed to determine compliance with the disinfection requirements. When using chloramines as the disinfectant, *CT* values for inactivation of viruses are much greater than values required for *Giardia*.

The *CT* values for inactivation of microorganisms are dependent on temperature and the residual disinfectant concentration. Therefore, to determine compliance with the *CT* requirements, the worst-case operating condition in regard to these parameters was evaluated. (For example, in Table 4.7, the worst-case conditions in the finished water were minimum temperature of 13°C, maximum pH of 8.2, chloramine residual of 3 mg/L). Using the worst-case operating condition scenario at HRWTP and assigned baffling factors, the calculated *CT* value for the clearwell with chloramines was 391.47 mg-min/L. The required *CT* value for a 2-log inactivation of viruses with chloramines is 363.8 mg-min/L. Therefore, HRWTP can achieve the *CT* requirements for the inactivation of *Giardia* and viruses in the clearwell using chloramines. Meeting the *CT* requirements under the worst-case condition assures that HRWTP can achieve compliance under normal operating conditions.

Implementational and Operational Issues

Operator Training and Start-Up

Training sessions on the new treatment strategy were given to the operating staff with technical assistance provided by the ferric sulfate vendor. It took a period of six months for the operational staff to be "comfortable" with enhanced coagulation and a period of one year for the treatment plant to operate like an integral unit. The treatment strategy implemented did not significantly change the operational needs of the plant and no additional staff were added.

Operational Issues

The only significant problem or "trade-off" experienced in implementing the treatment modification was the control of manganese. The enhanced-coagulation pH allowed manganese to pass through the filters. The situation was resolved by maintaining the pH on top of filters at >6.0.

Acid addition has not presented a safety problem. Tampa found that the addition of sulfuric acid did promote corrosion in their rapid-mix chamber at the feed diffuser. After the City found an acceptable coating system for their rapid-mix chamber, the corrosion problem was solved. The coating used in the rapid-mix basins where ferric sulfate and sulfuric acid are added is a two-part commercial membrane (CIM 1061, C.I.M. Industries, Inc., Peterborough, N.H.) applied at 60 wet mils, using an air-supplied mastic spray gun. The coating is approved for potable water systems. After application, the coating did require a seven-day cure period before Tampa could put the basin back in service. The settling basins were initially epoxy coated and Tampa did not experience corroding at this location. Tampa has had no problems since then and they clean and inspect treatment basins annually.

System Performance

Water Quality

The water quality data from the calendar year 1990 represents the treatment plant "before treatment modifications" and the "after treatment modifications" water quality data were collected in calendar year 1997 with additional supporting ICR data from July 1997 to December 1998 (Table 4.7). Care must be taken in comparing the year 1990 with the years 1997 and 1998, as Tampa finds not only seasonal variations in raw-water quality but also yearly cycles of variation. Nonetheless, the water quality parameters (e.g., TOC, temperature, pH, alkalinity, hardness, turbidity) of the influent to the plant were similar before and after treatment modifications.

TOC. The influent TOC varied from 4.3 to 26 mg/L, with an average of 13 mg/L. The seasonal rainfall in summer results in TOC levels increasing to a maximum. Finished-water TOC with alum coagulation ranged from 1.8 to 8.9 mg/L, with an average of 6.2 mg/L. TOC removal before modifications ranged from 21 to 77 percent, with an average value of 50 percent.

Finished-water TOC with enhanced coagulation (ferric sulfate) ranged from 1.6 to 5.1 mg/L, with an average of 2.9 mg/L. TOC removal ranged from 70 to 88 percent, with an average TOC removal of 81 percent. Thus, enhanced coagulation with ferric sulfate was much better than sweep coagulation with alum at removing TOC. Enhanced coagulation also allows for better compliance with the TOC-removal requirements in the D/DBP Rule.

UV. The plant influent had an average UV absorbance of 0.69 cm^{-1} (range = 0.15-1.27 cm^{-1}). Influent SUVA was 4.0 L/mg-m on average (range = 2.9-5.1 L/mg-m), which is indicative of a moderate-to-high level of humic substances in the raw water.

In pilot-plant testing—in which TOC removal was >75 percent—raw-water SUVA was reduced from a range of 3.4 to 4.3 L/mg-m to a range of 1.8 to 2.3 L/mg-m in the settled water (prior to the addition of a disinfectant/oxidant). In the Stage 1 D/DBP Rule, achieving a settled-water SUVA of <2.0 L/mg-m—prior to the addition of a disinfectant/oxidant that can destroy UV but not remove TOC—can be used as an alternative performance criterion for enhanced coagulation (USEPA 1998a).

Temperature. The City of Tampa is located in Florida, which is subject to a hot and humid climate in the summer months and balmy winters. The temperature of the raw water ranges from 15 to 29°C, with an average of 25°C. The humidity in the summer months results in rains, which results in higher levels of TOC and color in the raw water. This seasonal variation in the raw water characteristics is also reflected in the finished water quality.

Bromide. Influent bromide levels at HRWTP ranged from less than 20 μg/L to 113 μg/L, with an average of 57 μg/L. This may be of significance when HRWTP goes to ozonation in 2002. Bromide plays a significant role in compliance regulations due to the oxidation (by ozone) of bromide to bromate (Krasner et al. 1993), which is a regulated DBP (USEPA 1998a). In pilot-plant testing, bromate was not detected during the ozonation of settled water. This may be due to the acidic pH of the settled water, as pH depression prior to ozonation is a proven means of minimizing bromate formation (Krasner et al. 1993).

Turbidity. The raw water had turbidity levels ranging from 1.2 to 40 NTU, with an average of 2.1 NTU. (The maximum turbidity value of 40 NTU was an outlier.) Before modifications, conventional treatment with alum produced a finished water with an average turbidity of 0.3 NTU (range = 0.04-1.1 NTU). After modifications, enhanced coagulation with ferric sulfate produced a finished water with an average turbidity value of 0.1 NTU (range = 0.04-0.43 NTU). Enhanced coagulation was better than sweep coagulation at reducing the level of turbidity in the finished water.

TTHMs. Before modifications—for the year 1990—the finished-water TTHMs ranged from 27 to 111 μg/L, with an average of 59 μg/L (Table 4.8). (In 1990, Tampa did have a minimum daily finished-water TTHM level at 27 μg/L, however this value was considerably lower than normal for any year that they were utilizing alum.) After treatment modifications—for the time period July 1997 to December 1998—the TTHMs ranged from 47 to 67 μg/L, with an average of 60 μg/L. (The minimum TTHM level with ferric-sulfate treatment of 47 μg/L was a typical lower level found in Tampa's water under either treatment scenario.) Enhanced coagulation is better than sweep coagulation at removing DBP precursors (TOC) when the raw water is high in TOC (summer), which is reflected in the lower maximum level of TTHMs produced after treatment modifications (the maximum THM formation was reduced from 111 to 66 μg/L). The new treatment scenario reduced THM formation due to a combination of higher TOC removal and chlorination at a lower ambient pH. Implementing enhanced coagulation has enabled HRWTP to achieve compliance with

Table 4.8

Comparison of DBPs before and after enhanced coagulation at City of Tampa

DBPs	Before modifications* Finished water	*After modifications*† *Finished water*
TTHMs, μg/L		
Minimum	27	47
Average	59	60
Maximum	111	66
HAA5, μg/L		
Minimum		32
Average		47
Maximum		66
TOX, μg Cl^-/L		
Minimum		170
Average		212
Maximum		295
CNCl, μg/L		
Minimum		<0.25
Average		11
Maximum		21

*Data collected for calendar year 1990.

†Data collected for calendar year 1997; ICR data from July 1997 - December 1998 was used for organic analyses.

the Stage 1 TTHM MCL of 80 μg/L. However, enhanced coagulation will not be sufficient to meet the previously proposed placeholder Stage 2 TTHM MCL of 40 μg/L.

HAAs. There were no data collected in the before modifications period for the HAAs. After treatment modifications, HAA5 ranged from 32 to 66 μg/L, with an average of 47 μg/L. Because the coagulation pH was lower with enhanced coagulation and primary disinfection occurred before final pH adjustment, chlorination at a lower pH would also have some effect on potentially raising HAA formation. Nonetheless, enhanced coagulation has allowed HRWTP the ability to meet the Stage 1 HAA MCL of 60 μg/L (based on the average value) but does not meet the previously proposed placeholder Stage 2 HAA5 MCL of 30 μg/L.

TOX. After treatment modifications, the TOX level in the finished water ranged from 170 to 295 μg Cl^-/L, with an average of 212 μg Cl^-/L. There were no data available for TOX before modifications were implemented.

CNCl. CNCl—a by-product of chloramination (Krasner et al. 1989)—was detected in the finished water after modifications at an average concentration of 11 μg/L (Table 4.8). There were no data collected for CNCl before modifications were implemented.

Pathogens. Total coliforms in the raw water ranged from <10 to 6,000/100 mL, with an average value of 1,100/100 mL. Fecal coliforms averaged 152/100 mL (range <1 to 1,600/100 mL) in the raw water. Protozoans (*Giardia* and *Cryptosporidium*) and viruses were either not detected or present at very low numbers in the raw water (Table 4.7). The finished water meets current microbiological regulatory requirements.

Summary of System Performance

Enhanced coagulation with ferric sulfate at HRWTP produces a better quality and more consistent quality finished water than conventional coagulation with alum. It was better for controlling seasonal variations in water quality. Enhanced coagulation also resulted in a much higher removal of TOC than conventional coagulation, which resulted (in part) in lower THM formation. It enabled the plant to comply with the TTHM and HAA5 MCLs in Stage 1 of the D/DBP Rule. In addition, the switch to coagulation with ferric sulfate resulted in an iron solids with a beneficial reuse.

The current TTHM and HAA5 values do not meet the previously proposed placeholder Stage 2 TTHM and HAA5 MCLs. Future modifications (i.e., ozonation/BAF) should allow HRWTP to meet the previously proposed placeholder Stage 2 TTHM and HAA5 MCLs and an anticipated LT2ESWTR.

BIOLOGICALLY ACTIVE CARBON: HOPEWELL

Summary

The Hopewell (Va.) WTP is a 36-mgd treatment plant; it typically operates at a flow of 20 mgd. The WTP is divided into two sections to produce two grades of water: one for industrial customers and one for domestic customers. Approximately 4.1 mgd is delivered to the domestic customers and the remainder is supplied to industry.

The source water for the Hopewell WTP is the Appomattox River, with intakes located 1.5 miles upstream of the confluence with the James River. Several wastewater treatment plants that discharge to the James and Appomatox rivers are in the proximity.

At Hopewell, pre-chlorination after the application of coagulant is the primary factor in DBP formation and generation. An absence of a primary disinfectant/oxidant will contribute to increased algal growth in sedimentation basins. In addition, given the proximity to several wastewater treatment plants, the elimination of pre-disinfection is not a viable option.

A bacterial regrowth event occurred in the early 1990s, which ultimately was controlled by the use of post-chloramination. Post-disinfection using chloramines limited coliform regrowth in pipeline biofilms and the formation of DBPs in the distribution system.

GAC has been used since 1960 to prevent taste-and-odor problems in the finished water. A new filtration system that included GAC pressure filters became operational in 1990. Starting in 1994, the process of regenerating GAC was initiated. The GAC is also used for the removal of other micropollutants, but it was not installed for DBP precursor (TOC) removal. However, Hopewell is experiencing the benefit of GAC providing a better medium for biological activity, which is resulting in the control of HAA concentrations. During warm-water months, an applied HAA concentration of 100 μg/L will be reduced to <10 μg/L.

The GAC is initially effective in THM reduction, but this control measure lasts only 90 to 120 days. Seasonal control of DBPs, especially THMs, is being achieved using pre-chloramination. During warmer months, ammonia was added to convert the chlorine to chloramines after a 1-min chlorine contact time, whereas ammonia was not added until the plant effluent during colder months. In the former disinfection scenario, the amount of free chlorine contact time provided adequate inactivation of bacteria; between chlorination and subsequent chloramination, *CT* was achieved. The presence of either a chlorine or chloramine residual in the GAC influent has not negatively affected the biodegradation of the HAAs.

Background

Case History

Virginia American Water Company, Hopewell District, operates the Hopewell District Water Treatment Facility. The plant flow at design is 36 mgd and at typical flow is 20 mgd. The population served by this treatment plant is 28,000. Approximately 4.1 mgd is delivered to the residential community, including a military reservation, federal correctional institute, and a section of an adjoining county. The remainder is supplied to industry.

The source water for the Hopewell WTP is the Appomattox River, with intakes located 1.5 miles upstream of the confluence with the James River. This area is influenced by the tides and is prone to flow reversal at the intakes. As much as 70 percent of the source water can be contributed from the James River, especially during low-flow conditions.

The Hopewell WTP, a rapid filtration plant, is divided into two sections to produce two grades of water—one for industrial customers and one for domestic customers—served by separate distribution systems. Both plants provide full treatment—including pre-chlorination—but the domestic plant uses post-filtration GAC contactors. GAC reactivation frequency is once every three years, as these contactors are primarily used for the reduction of tastes and odors.

Hopewell had persistent coliform occurrences in 1991 (Norton and LeChevallier 1997). The average free chlorine residual in the distribution system was increased from 0.57 to 2.1 mg/L. However, coliform bacteria persisted. As a result of increased chlorine residuals, THM concentrations during the summer exceeded 100 μg/L. Continued coliform occurrences and excessive THM concentrations led Hopewell to convert its domestic distribution system to chloramines in July 1993.

The GAC is initially effective in THM reduction, but this control measure lasts only 90 to 120 days. However, the GAC has been effective in reducing HAA concentrations over extended periods of time. The reduction efficiency is temperature dependent, suggesting that the GAC is biologically active. During warm-water months, an applied HAA concentration of 100 μg/L will be reduced to <10 μg/L. Other research has demonstrated that HAAs can be removed by GAC filtration, presumably by biodegradation processes within the filter bed (Singer et al. 1999).

In 1998, WTP personnel initiated full-scale treatment for evaluating the effectiveness of pre-chloramination for minimizing DBP concentrations. In 1999, the application of an acidified aluminum sulfate was being evaluated to further reduce TOC and precursor concentrations.

Process Selection

At Hopewell, pre-chlorination after the application of coagulant is the primary factor in DBP formation and generation at this WTP. An absence of a primary disinfectant/oxidant will contribute to increased algal growth in sedimentation basins, resulting in increased TOC and precursor material. In addition, given the proximity to several wastewater treatment plants that discharge to the James and Appomatox rivers, the elimination of pre-disinfection is not a viable option.

A bacterial regrowth event occurred in the early 1990s, which ultimately was controlled by the use of post-chloramination (Norton and LeChevallier 1997). During the outbreaks, a treatment strategy was used to try and reduce AOC levels. These operational changes—i.e., the application of potassium permanganate, PAC, and change in chlorine application—were employed to reduce THM and AOC levels, with marginal effectiveness. Alternatively, post-disinfection using chloramines limited coliform regrowth in pipeline biofilms and the formation of DBPs in the distribution system.

GAC has been used since 1960. GAC was put in to prevent taste-and-odor problems in the finished water. A new filtration system that included GAC pressure filters became operational in 1990. The GAC is also used for the removal of other micropollutants, but it was not installed for DBP precursor (TOC) removal. (However, Hopewell is experiencing the benefit of GAC providing a better medium for biological activity, which is resulting in the control of HAA concentrations.) In 1994, the GAC in four contactors was replaced and regeneration was initiated. The frequency needed to regenerate the the GAC is three years. Every year, a portion of the GAC has been reactivated. The GAC is reactivated off-site.

In 1998, WTP personnel initiated full-scale treatment for evaluating the effectiveness of pre-chloramination for reducing DBP concentrations. For this disinfection scenario, a 3.0 mg/L free chlorine residual was achieved prior to the addition of aqua ammonia. A 15 to 20 second exposure to a free chlorine residual yielded a 100 percent inactivation of coliform bacteria. Aqua ammonia was added to interrupt the DBP formation reaction. Typically, the raw water is disinfected with free chlorine for a 1-min contact time before ammonia addition. The combined chlorine residual reduced the growth of algae in the sedimentation basin and the pre-oxidant maintained filter performance. Between chlorination and subsequent chloramination, *CT* is achieved.

In 1999, the application of an acidified aluminum sulfate was being evaluated to further reduce TOC and precursor concentrations. The aluminum sulfate contains a portion of free sulfuric acid.

Parameters that Influenced Process Design

The parameters that influenced process design were coliform bacteria, algae in sedimentation basins, tastes and odors, and DBPs.

Coliform control. Pre-chlorination has been used to inactivate coliform bacteria. Post-chloramination has been used to limit coliform regrowth in pipeline biofilms.

Algae control. Pre-disinfection (with free or combined chlorine) has been used to control algal growths in sedimentation basins.

Taste-and-odor control. Post-filtration GAC contactors have been used for the reduction of tastes and odors in the finished water.

DBP control. Post-chloramination has been used to minimize DBP formation in the distribution system. Biologically active GAC has been used to reduce HAA concentrations in the plant. Seasonal control of DBPs is also being achieved using pre-chloramination.

The Decision-Making Process

The American Water Works Company takes a proactive approach in meeting water quality regulations, planning several years in advance of the implementation of new regulations. The planning process involved personnel from the company, as well as corporate-level resources available in the American system.

The decision-making process considered a review of alternative treatment methods, experience in similar water systems, and previous operating experience at Hopewell.

Driving Forces

There were several driving forces that helped shape the selection of treatment options:

- The biggest driving factor was the variable raw water quality. The river-water intakes are located approximately 1.5 miles upstream of the confluence of the James River. This dynamic tidal estuary is prone to flow reversal. Although the intakes are located on the Appomattox River, up to 70 percent of James River water can be withdrawn. Two distinct watersheds, compounded by tidal influence, cause the river water quality to constantly change throughout the day. The plant receives an influent that is influenced by upstream and downstream industrial and wastewater discharges. Within a 25-mile radius, there are five major wastewater treatment plants that provide variable sources of organic nitrogen, as well as bacteria and pathogens. At times the source water quality is void of sufficient alkalinity to promote coagulation. This condition will occur during extreme rainfall events and requires the addition of a supplemental alkali in the form of sodium hydroxide or calcium oxide. The river water has a variable and moderate concentration of TOC. Tastes and odors and algae are issues.
- In 1990, completion of the second phase of a multiple-phase improvement program occurred. High-rate multimedia filtration with air scour (6.0 gpm/sf), followed in series with deep-bed GAC filtration/adsorption, became operative. The capacity of this portion of the treatment plant was 12 mgd. In 1991, the domestic distribution system experienced a bacterial regrowth problem. Ultimately, the problem was resolved by implementing a post-chloramine residual.
- In reviewing treatment options—in light of the experience with influent conditions—a primary concern arose over process control for assurance of filter and disinfection performance in the face of river water quality challenges. Operationally, there were concerns regarding the impact of solving one problem while creating others. Each alternative treatment technique was

scrutinized by addressing the impact on operations, the goal or outcome, ease of installation, and cost. There were concerns over algal growth in the sedimentation basins if chlorine were removed as a pre-disinfectant/oxidant, and sedimentation efficiency. There were issues of the formation of DBPs of alternative oxidants/disinfectants, and performance of the multimedia filters and GAC contactors. The final decision was to add aqua ammonia to the coagulated/chlorinated water. This addition was made after a brief period where the treated water was exposed to breakpoint chlorination. While being the most cost-effective and simplistic approach that would interrupt the formation kinetics of the DBPs and maintain sedimentation-basin water quality, pre-chloramination could have several major impacts on the successive treatment processes. Of primary concern was ammonia toxicity to the bacterial population on the GAC and nitrification within the GAC contactors.

- Cost was also a driver. This was made especially acute by the fact that a majority of utility revenues came from large industrial customers. The existing utility provided two different grades of water for domestic and industrial purposes. The different processes and resulting costs presented a challenge to preserve cost advantages for industrial users while meeting higher standards for domestic customers.

Constraints

The major constraint affecting the decision process was the existing plant design. The plant consists of very differing process trains, including an entire set of high-rate wooden tub filters on the industrial side. Solutions that would fit the existing processes, both physically and operationally, were required in order to avoid major changes in rates to industrial customers. The chloramination approach met this requirement, but raised a series of issues regarding pH and alkalinity balancing; interference from influent ammonia; and ammonia chemistry throughout the process.

Technology Checklist – Anticipated Factors

- The need for enhanced attention to process control (pH, alkalinity, ammonia) was correctly anticipated. And, the measures employed to address these concerns (e.g., online SCADA, aqua ammonia) have performed to meet the challenge. Operator training in the new regime was also employed as a strategy to assure success.

- It was confirmed that an amount of free chlorine contact time could be maintained to provide assured inactivation of bacteria and algal control while still meeting Stage 1 DBP levels after quenching with ammonia.

Technology Checklist – Less Well Anticipated Factors

- Although control of THMs was achieved with pre-chloramination, control of HAAs would not have been obtained were it not for the fact that GAC was already in place. Biological activity in the GAC filters achieved sufficient HAA reduction to meet Stage 1 goals.
- Despite efforts to comply with new regulations via incremental modifications to the existing split treatment works providing both domestic and industrial supplies, some industrial customers subsequently declined to make the additional investment that would be required of them to install dual plumbing systems at their facilities. As a result, the wheels are now in motion to rethink the solution for Hopewell in terms of overhauling the entire plant to meet new DBP standards at significantly increased cost.

Compliance Issues

The biologically active GAC has been effective in reducing HAA concentrations to below the previously proposed Stage 2 "placeholder" MCL (of 40 μg/L) (USEPA 1994a). Seasonal usage of pre-chloramination, along with post-chloramination, is being used to control THM concentrations. Between chlorination and subsequent chloramination, *CT* is achieved.

Key Players

Major department heads had input into the decision-making process. Because cost is a very sensitive issue at this location, the community (City of Hopewell) and industrial customers have actively intervened at rate hearings.

Demographic information from 1990 was obtained for the City of Hopewell. The median household income (in 1989) was $26,934. The median cost of housing was $54,500. The median age of the population was approximately 30 years.

Selection Issues and Future Strategies

The major new stategy is the use of seasonal control of DBPs using pre-chloramination.

Economic Analysis

The cost of the research project addressing biofilm bacteria was $60,000.

The chemical cost increased $6.00 to $10.00 per MG for post-chloramination. For pre-chloramination, the chemical cost increased by $6.00 to $8.00 per MG. For acidified aluminum sulfate, the increased chemical cost (for coagulant only) is nearly $20.00 per MG.

For the pre-chloramination system only, the total cost was $4,765:

- Total material: 350 ft of polyvinyl chloride (PVC) schedule 80, 2-inch pipe @ $240;
- Metering pump for aqua ammonia @ $1,200;
- Service saddle for diffuser @ $375;
- 1,500-gallon polypropylene tank (contributed by chemical supplier) @ no cost;
- Concrete tank pad @ $450;
- Copper line @ $300;
- Water supply booster pump @ $1,200; and
- Labor for installation @ $1,000.

The additional annual cost of operations due to the pre-chloramination modifications are $2,800 for aqua ammonia and $3,300 for liquid chlorine (testing materials were $200).

Because GAC was installed for taste-and-odor control and is only regenerated once every three years—where biodegradation of HAAs has been an additional benefit—there have been no additional costs for GAC that are directly related to DBP control at Hopewell. GAC is now part of Hopewell's DBP-control strategey, so a portion of the cost could be attributed.

Treatment Facilities and Operations

Water Supply

The source water for the Hopewell WTP is the Appomattox River, with intakes located 1.5 miles upstream of the confluence with the James River. This area is influenced by the tides and is prone to flow reversal at the intakes. As much as 70 percent of the source water can be contributed from the James River, especially during low-flow conditions. In addition, several wastewater treatment plants that discharge to the James and Appomatox rivers are in the proximity.

Typical water quality parameters for the plant influent are presented in Table 4.9. These data are from 1980 through the present. The influent to the plant was moderate in TOC (3.0-4.8 mg/L,

Table 4.9

Typical raw water quality at Hopewell WTP

Parameter	Minimum	Average	Maximum
TOC (mg/L)	3.0	3.9	4.8
UV (cm^{-1})	0.020	0.128	0.245
Temperature (°C)	3.9	18	30
pH	6.3	7.2	9.4
Alkalinity (mg/L as $CaCO_3$)	11	45	75
Total hardness (mg/L as $CaCO_3$)	23	43	115
Turbidity (NTU)	10	26	104
Total coliforms (#/100 mL)	ND	898	6,100
Fecal coliforms (#/100 mL)	ND	371	4,500
Giardia (# cysts/100 L)	ND	19	93
Cryptosporidium (# oocysts/100L)	ND	0.3	4
Viruses (pfu/100L)	1,840	2,883	4,280

average = 3.9 mg/L) and UV (0.020-0.245 cm^{-1}, average = 0.128 cm^{-1}). The alkalinity (11-75 mg/L, average = 45 mg/L) and hardness (23-115 mg/L, average = 43 mg/L) were low to moderate.

The variability in microbial loadings (e.g., the concentration of fecal coliforms ranged from ND to 4,500/100 mL, with an average of 371/100 mL) probably reflects the influence of wastewater discharges in the source waters. *CT* is very important to achieve for this facility.

Hopewell WTP: Current and Seasonal Treatment Operations

The Hopewell WTP is a 36-mgd treatment plant; it typically operates at a flow of 20 mgd. The WTP is divided into two sections to produce two grades of water: one for industrial customers and one for domestic customers. Approximately 4.1 mgd is delivered to the domestic customers and the remainder is supplied to industry.

Water treatment processes at the domestic plant included prechlorination, flocculation, clarification, multimedia filtration, GAC filtration, and post-chloramination (Figure 4.11). Currently, Hopewell uses pre-chloramination for seasonal control of DBPs (Figure 4.12). Pre-chloramination is used during the warmer months (e.g., during summer and fall).

Raw water (pH varies between 6.3 and 9.4, average = 7.2) was treated with alum (range = 25-120 mg/L, average = 35 mg/L) at a pH of 5.9 (range = 4.9-6.5). In 1999, acidified aluminum sulfate was used for coagulation (at a pH of 5.5 to 5.9). The settled water was filtered with multimedia: 17 inches of anthracite, 12 inches of silica sand, and four inches of garnet.

The filtered water then passed through a GAC contactor. The carbon installed in 1990 was supplied by ATOCHEM (CECA Carbon 30S) (a total of 11,020 cu ft). In 1994, four contactors were replaced with American Norit carbon. Starting in 1994, regeneration of GAC was initiated. Every year thereafter a portion of the GAC has been reactivated. The makeup carbon has been supplied either by Calgon or has been imported. The GAC specifications are as follows: apparent density = 0.36 to 0.60 g/mL; effective size = 0.8 to 1.0 mm; uniformity coefficient = 1.6 to 1.9; iodine number = not less than 800.

The depth of the GAC is 12 ft, 4 in. The maximum loading rate of the GAC filter is 9.2 gpm/sf (EBCT = 10 min). The normal loading rate is 6.1 gpm/sf (EBCT = 15 min). The backwash frequency of the GAC filter is after 200 hours of continuous operation. The criteria for

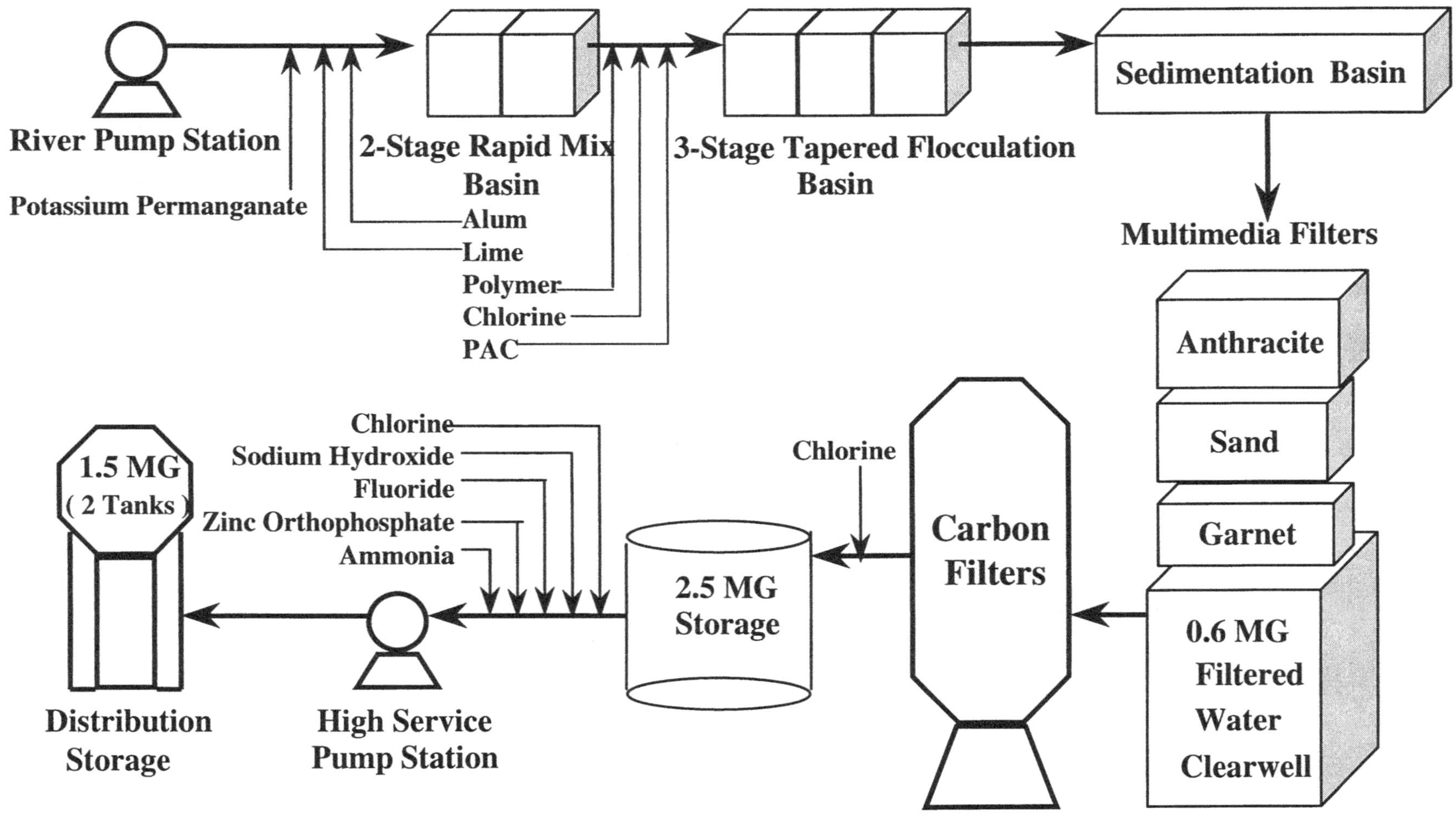

Figure 4.11 Flow schematic diagram at Hopewell with pre-chlorination

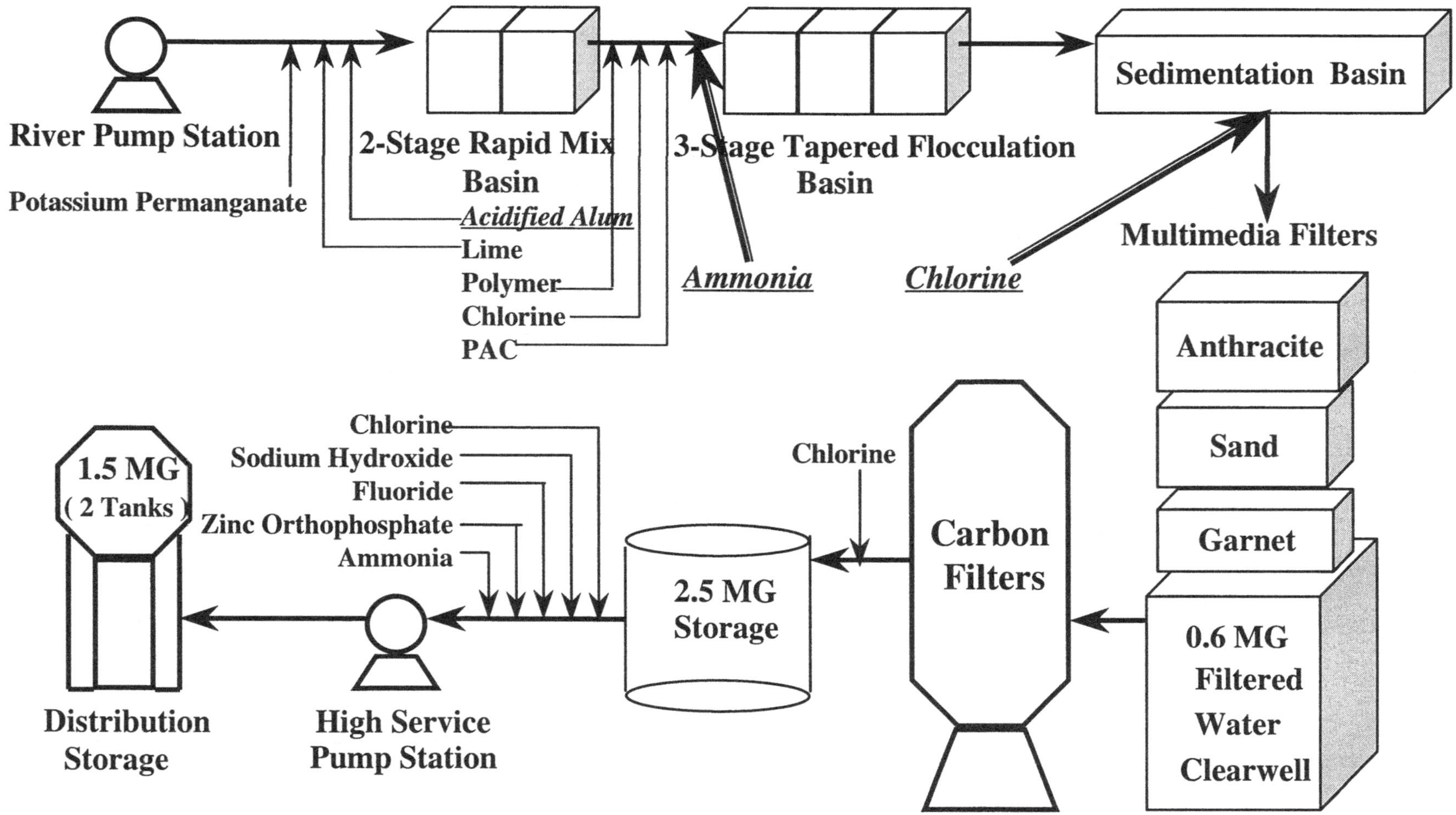

Figure 4.12 Flow schematic diagram at Hopewell with pre-chloramination

backwashing the GAC filter are based on time, pressure differential, and turbidity. The GAC is regenerated once every three years. A portion (1/3) of the total volume is reactivated yearly. The criterion for regenerating the GAC is based on iodine number.

In terms of the disinfection process, a free chlorine residual was maintained in the coagulated water that exited the rapid-mix basin. Chlorine was applied after the coagulant and flocculent. To achieve a 3.0 mg/L free chlorine residual at the exit of the mix tank, the applied chlorine dose rate varied from 3.5 to 10.0 mg/L. Because the coagulant was alum or acidified aluminum sulfate, the coagulated-water pH was depressed to approximately 5.9, depending on the raw-water alkalinity. This low pH was maintained through the treatment process, with final adjustment at the suction of the high service distributive pumps.

With pre-chlorination (Figure 4.11), the chlorine concentration at the exit of the sedimentation basin was not less than 0.75 mg/L as a free chlorine residual. This equated to an inactivation rate of 5 to 7 logs of *Giardia*. After filtration and subsequent GAC filtration/adsorption, the filtered water was chlorinated again to a 1.0 mg/L free residual and entered a 2.5 MG storage (clearwell). The tank was not baffled, having a baffling factor of 0.10. A 0.5 log inactivation rate was maintained, based on pH, temperature, chlorine concentration, and time. Post-chloramination was performed immediately prior to the distributive pumps.

With pre-chloramination (Figure 4.12), chloramination was conducted from 15-20 seconds to one minute after chlorine application. For pre-chloramination, a 3.0 mg/L free chlorine residual was achieved prior to the addition of aqua ammonia. The ratio of chlorine to ammonia was approximately 3:1. This resulted in a slight excess of free ammonia (the stoichiometric optimum for monochloramine formation is a 5:1 ratio; i.e., no excess ammonia is formed at that ratio). The predominate chloramine species was monochloramine (approximately 75 percent); the other chloramine species was dichloramine. The pH of coagulation was 5.5 to 5.9, which drove the formation species. At the exit of the sedimentation basin, chlorine can be added. Between chlorination and subsequent chloramination, *CT* is achieved.

In terms of corrosion control, prior to 1994, a zinc orthophosphate corrosion inhibitor was used having a 1:1 zinc to orthophosphate ratio, which was applied at a maximum rate of 0.75 mg/L as zinc during warmer water months (70°F). After 1994, the corrosion inhibitor was changed to a 1:11 zinc to orthophosphate ratio, applied at a maximum rate of 2.0 mg/L as orthophosphate. The finished-water pH was maintained at pH 7.4 using caustic soda addition prior to ammonia, fluoride, chlorine, and corrosion inhibitor application. The point of post-chemical addition was into an "in-line" Komex mixer located upstream of the distributive high service pumps.

Implementational and Operational Issues

Operator Training and Start-Up

With the recent seasonal (warm months') switch to pre-chloramination, operator training was conducted before implementation; therefore the transition was one to two weeks. In addition, analytical techniques were already in-place, which allowed for a confidence factor. It took three weeks maximum for this implementation. Furthermore, continued reinforcement was provided.

System Performance

Water Quality

The water quality data from January 1989 though December 1990 represents the treatment plant "before treatment modifications" and the "after treatment modifications" water quality data were collected from January through December 1998 (Table 4.10).

TOC. Historical TOC removal ranged from 16 to 57 percent, with an average value of 41 percent (Table 4.10). In 1998, TOC removal ranged from 33 to 66 percent, with an average TOC removal of 52 percent (Table 4.10). From 1996 through 1998, the coagulant dose was sufficient in meeting enhanced coagulation requirements for TOC reduction (USEPA 1998a).

Turbidity. Historically, Hopewell's WTP produced a finished water with an average turbidity of 0.37 NTU (range = 0.03-1.53 NTU). In 1998, the WTP produced a finished water with an average turbidity value of 0.07 NTU (range = 0.05-0.12 NTU). Hopewell's filtered-water turbidity goal is 0.10 NTU. At the time of the preparation of this case study, the WTP had met the "Partnership for Safe Water" (AWWA 2001) goal for 35 consecutive months (99 to 100 percent $\leq$0.10 NTU).

Table 4.10

Comparison of water quality and DBPs before and after treatment modifications at Hopewell

Water quality parameter	Before modifications* Finished water	After modifications† Finished water
pH		
Minimum	6.8	7.0
Average	7.2	7.4
Maximum	7.6	7.8
Alkalinity, mg/L as $CaCO_3$		
Minimum	14	11
Average	43	37
Maximum	78	59
Total hardness, mg/L as $CaCO_3$		
Minimum	45	39
Average	68	54
Maximum	134	119
Turbidity, NTU		
Minimum	0.03	0.05
Average	0.37	0.07
Maximum	1.53	0.12
TOC, mg/L		
Minimum	1.9	1.2
Average	2.6	1.9
Maximum	3.6	2.6
TOC removal, %		
Minimum	16	33
Average	41	52
Maximum	57	66

(continued)

Table 4.10 (Continued)

Water quality parameter	Before modifications* Finished water	After modifications† Finished water
TTHMs, μg/L		
Minimum	44	52
Average	62	72
Maximum	93	100
HAA5 (HAA6), μg/L		
Minimum	7.0	3.3 (4.3)
Average	24	9.6 (12)
Maximum	45	17 (23)

*Data collected January 1989 though December 1990.

†Data collected January through December 1998.

THMs. From January 1991 through June 1993, with free chlorine in the distribution system, the six-month averages of THMs in the distribution system ranged from 63 to 69 μg/L for January through June and from 101 to 130 μg/L for July through December. From July 1993 through December 1994, with chloramines in the distribution system, the six-month average of THMs in the distribution system was 48 μg/L for January through June and was 68-87 μg/L for July through December. The use of post-chloramination minimized THM formation relative to that formed with free chlorine in the distribution system, especially in the latter six months of the year.

Figure 4.13 shows some examples of the effect of GAC on THM control at Hopewell. After a carbon change out in May 1993, the June and July carbon-contactor effluents had 1 and 60 μg/L TTHMs, respectively (the filter effluent upstream of the GAC had 124 and 138 μg/L TTHMs, respectively). By August, the adsorptive capacity of GAC was exhausted. Likewise, after a carbon change out in April 1994, the July and October carbon-contactor effluents had 20 and 19 μg/L TTHMs, respectively (the filter effluent upstream of the GAC had 93 and 74 μg/L TTHMs, respectively). By the December sampling, the adsorptive capacity of GAC was exhausted. However, because this mode of GAC operation did not remove DBP precursors, THMs were formed again during post-GAC chlorination.

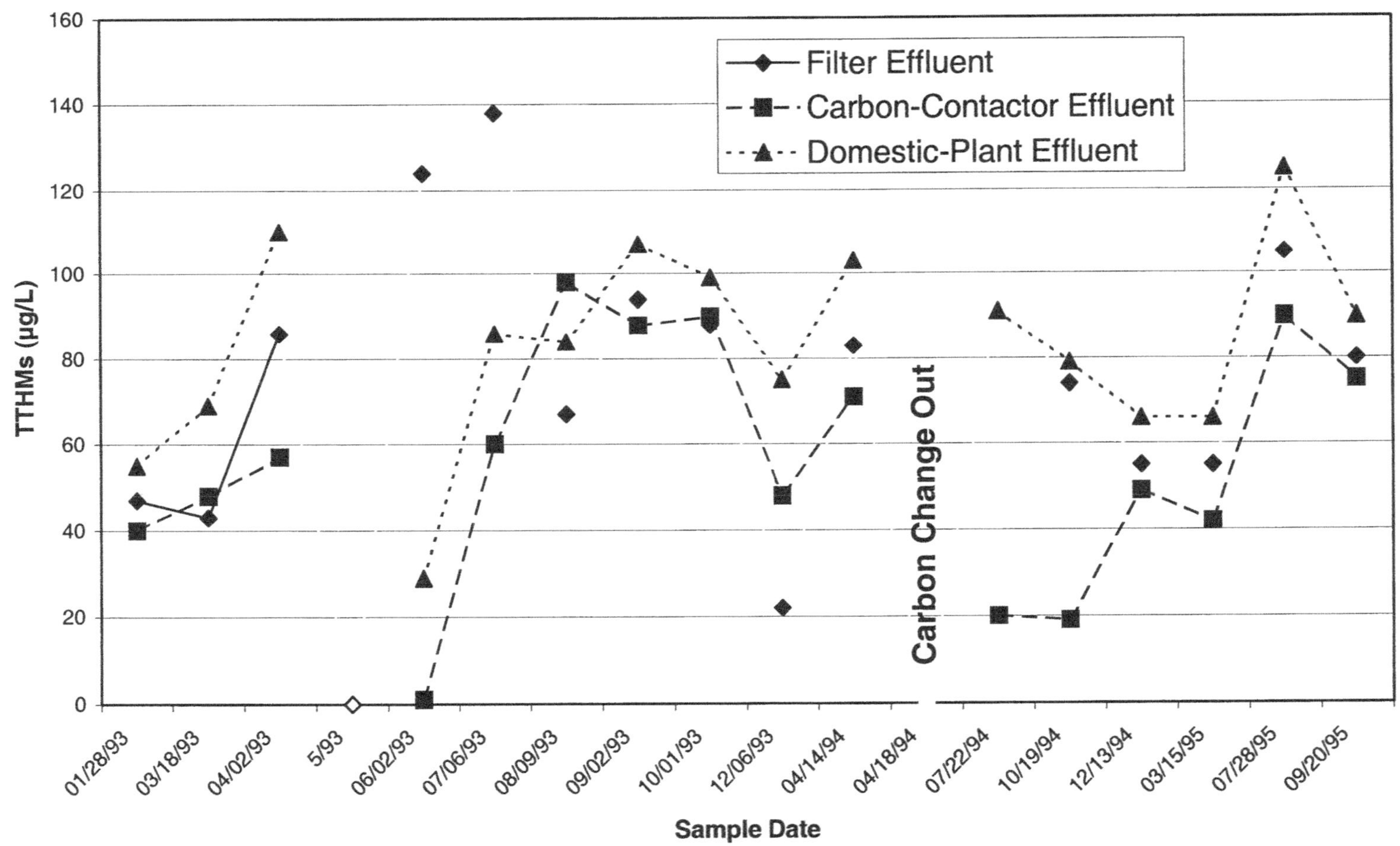

Figure 4.13 Effect of GAC on THM control at Hopewell: January 1993 - September 1995

Because the THMs continue to remain high at Hopewell (1998 average concentration similar and somewhat higher than the 1989-90 average level) (Table 4.10), seasonal pre-chloramination is being used to better minimize THM formation. For example, in November 2000, pre- and post-chloramination resulted in TTHMs of 36-45 µg/L in the domestic distribution system, whereas the distribution system of the industrial plant (with chlorination, no chloramination) had 72-81 µg/L TTHMs (Figure 4.14). In the November sampling, the short free chlorine contact time prior to pre-chloramination produced 11 µg/L TTHMs at the domestic plant, whereas pre-chlorination through the sedimentation basins (with no ammonia addition to quench THM formation) at the industrial plant produced 53 µg/L TTHMs. In February 2001, pre-chlorination and post-chloramination resulted in TTHMs of 40-44 µg/L in the domestic distribution system. In February, pre-chloramination was not needed, as THM formation with pre-chlorination was lower in the winter (pre-chlorination through the sedimentation basins at the domestic plant produced 18 µg/L TTHMs). Thus, pre-chloramination is being used at the domestic plant at a time of the year in which THM formation would be too high with pre-chlorination.

HAAs. Figure 4.15 shows an example of the seasonal variations in HAA5 formation and control at Hopewell. In February 1996, when the water temperature was 39°F, there were 56 and 43 µg/L HAA5 at the filter and plant effluents, respectively. In August 1996, when the water temperature was 80°F, there were 105 and 8 µg/L HAA5 at the filter and plant effluents, respectively. Although there was twice as much HAA formation (by the filter effluent) in the summer than in the winter sampling, during the warm-water months there was almost completion degradation of the HAAs across the carbon contactor. As has been observed in other research, the biodegradation of HAAs can be pronounced under warm-water conditions and absent during cold-water periods (Baribeau et al. 2000).

When pre-chlorination is practiced at the domestic plant, the chlorine residual is reduced through the sedimentation basins and is completely removed in the GAC contactor (e.g., in February 2001, the chlorine residuals of the rapid mix effluent, GAC influent, and GAC effluent were 3.5, 0.5, and 0 mg/L, respectively). Coffey and colleagues (1996) showed that biodegradation in a GAC filter can be maintained in spite of the presence of a chlorine residual in the GAC influent due to the reduction of chlorine on the surface of the GAC. Although GAC may not be as effective in removing chloramines (in November 2000, the chloramine residuals of the GAC influent and

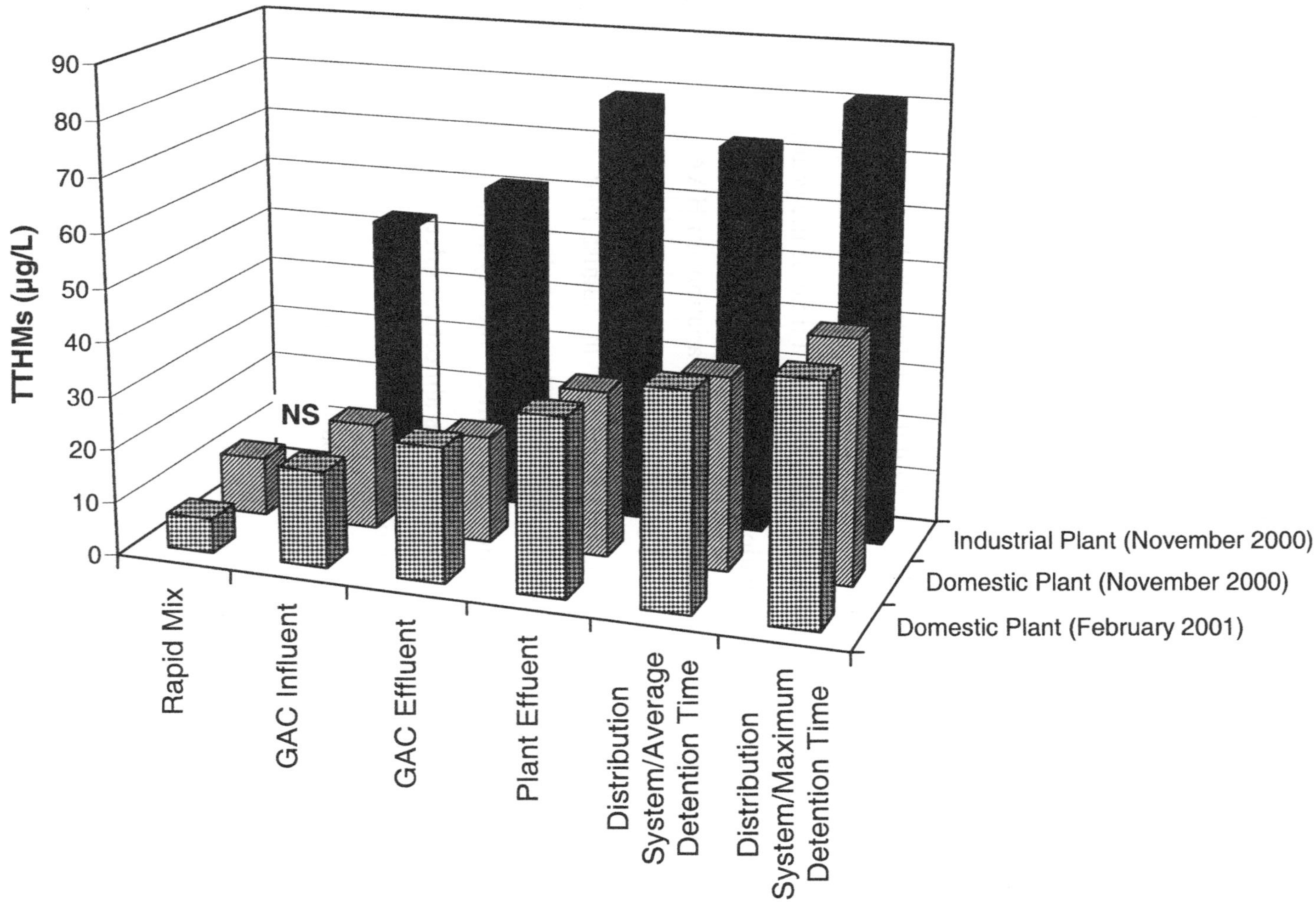

Figure 4.14 Seasonal control of THM formation at Hopewell domestic plant with pre-chloramination: domestic plant used pre- and post-chloramination in November 2000 and pre-chlorination/post-chloramination in February 2001; industrial plant used pre- and post-chlorination

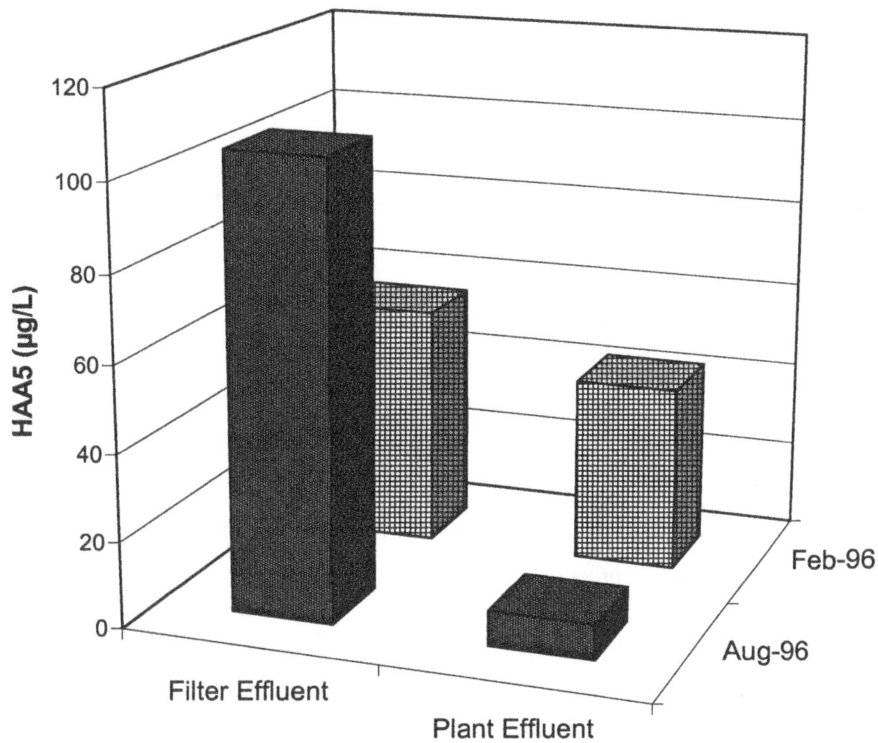

Figure 4.15 HAA5 seasonal variations at Hopewell

effluent were 0.9 and 0.1 mg/L, respectively), good biodegradation of HAAs can still be achieved (e.g., 91-percent removal of the HAAs in November 2000) (Krasner et al. 2001).

Historically, HAA5 formation in the finished water had ranged from 7.0 to 45 µg/L (average = 24 µg/L) (Table 4.10). In 1998, HAA5 formation ranged from 3.3 to 17 µg/L (average = 9.6 µg/L) (Table 4.10). The biologically active GAC has been effective in reducing HAA concentrations to well below the Stage 1 MCL of 60 µg/L (even well below the Stage 1 MCL with a 20-percent safety factor, i.e., <48 µg/L). In addition, HAA5 concentrations are below the previously proposed Stage 2 "placeholder" MCL (of 40 µg/L) (USEPA 1994a).

Summary of System Performance

At Hopewell, a combination of post-chloramination and seasonal pre-chloramination is being used to control THM formation. Pre-chloramination is used during the warmer months (e.g., during summer and fall). Thus, pre-chloramination is being used at a time of the year in which THM formation would be too high with pre-chlorination. With seasonal pre-chloramination, TTHMs have been less than 80 μg/L. Between chlorination and subsequent chloramination, *CT* was achieved. Biologically active GAC, especially in the warm-water months, was able to effectively degrade HAAs produced by pre-chlorination. The presence of either a chlorine or chloramine residual in the GAC influent has not negatively affected the biodegradation of the HAAs.

MEMBRANE TREATMENT: CITY OF PLANTATION

Summary

The City of Plantation switched from lime softening to membrane softening for the treatment of a hard, colored groundwater. The driving forces in the decision to install membranes included: (1) The condition and age of the existing softening plant. (2) The expanding growth of the community made it imperative to have increased plant capacity. (3) The membrane plant would satisfy anticipated or contemplated future water quality regulations. An upgraded softening plant would have been more expensive than a membrane plant due to the extreme treatability issues of the groundwater (i.e., hard and highly colored). In addition, a membrane plant delivered additional financial benefits by enabling incremental expansion to meet demand growth in the service area.

The process design was influenced by the hardness, color, and iron in the groundwater. In addition, fouling-rate potential and permeate control were evaluated for different membrane types. A combination of reverse osmosis (RO) and nanofilters (a mixture of thin film composite [TFC] and cellulose acetate [CA] membranes) were installed (in series, with the nanofilters as the lead elements). The CA membranes were later replaced by TFC membranes. The cost of the membrane plant (12-mgd design flow) was $14 million and the annual operating cost (for labor, electricity, and chemicals) is $1.6 million.

The case study highlighted key facts about membranes and their implementation: (1) Membranes are not a "black-box" technology; there can be significant operational problems with a membrane system. (2) Research and testing is essential to select the right membrane type for cost considerations and reliability concerns. (3) Disposal options for membrane concentrates may be limited by state requirements. (4) Membranes may not be suitable for some utilities when the system can not afford to lose 15-30 percent of the influent flow to reject water.

Safety considerations for a membrane plant are much greater than for a lime softening plant, especially with regards to proper handling of chemicals. The response time in the operation of a membrane plant is much less forgiving than in the operation of a lime softening plant. Special emphasis has been placed on staff training, automation, maintenance of instrumentation, chemical handling, and membrane cleaning. It took a period of one year for the staff to feel "comfortable" with the new technology.

Prior to the use of membranes, Plantation—utilizing lime softening and chloramination—had a range of TTHMs in the distribution system of the two plants from 28 to 58 µg/L. With membranes and chloramines, Plantation has a range of TTHMs from 1.0 to 3.5 µg/L. The membrane system is more effective at removing the hardness of water (the influent concentration has ranged from 240 to 370 mg/L as $CaCO_3$) as well as producing a more consistent product water quality (effluent hardness of 16 to 27 mg/L) than lime softening (effluent hardness of 48 to 167 mg/L). The membrane system removed color (the influent color has ranged from 45 to 150 pcu at the two plants) to a larger degree (down to 3-4 pcu) as compared to the lime softening/chlorination process (down to 4-13 pcu). The stability index of the lime-softened water was 0.22 on average, whereas it was -1.69 on average for the membrane-softened water (negative values indicate a potential for corrosion). Zinc orthophosphate is used at the membrane plant to control corrosion and the City has future plans for post-treatment lime addition.

The Plantation case study demonstrated that membranes can be a cost-effective alternative to lime softening for a very difficult-to-treat water. The quality of the finished water from the membrane plant far exceeded expectations and meets all regulatory goals. In addition, price competition and technical improvements from the manufacturers have been facilitating the implementation of membrane filtration.

Background

Case History

The City of Plantation operates two treatment plants—the East Water Treatment Plant (EWTP) and the Central Water Treatment Plant (CWTP)—and serves a population of 78,000. Both plants use groundwater as their source of water supply with wells connected to the Biscayne Aquifer. The groundwater is hard and highly colored, with the color indicative of large amounts of NOM. Both plants had been designed for lime softening

In late 1987, faced with expanding growth in the community, the City realized it needed to increase the capacity of its water treatment system (Figure 4.16). At the time, the EWTP was the baseload plant with a total plant capacity of 13 mgd. (Although the EWTP had a rated capacity of 10 mgd, a higher capacity could be provided based upon actual operational experience.) The peaks were made up by the CWTP. The CWTP had a capacity of 3 mgd. A master plan was prepared that showed the future capacity needs (e.g., maximum daily demand of 24.5 mgd by the year 2005) and evaluated the alternatives for treatment. Of the two existing WTPs, the CWTP was older and near its effective useful life, therefore construction of a new facility at the CWTP site was chosen as the best alternative to meet future capacity needs.

Given the organic color and the hardness of the water, a decision was made—with information provided by the membrane manufacturer's specific computer projections (using "ROPRO" (c) version 3.22, Fluid Systems Corp.)—for the new facility to be a membrane softening plant. In addition, amendments to the SDWA in 1986 were requiring utilities to modify their existing treatment plants in order to meet the new regulations. The City selected process modifications based on (1) the fact that construction and operating and maintenance costs for a low-pressure membrane softening plant in lieu of a modified lime softening plant would prove to be cost effective and that (2) the product water from the membrane plant would satisfy all current and future contemplated water quality regulations.

The City decided to make the new CWTP operate as the baseload plant at a capacity of 12 mgd (because membranes are best operated at a constant rate) and to meet any additional water demand with the EWTP (currently operating at 3 mgd instead of 10 to 13 mgd). In this manner the role of the two plants was reversed. Consequently, the EWTP has been chosen as the "before treatment modifications" plant and the new CWTP as the "after treatment modifications" plant for this case study.

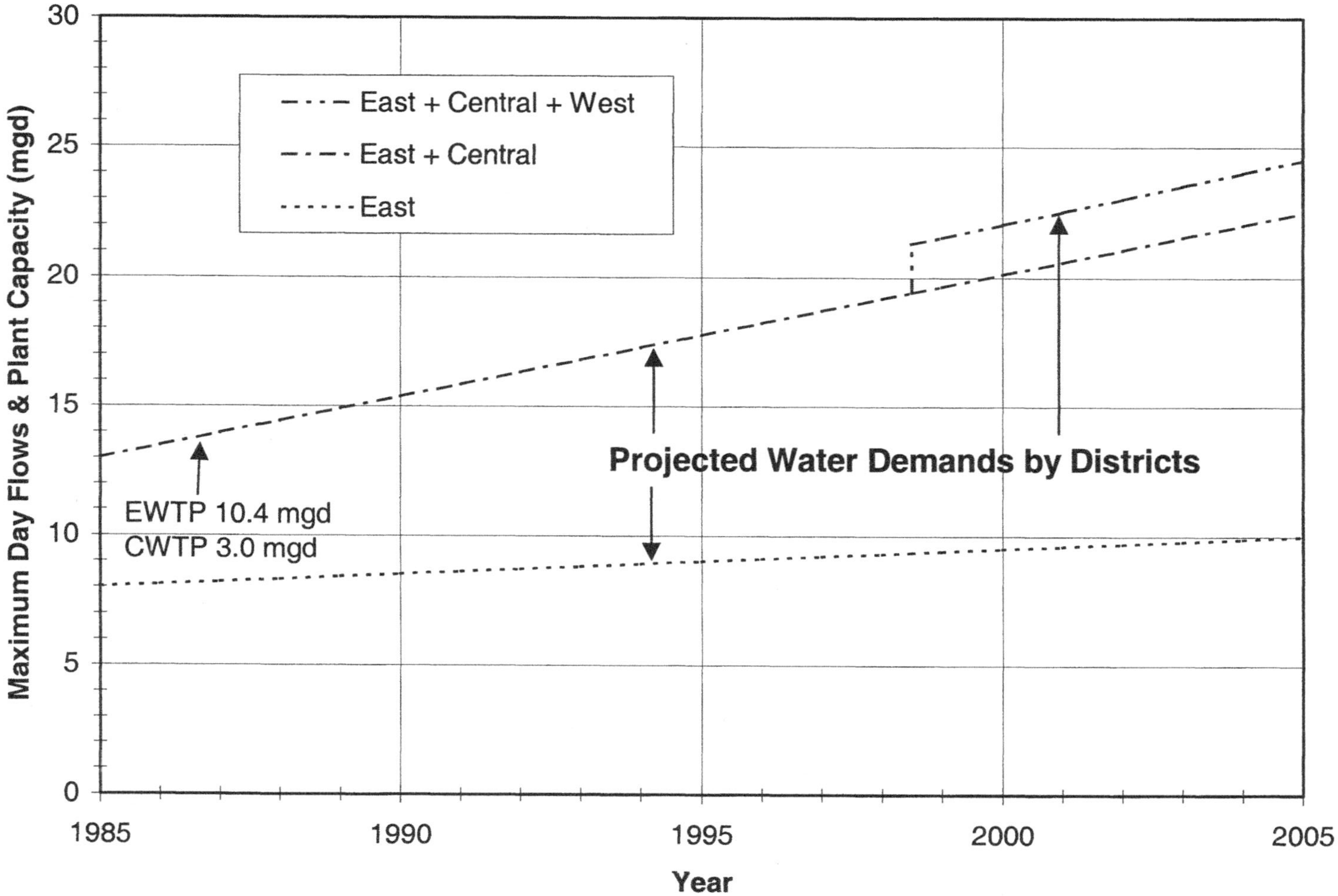

Figure 4.16 City of Plantation water system demand (*Source*: Adapted from Malcolm Pirnie, Inc. 1988)

Process Selection

In the 1980s, membrane softening was a modification of the "traditional" RO process used for saline- and brackish-water desalination. In the membrane-softening process, semi-permeable membranes were specifically formulated to retain calcium, magnesium, and other multi-valent ions that comprise water hardness while allowing suitable water flux at relatively low pressures (90-150 psi). The City of Plantation uses RO membranes with an operating pressure <150 psi as opposed to traditional RO membranes where the operating pressures range anywhere between 250 and 800 psi.

Softening membranes can include nanofilters with molecular-weight cutoff (MWCO) values of 200-500 daltons. If greater removal of dissolved solids are needed than can be achieved by nanofiltration alone, RO membranes (e.g., MWCO <10 daltons) are also included in the membrane-softening system. The effective pore size for the membranes also retains humic and fulvic acids and other long-chained organic molecules, thus removing color and DBP precursors to a large degree. In addition, the nearly complete retention of bacteria, viruses and protozoans allows the use of chlorine (or chloramines) at relatively low doses to satisfy disinfection requirements.

The City of Plantation authorized Malcolm Pirnie, Inc., to provide engineering design and construction services for the membrane softening plant. The other option that had been considered was the traditional lime-softening process. Process selection was based on an in-house paper evaluation conducted with the full participation of the plant staff. The evaluation was conducted in 1988 for a period of six months. A full evaluation was performed at the design stage.

Two types of membrane materials (i.e., CA and TFC) were evaluated in a design-criteria computer analysis. CA membranes have inherently lower water flux rates when compared to TFC membranes; therefore, to produce an equivalent flow of water at equivalent pressures, a greater CA membrane area is required necessitating greater numbers of elements and pressure vessels. On the other hand, CA membranes have a lower fouling-rate potential. TFC membranes have a higher water flux rate but tend to foul more rapidly. Fouling results in an increase in the necessary feed water pressure to achieve target flows. In order to reduce fouling, water flux must be limited for TFC membranes by permeate water flow control, which increases the instrumentation and control system costs. A combination of CA and TFC membrane materials was chosen for use in the plant.

Membrane-specific computer programs were used to provide preliminary design criteria and estimates of product water quality. Pilot testing by the membrane manufacturer was used in developing final design criteria. A combination of nanofilters and RO membranes were chosen: (1) nanofilters with an average MWCO of 100 daltons and (2) RO membranes—which operate at low pressures—with an average MWCO of 10 daltons.

Parameters that Influenced Process Design

The most significant inorganic parameter in terms of design was iron. The raw water has an iron concentration ranging between 0.7 and 2.5 mg/L (Table 4.11), which is in the form of ferrous ion. Because the membrane system is maintained under reducing, anaerobic conditions, iron removal is primarily dependent on membrane retention. Thus, up to 88 percent of the iron must be removed to meet the 0.3 mg/L secondary MCL. Thus, a combination of nanofilters and RO filters were ultimately selected.

Further iron reduction can occur in the degassing step where soluble ferrous iron can be oxidized to insoluble ferric iron, which can deposit out in the chlorine contact chamber or the storage tanks. However, the relatively light density nature of the iron floc could pass through the service pumps causing finished-water turbidity problems. This can be prevented by providing pressure sand filters between the clearwell transfer pumps and the storage tanks. Additional steps were taken to prevent iron breakthrough by adding an inorganic phosphate to sequester the iron present in the membrane effluent before it went to the degasser.

The primary organic parameter in terms of design was color. Color can range between 100 and 150 pcu, which represents a significant humic and fulvic acid content that can impact the organic fouling rate of the membrane. This can be controlled by periodic flushes of the membrane with product water.

Membranes can also form a biological slime layer over time due to the presence of iron bacteria (that can be present on the well casings, which are fabricated from unlined carbon steel) and other biological life. These organisms can be controlled by a low feedwater pH and high pH cleaning and/or high feedwater velocity.

Table 4.11

City of Plantation water quality:

comparison of lime softening (EWTP) and membrane softening (CWTP)

	EWTP			CWTP	
Water quality*	Influent	Settled water	Effluent	Influent	Effluent
pH					
Minimum	6.8	8.3	7.4	7.0	7.3
Average	6.9	9.2	8.5	7.1	7.4
Maximum	7.1	10	9.4	7.4	8.0
Alkalinity, mg/L					
Minimum	224	30	24	242	27
Average	244	44	37	272	28
Maximum	260	115	82	316	31
CO_2, mg/L					
Minimum	34				
Average	57				
Maximum	78				
Total hardness, mg/L					
Minimum	244	61	80	240	16
Average	289	84	NA	293	18
Maximum	314	151	167	370	27
Ca hardness, mg/L					
Minimum	242	48	70	215	15
Average	261	68	NA	270	17
Maximum	296	110	141	356	25
Mg hardness, mg/L					
Minimum	6				
Average	28				
Maximum	78				

(continued)

Table 4.11 (Continued)

	EWTP			CWTP	
Water quality	Influent	Settled water	Effluent	Influent	Effluent
Chloride, mg/L	*52*†		*49*	*62*	*16*
Minimum	45			44	12
Average	56			58	14
Maximum	89			82	16
TDS, mg/L	*400*		*170*	*400*	*68*
Minimum				320	56
Average				355	59
Maximum				407	62
Iron, mg/L					
Minimum	0.3		0.01	0.7	0.05
Average	0.9		0.02	1.4	0.05
Maximum	1.2		0.04	2.5	0.1
Color, pcu	*45*		*10*	*120*	*5*
Minimum	45		4	100‡	3
Average	69		7		4
Maximum	85		13	150‡	4
Turbidity, NTU	*1.7*		*0.06*	*0.60*	*0.07*
Minimum			0.03		0.09
Average			0.11		0.12
Maximum			0.28		0.25
Coliforms (MF)	Absent§		Absent§	Absent§	Absent§
Total, col/100 mL	*<1*		*<1*	*<1*	*<1*
Total chlorine, mg/L					
Minimum			1.9		3.1
Average			2.9		3.5
Maximum			4.1		4.1

(continued)

Table 4.11 (Continued)

	EWTP			CWTP	
Water quality	Influent	Settled water	Effluent	Influent	Effluent
Stability index					
Minimum			-1.52		-1.80
Average			0.06		-1.69
Maximum			1.17		-1.02
TOC,** mg/L	11.3		4.43††	17.8	0.23‡‡
UV,** cm^{-1}	0.328		0.132††	0.635	0.01‡‡
SUVA,** L/mg-m	2.9			3.6	
Bromide,** mg/L	0.09			0.14	0.02‡‡
Ammonia,** mg/L as N	0.86			1.57	0.87§§

*Water quality from June 1996 to May 1997 unless noted otherwise.

†Italicized values for samples collected September 1996.

‡Range of values for wells sampled July 1996.

§Coliforms absent in wells, plant effluents, and distribution system sampled May 1997.

**Water quality data sampled on September 30, 1997.

††Lime softening filter effluent before secondary chlorination.

‡‡Membrane effluent before chlorination.

§§Membrane effluent after degasser.

The Decision-Making Process

The City of Plantation, in order to meet the increasing growth in population acquired a private utility (Gulfstream), which operated a lime softening plant (i.e., CWTP). The existing plant was not capable of meeting the new water quality regulatory requirements and was not sufficient in capacity to meet the expected future population growth. An in-house decision was made in conjunction with the city officials to build a state-of-the-art membrane softening treatment plant.

Interviews were conducted with the plant manager and utility manager to evaluate the decision process that led to the construction of a membrane plant. Although the interviews were very wide-ranging, the results of these interviews are grouped into four categories: driving forces, constraints, technology checklist/anticipated factors, and technology checklist/less well anticipated factors.

Driving Forces

Three factors were identified as driving forces in the decision (Figure 4.17). These are the compelling factors that steered the process in the direction of a membrane plant.

- First, the condition and age of the existing softening plant was such that an infrastructure replacement decision was imminent regardless of technology or regulatory issues.
- Second, the capacity of existing facilities was being strained by growth pressures, making it imperative that treatment decisions be made so that capacity could be made available as needed.
- Third, new regulations on DBPs (anticipated stringent new MCLs) made it clear that an upgraded softening plant would be more expensive than a membrane plant due to the extreme treatability issues of the source water.

The respondents were asked what they might have done if the first two conditions had not existed, but the third still existed. They indicated that they probably would have adopted a wait-and-see approach, delaying action until the D/DBP Rule was finally defined. Conversely, if the first two conditions were assumed to be true, but the third was lacking, they indicated they would have gone ahead with a membrane plant. Replacement and growth were more compelling drivers than regulation.

The decision was made easy by the fact that the Gulfstream plant was a steel-erected, lime-softening plant whose useful life was at an end. Continuing with lime softening and addressing the problems of how to achieve better color and TOC removal would have required extensive modification and addition of new equipment to the old equipment. New plant facilities were required for a plant with a 50-year lifespan that would meet all the SDWA requirements. The choice of membranes and new plant facilities was easier given the problems with the older plant.

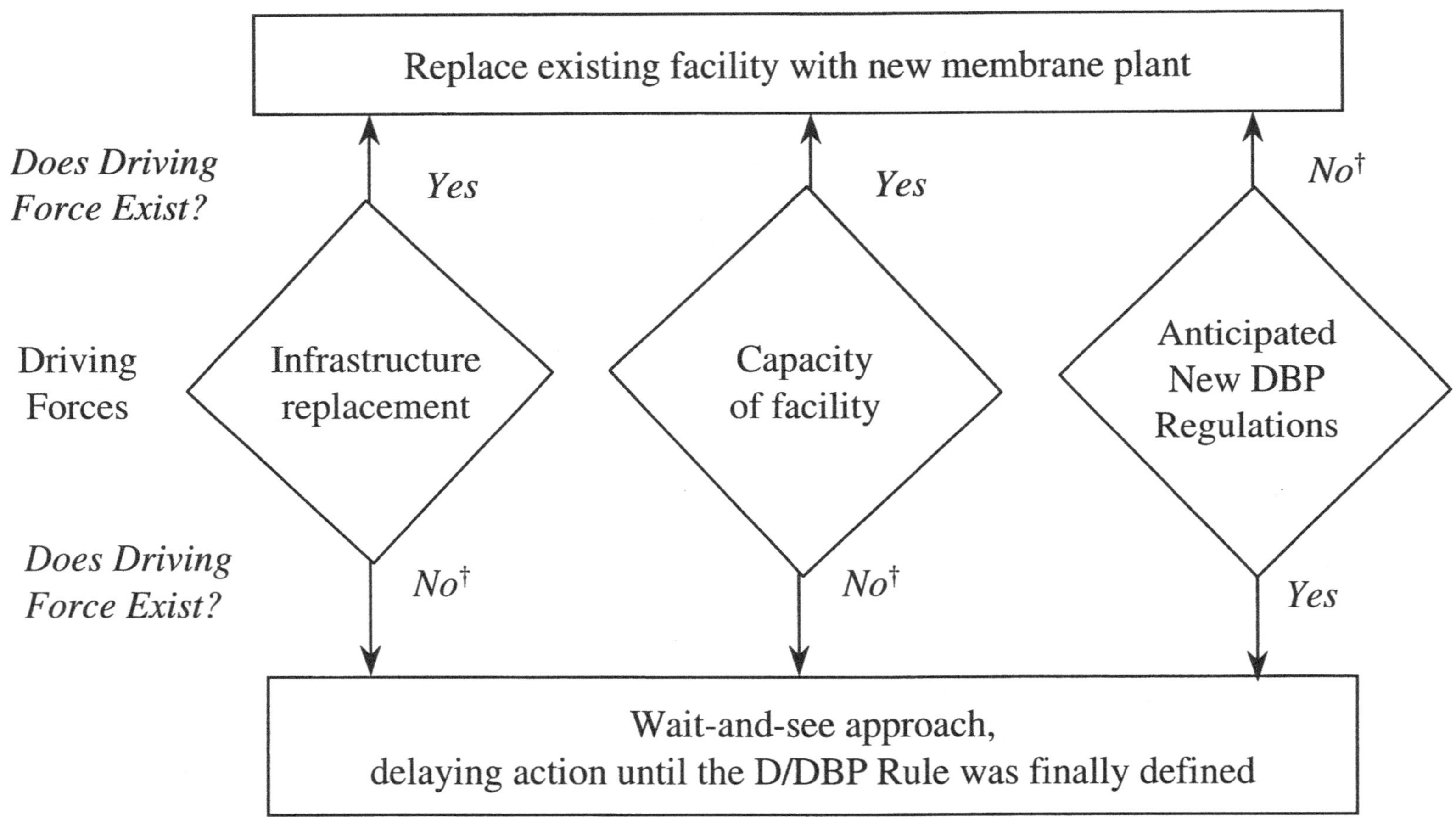

*1988 timeframe. † In actuality, this driving force existed as well.

Figure 4.17 The decision-making process* at the City of Plantation: replacement of existing lime softening plant (CWTP) with membrane plant

Constraints

A series of questions were asked regarding typical types of constraints encountered in a utility's decision process. In this unique case, the general conclusion is that there were no binding constraints from the usual sources.

- Access to capital was not a problem. The plant was paid for from a $20 million cash reserve. A bond was floated to replenish the reserves, but they were able to pay cash. The service area benefits from a strong economy and vigorous population growth, as reflected in a strong financial position.
- Rate impacts were not a problem. Gulfstream Utilities had been charging higher rates before the acquisition than those needed to recover the cost of the membrane plant.
- Equity in cost allocation was not a problem because the Gulfstream Utilities district is a separable service area that could be separately assessed for the cost of the new plant, leaving other ratepayers on the system unaffected.
- Public attitudes and board politics were not constraining because the non-regulatory water quality benefits of the proposed plant were sufficiently well communicated and understood that there was general support for the project.
- Internal communications did not present a barrier because the utility staff had developed a shared understanding of the project over the course of a lot of study and analysis of membrane technology.

Technology Checklist – Anticipated Factors

Our interviews revealed the following technology factors were well anticipated by the utility and plant staff as part of their decision process.

- They correctly anticipated that membranes would be cost competitive with lime softening for treatment of their unique source water.
- They correctly anticipated price competition among membrane manufacturers, resulting in reduced costs.

- They correctly anticipated technological advancements that produced low-pressure membranes and lower energy costs.
- They correctly assessed that a membrane plant would deliver additional financial benefits by enabling incremental expansion to meet demand growth in the service area.

Technology Checklist – Less Well Anticipated Factors

Our interviews revealed the following technology factors were less well anticipated by the utility and plant staff as part of their decision process.

- They under-estimated the extent of operational and regulatory issues associated with deep-well injection of wastes, resulting in additional expenses and implementation hang-ups.
- They initially under-estimated the operational demands of a membrane plant. Contrary to the popular perception that membrane plants are simpler to run, they discovered the need for an extremely high degree of staff training and reliability assurance. The response time in the operation of a membrane plant is much less forgiving than in the operation of a softening plant. Missed-steps are potentially costly in terms of both damage to membranes and personnel safety. The safety considerations of a membrane plant are much greater than a lime-softening plant, especially in regards to proper handling of chemicals. Special emphasis had to be placed on staff training, automation, maintenance of instrumentation, chemical handling, and membrane cleaning.
- They under-estimated the extent to which the membrane plant would improve finished water quality and the broad community support the improvement would generate. It is largely as a result of this visible success that they are moving forward with plans for a second facility. (At the time this case study was prepared, the City of Plantation was involved in the construction of a 6-mgd membrane softening plant at the EWTP site to replace that lime softening facility.)

In 1988, membrane softening technology was not a standard method for water treatment and was, in fact, cutting-edge technology at that time. There were many uncertainties, including lack of knowledge and experience with contractors and operations and maintenance for the membrane equipment. A major tripwire in terms of operations and system reliability was membrane life and replacement cost. Five years was the estimated lifespan for a membrane, and this greatly affected operations and maintenance costs. As the technology was new and there were no plants at that time with a membrane process greater than five years, there was no way to know if this was a good assumption. It has now been determined that the membrane life of five years is a conservative estimate.

Additional research that would be worthwhile to improve the decision-making process includes research or testing that needs to focus on membrane types for cost and reliability concerns. Pilot testing of the membranes was done by the membrane manufacturers to screen the membranes for water treated by the City and to develop the design to meet the performance requirements in the contract, as outlined by the City of Plantation and Malcolm Pirnie.

Compliance Issues

The decision to implement a membrane softening plant was taken with the knowledge that the City of Plantation would, with this strategy, be able to meet all current and future contemplated regulatory requirements and would thus avoid the need for future modifications. The treatment modification addressed prevalent issues, such as TOC removal, DBP control, and color removal. In addition, the high removal of TOC increased the stability of chloramine residuals and decreased distribution-system maintenance issues.

Originally, two membrane types were specified because of the iron in the raw-water supply. In 1989-1990, there were limited choices than now exist. Therefore, an iron sequestration agent has been used as part of the iron-control strategy. Iron will be further resolved with the next membrane element replacements, which are low-pressure RO membranes that were not available in 1990. Originally, the first bank of membranes, which were nanofilters, were intended to leave some calcium in the water because of corrosion concerns. Because RO elements were subsequently installed in part of the first bank of membranes (see "Implementation and Operational Issues" section below), a future modification will be post-treatment lime addition.

A compliance issue, which is a natural consequence of the membrane process, is membrane-concentrate disposal. This was resolved by using deep-well injection, which is the only permitted disposal option in South Florida.

Key Players

The key players involved in the decision-making process were the City of Plantation's Utility Department staff, Malcolm Pirnie engineers, and the City Council. The Utilities Department staff and the consultants from Malcolm Pirnie kept the City Council informed during the decision-making process and the council fully supported the modifications undertaken.

Customers were informed of the change in the treatment process and extensive efforts were made by the staff to "educate" the public about the new water treatment system. In addition, an article in the *Sun Sentinel* in 1988 indicated that "A $9 million water treatment plant will be built within two years to keep pace with a growing city and stringent clean-water standards." The average consumer in the City of Plantation is young (median age = 37 years), with a median household income of $41,832, and lives in an upscale community where the cost of housing ranges from $70,000 to over $750,000. The customers were very supportive of the new system.

Selection Issues and Future Strategies

The success of the membrane softening system in meeting the regulatory goals has affirmed the faith of the decision-making team in the process. Key individuals involved in the selection process indicated that that they would stand by their original choice if given the opportunity to make the decision again. The only decision they would have changed would be their selection of membrane material; given the choice again they would have chosen to implement TFC membranes instead of a combination of CA/TFC membranes in the entire plant.

In addition, when the time comes for the replacement of existing RO membranes, the installation of more efficient, low-pressure RO membranes will make the system more selective for iron and increase the overall efficiency of the system. Moreover, at the time this case study was prepared, the City of Plantation was involved in the construction of a 6-mgd membrane softening plant at the EWTP site to replace that lime softening facility.

Economic Analysis

It cost the City of Plantation approximately $14 million to build the 12-mgd membrane plant. Water production costs for 1996 for the CWTP included $654,123 for labor (10 operators, 1 lead operator, 1 supervisor, and 1 maintenance worker), $529,994 for electrical, and $388,085 for chemicals. The total water production (on-line capacity factor = 100 percent) in 1996 for the membrane plant was 4123.75 MG (if the daily water production in a year [i.e., 365 days] was always exactly 12.0 mgd, the total annual water production would be 365 x 12.0 = 4380 MG). The cost per MG—not including capital recovery costs—was $376.41. The membrane replacement cost is estimated at $607,000, and each membrane element is expected to have a lifespan of 5 to 7 years.

Additional information on membrane costs are available elsewhere (Bergman 1996). This article includes a discussion of costs at the Plantation facility.

Treatment Facilities

Water Supply

The EWTP treated groundwater from up to eight wells on its site and the CWTP from eight wells on its site. The raw water quality data (Table 4.11) were very similar for both sites.

EWTP (Conventional Lime-Softening Plant)

The EWTP has a potential total production rate of 13 mgd. Before the introduction of the membrane plant, it was producing water at the rate of 10 mgd. Its current average water production varies between 1 and 3 mgd, as it functions as a secondary source of water supply making up the peaks in the daily water demand. (The finished water from both treatment plants are delivered to a common distribution system.) However, it is anticipated that the projected future water demand (Figure 4.16) would make the EWTP function at its full capacity in the future.

The influent to the EWTP contained no coliforms, but was high in color (45-85 pcu), alkalinity (average = 244 mg/L as $CaCO_3$) and hardness (average = 289 mg/L as $CaCO_3$) (Table 4.11).

There were four operational treatment units in the plant (Figure 4.18), which consisted of multi-cone aerators, Infilco accelerators, rapid-sand filters, and clearwells. The raw water (average pH = 6.9) could be pretreated with chlorine to remove color and was then lime softened (average softening pH = 9.2) and filtered. Ammonia, chlorine, and fluoride could be added to the filtered water prior to storage in the clearwell. The finished water was distributed at a pH of 8.6 (on average).

Table 4.12 lists the dosage of the chemicals added and their usage.

Finished-water ammonia addition for the formation of chloramines was not needed as the raw water typically has approximately 2 mg/L of ammonia. Even with chloramination, chlorine demands in a colored groundwater can be high. The organic chlorine demand is due to TOC (including the color) and a significant source of inorganic chlorine demand—in addition to ammonia—is hydrogen sulfide (White 1999).

CWTP (Membrane Softening System)

The CWTP was designed to produce 12 mgd. The plant was started in May 1991, then on-line at full-scale production in July 1994 (May 1991, 6.0 mgd; September 1991, 9.0 mgd; July 1994, 12.0 mgd). The membrane plant replaced a 3-mgd lime-softening facility, which had been originally owned and operated by Gulfstream Utilities and was purchased by the City of Plantation in February 1985. The membranes are operated on a consistent, steady-state basis for process optimization. It should be noted that membranes can not be peaked to produce additional flows. As such, the maximum finished-water flow available on a daily basis is the typical flow of 12 mgd. The typical raw-water intake to produce such a flow based on an 85-percent permeate water recovery rate is 14 mgd.

The influent to the CWTP was very similar in its characteristics to the EWTP. It is high in color (100-150 pcu), TOC (18-29 mg/L), alkalinity (average of 272 mg/L as $CaCO_3$) and total hardness (average of 293 mg/L as $CaCO_3$) (Table 4.11).

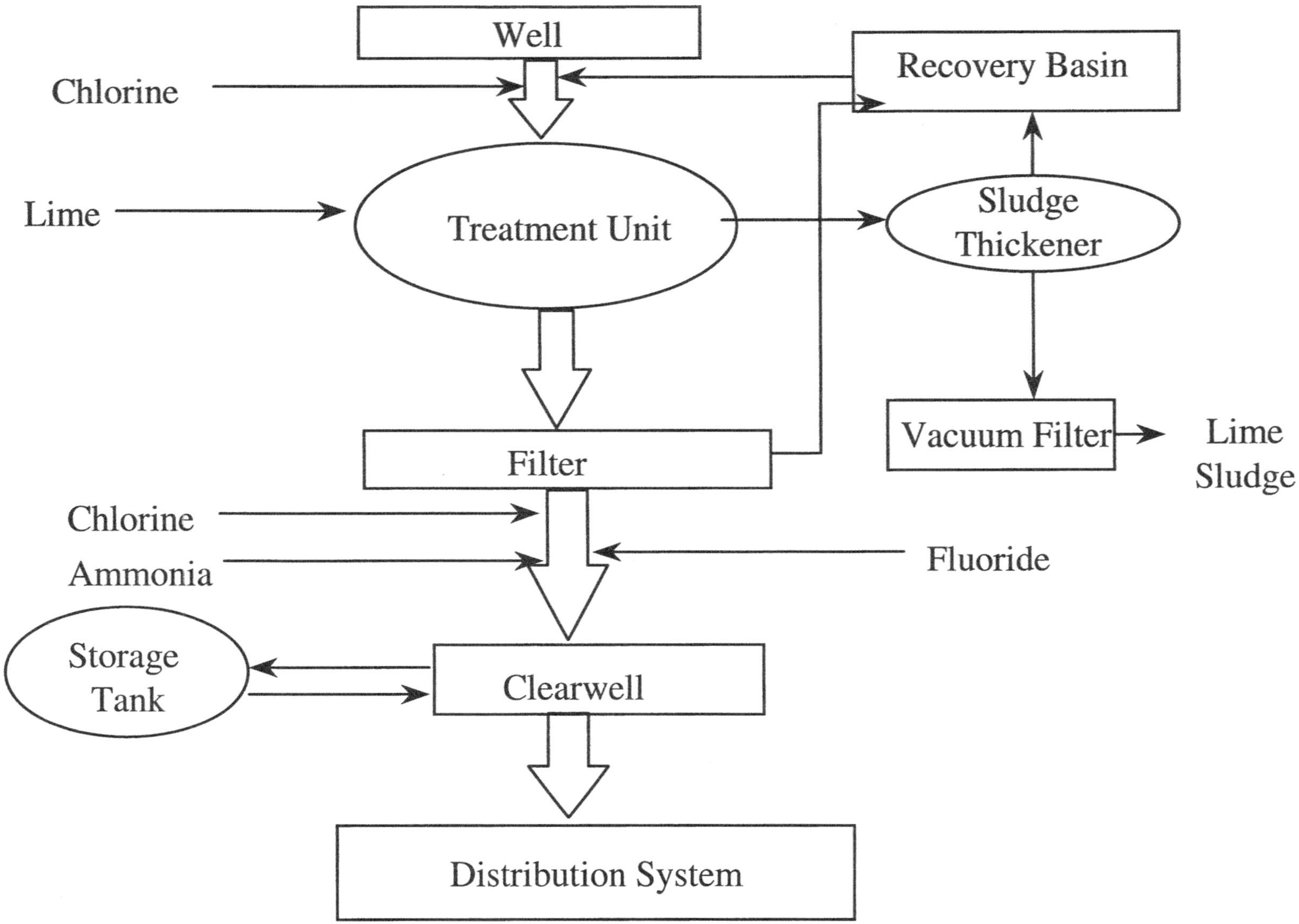

Figure 4.18 EWTP (lime-softening process)

Table 4.12

Chemical dosage and usage at EWTP (lime-softening process)

Chemicals added	Dosage*	Chemical usage
Lime as CaO, mg/L		Softening
Minimum	145	
Average	194	
Maximum	250	
B-floc coag (nonionic polymer), mg/L		Coagulant aid
Minimum	0.01	
Average	0.10	
Maximum	0.17	
Post chlorine, mg/L		Disinfection, color removal
Minimum	7.9	
Average	11	
Maximum	19	
Calgon RS-2 (inorganic phosphate)	†	Iron sequestration

*Range of dosages used from June 1996 to May 1997.

†Stopped using in December 1996.

Figure 4.19 shows a schematic of the membrane softening plant. Acid and scale inhibitors are injected into the feedwater line (average raw-water pH = 7.1). The pH of the feedwater is adjusted to approximately 6.2 to minimize membrane scaling due to $CaCO_3$. The acid addition reduces the bicarbonate ion concentration by converting it to sulfates and/or carbon dioxide, and the carbon dioxide is later removed from the product water by degassification. The scale inhibitor handles other inorganic scale problems. The water then passes through cartridge filters (5 μm), which serve as a barrier to particles that can damage the membranes. Inorganic phosphate is added after the membranes and prior to the degasser to sequester any iron that is not removed by the membranes.

The membrane arrays—four, independently operable 3-mgd product trains arranged in parallel—are fed from a common header. (Membrane plants utilize separate parallel trains, as is

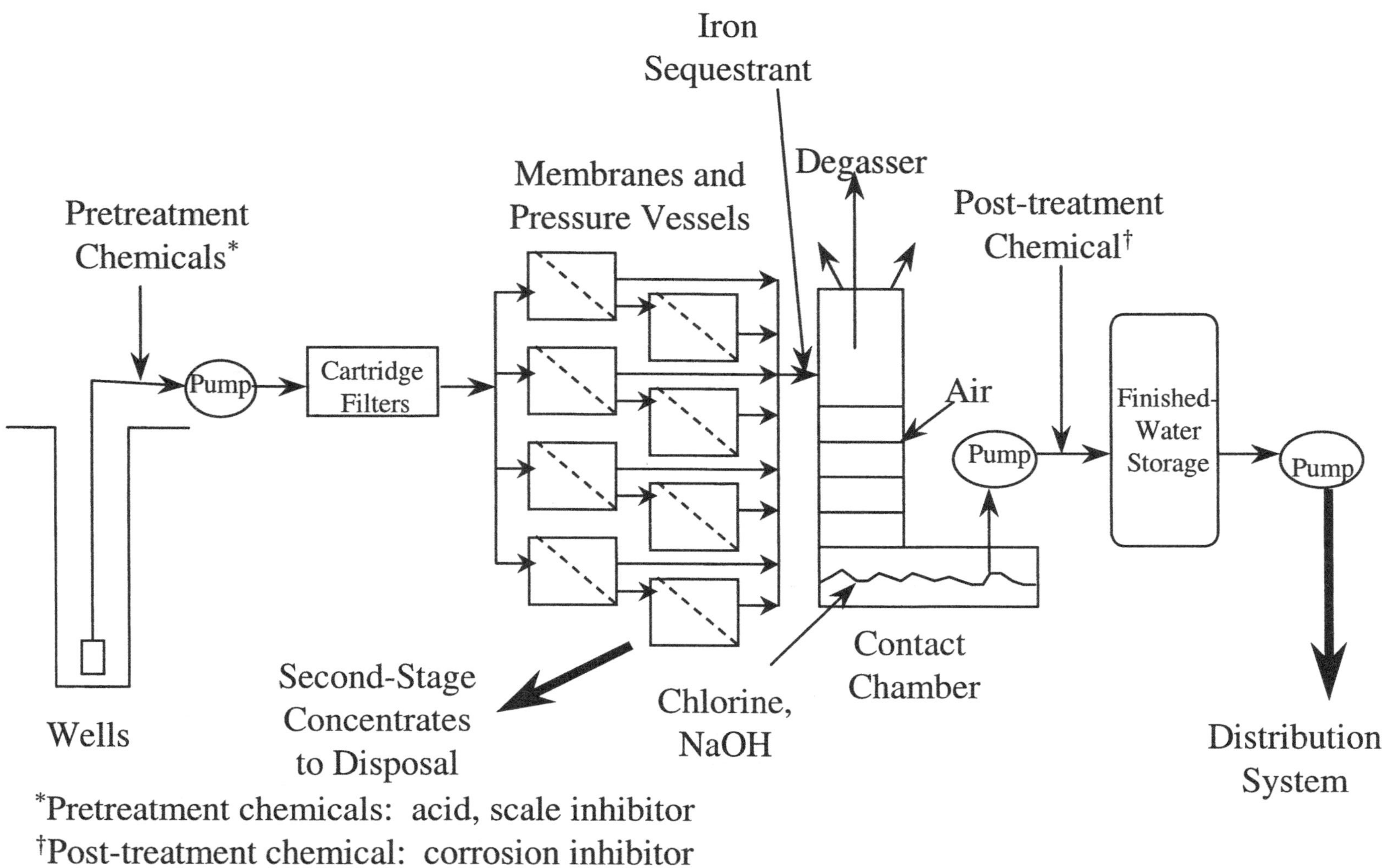

Figure 4.19 CWTP (membrane softening process)

done in conventional treatment plants.) Each membrane array consists of two banks/stages of TFC membranes arranged in series. Reject flow from the first stage serves as the feed flow to the second stage (Figure 4.19).

The first bank is composed of 61 pressure vessels, each of which consists of seven spiral-wound polyamide elements that replaced the CA elements in July 1994. The four lead elements (8921S, Fluid Systems Corp., San Diego, Calif.) are nanofilters with a MWCO <100 daltons with an operating pressure <150 psi. The three tail end elements are RO filters (8821LP, Fluid Systems Corp.) with a MWCO <10 daltons with an operating pressure <150 psi. The second bank of membranes is composed of 45 vessels, each consisting of seven spiral-wound polyurea RO elements (8021LP, Fluid Systems Corp.) with a MWCO <10 daltons and an operating pressure <150 psi. The membranes are operated at 124 psi. A few 8822HR elements (Fluid Systems Corp.)—which did not completely conform in specifications (i.e., exhibited 20- to 30-percent lower flow than rated design)—were scattered among some vessels to replace a few scaled 8021 elements.

The softened/permeate water is collected and directed to the degasser towers. The product water is aerated and stripped of carbon dioxide and hydrogen sulfide. The carbon dioxide stripping raises the product water pH (with an increase of over 1.0 pH unit), which decreases the necessary downstream pH adjustment chemical dose. The hydrogen sulfide stripping eliminates an odorous contaminant that has a high chlorine demand. The degasser efficiency at the CWTP is approximately 97 percent.

The water is then chlorinated for primary disinfection. Typical contact time is approximately 35 to 40 minutes. A certain proportion of the raw-water ammonia is oxidized to nitrate in the degassing tower. (It is questionable how much the ammonia is reduced by degassification only. For example, during sampling in September 1997, the raw-water ammonia was reduced in concentration from 1.6 to 0.87 mg/L after membrane treatment and degassification plus chlorination. It is doubtful that the permeate ammonia is in a very volatile state.) The residual ammonia reacts with free chlorine to form chloramines. Although the degassification process is not controlled for chloramine production, there is a sufficient concentration of raw-water ammonia in the source water to ensure that there will be an adequate amount of residual ammonia after degassification for the formation of chloramines.

Baffle design inside the storage tanks allows for additional *CT* and full formation of chloramines. The City treats a groundwater source and chloramines as the primary disinfectant has been the approved disinfection practice at this plant. Disinfection requirements for groundwaters under the Groundwater Rule had not been established at the time this case study was prepared. The finished water is stabilized by the addition of caustic soda to (achieve an average distribution pH of 7.4) and a superphosphate-type corrosion inhibitor.

The concentrate water (approximately 2 mgd) is discharged into a deep-injection well.

The dosage of the chemicals added and their usage are listed in Table 4.13. The chlorine demand of the membrane-treated water is significantly lower than that of the lime-softened water. On average, the post-chlorine doses at the CWTP and EWTP were 6.6 and 11 mg/L, respectively (Tables 4.12 and 4.13). The membrane-treated water is lower in chlorine demand because (1) the TOC is lower (e.g., during sampling in September 1997, the TOC of the membrane- and lime-softened waters was 0.2 and 4.4 mg/L, respectively [Table 4.11]); (2) the color is lower (the membrane process is highly effective at removing color [Table 4.11], whereas a combination of softening and chlorination are required to remove color at the EWTP); and (3) most of the hydrogen sulfide is stripped out of the water during degassification.

Operations

Instrumentation and Control

The membrane process monitoring and control system consists of PLCs linked to a process computer. The PLC system contains the timing, sequencing, and logic functions for control of the membrane units, which includes feed, permeate, bypass, and concentrate valve operation, chemical feed systems, sequential train startup, control of permeate flow and permeate recovery recovery rate, and normal and emergency shutdown. The PLC system collects data from the process, which is transmitted to a PC in the main control room to which the operator can interface. Instrumentation in the main process floor area includes a feedwater panel, permeate panel, concentrate panel, and a train instrument panel.

Table 4.13

Chemical dosage and usage at CWTP (membrane softening process)

Chemicals added	Chemical formula	Dosage*	Chemical usage
Acid, mg/L	93% H_2SO_4†		Lowers pH to control membrane scaling due to $CaCO_3$
Minimum		117	
Average		138	
Maximum		154	
Anti-scale, mg/L			Handles other inorganic scale problems
Minimum		1.21	
Average		1.49	
Maximum		1.66	
TPC-556, mg/L	Inorganic phosphate		Iron sequestration
Minimum		0.14	
Average		0.39	
Maximum		0.53	
Chlorine, mg/L	Cl_2		Disinfection
Minimum		4.8	
Average		6.6	
Maximum		8.3	
Superphosphate, mg/L	Zinc orthophosphate		Inhibits corrosion
Minimum		0.58	
Average		1.57	
Maximum		2.01	
Caustic soda, mg/L	NaOH		Raises the pH for distribution
Minimum		17.2	
Average		21.1	
Maximum		27.2	

*Range of dosages used from June 1996 to May 1997.

†Contaminants: 28.3 ppm Fe, 0.7 ppm $KMnO_4$.

Control of Membrane Fouling

Fouling is the progressive deposition of inorganic, organic, and biological materials onto the membrane surface to form a secondary hydraulic barrier, thus decreasing membrane flux. Periodic membrane cleaning is a standard operational and maintenance factor. Two separate cleaning systems are provided for the facility: a Flush System and a Clean-in-Place System. The Flush System flushes the feedwater and concentrate water pressure tubes with permeate water. This prevents fouling by (1) reducing stagnation and deposition of chemicals from the high-TDS water and by (2) inducing back diffusion of materials from the membrane surface. The Flush System is used after the shutdown of a train. A permeate flush is performed on every train shutdown. No chemicals are added. Each train typically averages one to two shutdown flushes per week. The Clean-in-Place System is utilized for deep cleaning of the membranes (e.g., once or twice a year), using chemicals such as detergents, citric acid and EDTA in combination with soda ash. The City of Plantation has been averaging one cleaning per year per train, using a three-day process of pumping/soaking, first with low pH, then high (citric acid and soda ash).

Spent recirculating chemical solutions used for membrane cleaning are wasted to a large on-site wastewater master pump station, not to a deep well. Cleaning chemicals go to the on-site master pump station with no noticeable problems to the wastewater plant.

Challenges

Corrosion Control

The City of Plantation has met the challenge of corrosion control in the CWTP effluent water by the addition of corrosion inhibitors such as zinc orthophosphate and by maintaining the effluent pH at 7.4 (on average). In addition, they have future plans for post-treatment lime addition to meet their corrosion-control goals.

Membrane Fouling

During the operation of the CWTP, the second-stage of membranes were subjected to organic and iron fouling. The City met the challenge of preventing membrane fouling by periodically cleaning them (i.e., they have been averaging one cleaning per year per train, using a three-day process of pumping/soaking, first with low pH, then high [citric acid and soda ash]).

Concentrate Disposal

The issue of concentrate disposal is a critical one, as a constant concentrate flow of 1.3 to 2.1 mgd has been produced. The concentrate could contain >2000 mg/L of TDS and approximately 300 mg/L of chlorides.

Four possible routes of disposal were considered:

1. Discharge to a surface-water canal.
2. Irrigation.
3. Deep-well injection on-site.
4. Routing to a sewage treatment plant for treatment and on-site deep-well injection using the wastewater disposal well.

Of the four choices, the City preferred to implement a discharge into the surface-water canal, as (1) it is a less expensive alternative, (2) a portion of the water supply would not be lost, and (3) the water upon dilution with the canal water would be suitable for irrigation. The major challenge seen by the City to this proposal was that the level of TDS in the concentrate was higher than the amount in the canal (550 mg/L). However, the relatively small flow of the concentrate and the absence of any constituents other than those found naturally in the groundwater suggested that this would be an acceptable option.

The preferred discharge option was into the surface-water canal (item #1 above). When the decision was made by the City to construct the membrane plant, it was assumed (in 1988) that they would be able to discharge the reject water into the adjacent canal. The proposed discharge was believed to pose no threat to the receiving waters and that it would not cause any undue, adverse human health or natural ecosystem effects. The concentration of salts in the waste stream of a membrane softening plant is considerably lower than in the concentrate from a brackish-water RO plant (Conlon and McClellan 1989).

In the late 1980s, the issue of concentrate disposal was not well resolved. Over objections by Florida officials, the USEPA considered the concentrates as an industrial waste, which compounded disposal problems (Dykes and Conlon 1989). The City asked the Broward County Environmental Quality Control Board for a variance to discharge the reject water from the membrane plant to the adjacent canal. Other permits that were required for a discharge included an Industrial Waste Discharge Construction Permit and Industrial Waste Discharge Operations Permits from the Florida Department of Environmental Regulation and a National Pollutant Discharge Elimination System Permit (Form 2A) from the USEPA. The final issuance of these permits defined the treatment necessary for the discharge of the concentrate waters and other waste streams. The City ultimately ended up going with the deep-injection wells (item #3 above) as this was the only permissible option in South Florida.

Changes in Plant Operations/Staffing

There were major changes in plant operations, with the membrane plant becoming highly automated, however there were no changes in plant staffing. There is a larger degree of automation, which requires increased maintenance of instrumentation and control units. The membrane system also requires a high level of operator training and knowledge, due to a low tolerance by the system for mistakes. In comparison, lime softening is much easier to operate, but that process requires a lot of physical labor for maintenance. Membrane plants require some physical maintenance as well (e.g., the changeout of membranes and pre-filters). In addition, there is a much shorter detention time for the water in the membrane plant in comparison with the lime softening plant.

Membrane processes are operated on a consistent, steady-state basis, and the four membrane units are independently operable. Thus, by operating each unit for different lengths of time, flow can be matched to demand, resulting in much shorter detention times in the plant. However, the membranes can not be peaked to produce additional flows (unless additional units are brought on-line).

Implementation of the new technology made operator training become very important. A membrane plant is a highly automated and complex water-treatment system, whereas the operational staff was familiar with the simpler and more flexible lime-softening systems. Training was required. Training was included in the initial contract. Operational staff also had to be trained in the use of different kinds of test equipment (e.g., conductivity meters, etc.). In spite of the complexities of the membrane process, there was no resistance by the operating staff to the treatment modifications. Although the operational staff were to undergo training before the start-up of the plant, due to some major problems getting on-line, the operations staff was trained primarily by "fire." It took a period of one year for the operational staff to feel "comfortable" with the new technology and for the plant to operate as an "integral unit."

Implementation and Operational Issues

There were two major unanticipated problems that occurred during plant start-up. In the original plant design, the first bank of membranes was composed of 64 pressure vessels with CA nanofilters and the second bank was composed of 42 vessels with TFC membranes, all of which were RO filters. CA membranes are highly pH sensitive and can only tolerate pH levels between 3 and 6, whereas TFC membranes have a higher range of pH tolerance (pH limits = 2-11). Therefore, the CA membranes were cleaned by glutaradehyde—that controlled the main problem, which was biological—whereas the TFC membranes were cleaned with soda ash.

Approximately 12 days after the start-up, some of the CA membranes were inadvertently exposed to caustic soda, which hydrolyzed the membranes. The error was attributed to the common piping between the flush system and the post-treatment caustic system. A field change was made to disconnect the common piping. This is when all chemical addition piping (caustic and sulfuric) was removed from the flush system, leaving the capability of using only permeate flushes. On the cleaning system, all chemicals are manually batch mixed and manually pumped into a cleaning tank.

Of the four trains, train B received the most damage. It appeared that train D was also damaged, possibly from a pocket of caustic in the flush line, but not from a full flush. (Conductivity is measured to monitor membrane performance [fouling] or detect membrane failure. Conductivity confirmed the damage.) In the next few months after start-up, train B was taken out of service. Trains C and D showed continued deterioration of the CA membranes, possibly from biological attack. In September 1991, TFC membranes from train B were "cannibalized" and mixed with elements from trains A and D (first-stage CA membranes that were still functional). Train C was not changed. The plant was 75-percent operational by September 1991. Other membranes were 100-percent replaced by July 1994. The trains were reconfigured to a 61/45 array of vessels (first/second stage) at that time.

Although the membrane process represents a highly automated system, operators must be made aware of the potential for catastrophic results, such as the inherent danger of damaging membranes beyond their usefulness by exposure to the wrong chemical solutions. In addition, by ultimately converting all of the membranes to TFC membranes, this should no longer be a problem.

In terms of other membranes issues, as discussed above, some elements in the second stage were replaced with nonconforming 8822HR membranes. During the manufacturing process, some problems were experienced with the first manufacturing run, resulting in some elements that exhibited 20- to 30-percent lower flow than the rated design. Although the 8822HR membranes did not completely conform in specifications, the 8822HR elements conformed well with the original TFC membranes, so they were offered to the City.

The second problem area had to do with the post-treatment precipitation of iron on the degasser media. This was ascribed to precipitation of iron on the degasser media resulting from high concentrations of iron passing through the initially damaged membranes, and the resulting low iron rejection by the membranes. Most of this iron was being oxidized in the degassers. The iron sequestrant helped initially, then the element replacement in July 1994 reduced the iron content by two thirds.

System Performance

Data from the two plants were collected for a period of one year (June 1996 - May 1997). In addition, historical data for THMs collected in 1989 were also submitted. Furthermore, sampling for selected parameters was performed on September 30, 1997 to fill some data gaps. These data have been tabulated and summarized in Table 4.11. Selected data have been depicted in graphs to show the degree of retention by the membranes and the corresponding degree of removal by the conventional lime softening process (Figures 4.20-4.22).

Water Quality

Hardness. Figure 4.20 and Table 4.11 show the effect of membrane softening and lime softening on reducing total hardness and calcium hardness, respectively. The magnesium hardness data for the EWTP influent is given in Table 4.11. Approximately 90 percent of the hardness of the influent water (on average) is calcium hardness. The membrane process was more effective at removing the hardness of the water, as well as producing a more consistent product water quality—total hardness = 16-27 mg/L and calcium hardness = 15-25 mg/L—than lime softening—total hardness = 48-167 mg/L and calcium hardness = 35-141 mg/L. The minimal difference in total and calcium hardness in the membrane product water suggests that magnesium hardness was well removed, whereas the lime-softened water appears to still retain a significant portion of the magnesium hardness. Lime softening at a higher pH (e.g., $\geq$10.5) would be required to remove magnesium hardness (Thompson et al. 1997).

TDS. Membrane softening reduced the TDS in the influent from an average concentration of 355 mg/L to an average value of 59 mg/L (approximately 80-percent removal) (Table 4.11). This is much higher than the reduction achieved by lime softening (based on a single data point, Table 4.11), which reduced the TDS from 400 to 170 mg/L (58-percent removal).

Chloride. Table 4.11 illustrates the efficiency of the nanofilter/RO membrane combination in reducing the chloride content from an average of 58 mg/L to an average of 14 mg/L. A single measurement at the EWTP plant (Table 4.11) shows that lime softening did not reduce the chloride concentration (influent = 52 mg/L, effluent = 49 mg/L).

Bromide. As shown in Table 4.11 (September 30, 1997 sampling), the nanofilter/RO membrane system was very efficient (86 percent) at removing bromide from the water (influent = 0.14 mg/L, effluent = 0.02 mg/L).

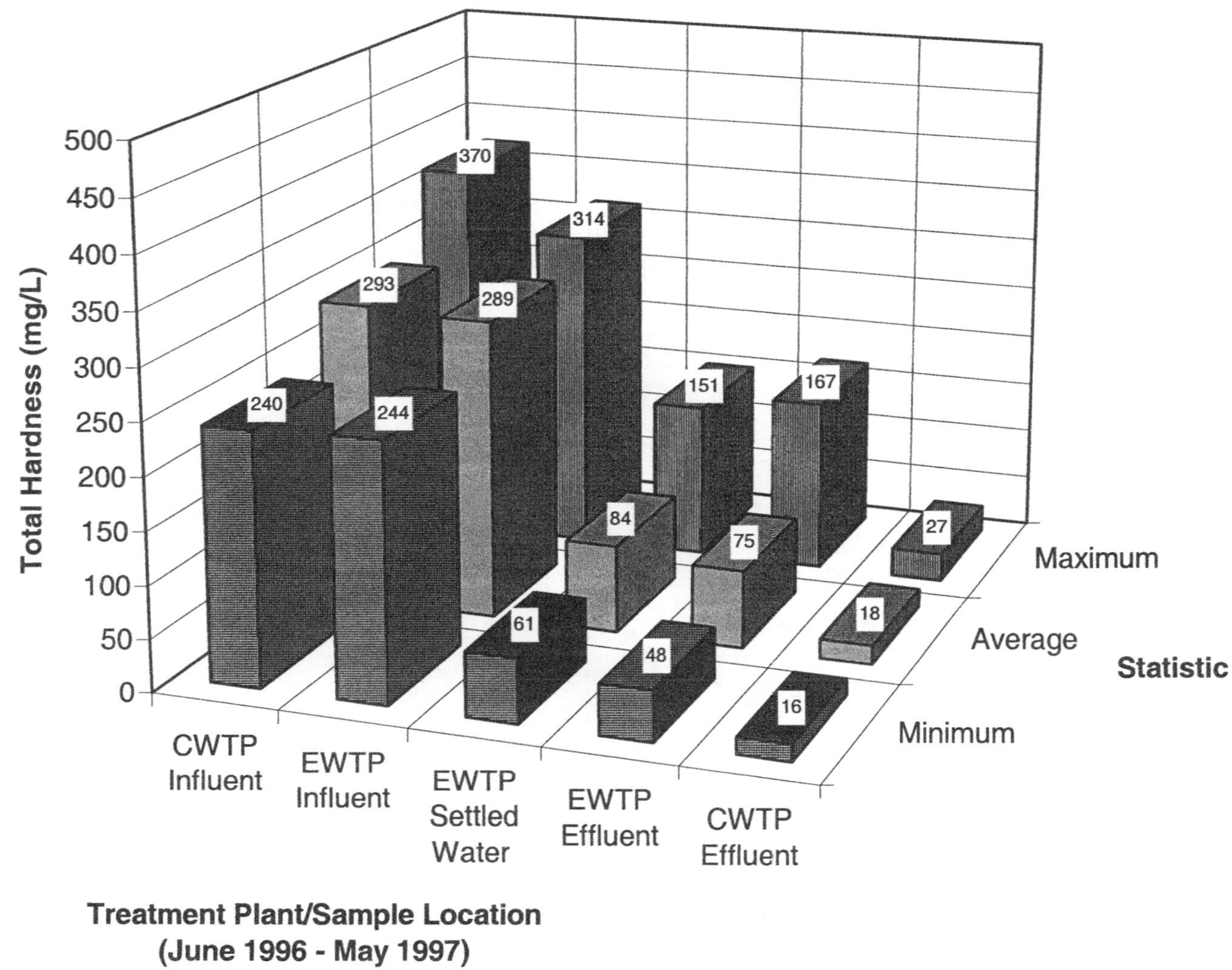

Figure 4.20 Effect of membrane treatment at CWTP and lime softening at EWTP on reducing the hardness of the groundwater treated by City of Plantation

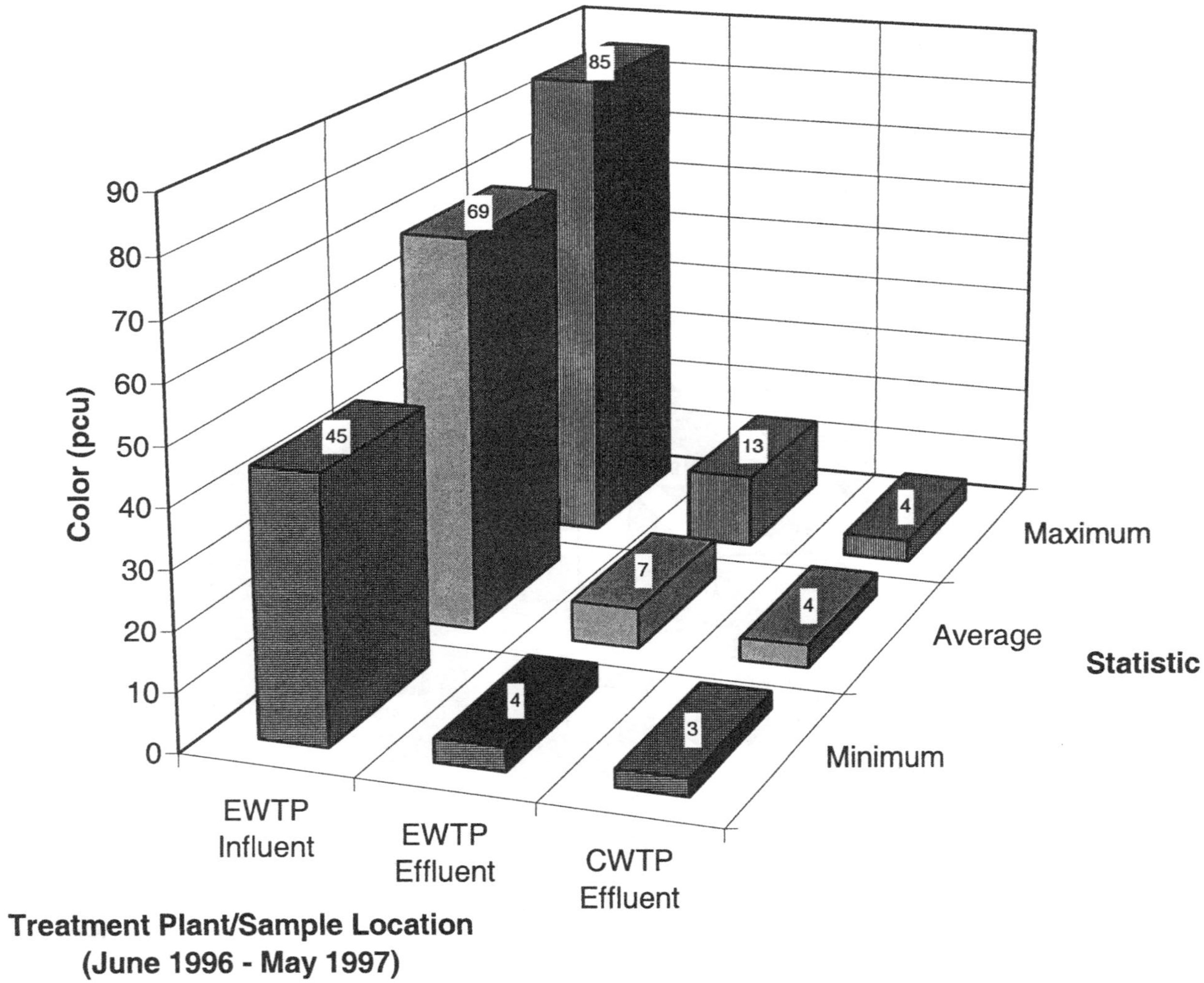

Figure 4.21 Effect of membrane treatment at CWTP and lime softening at EWTP on reducing the color of the groundwater treated by the City of Plantation

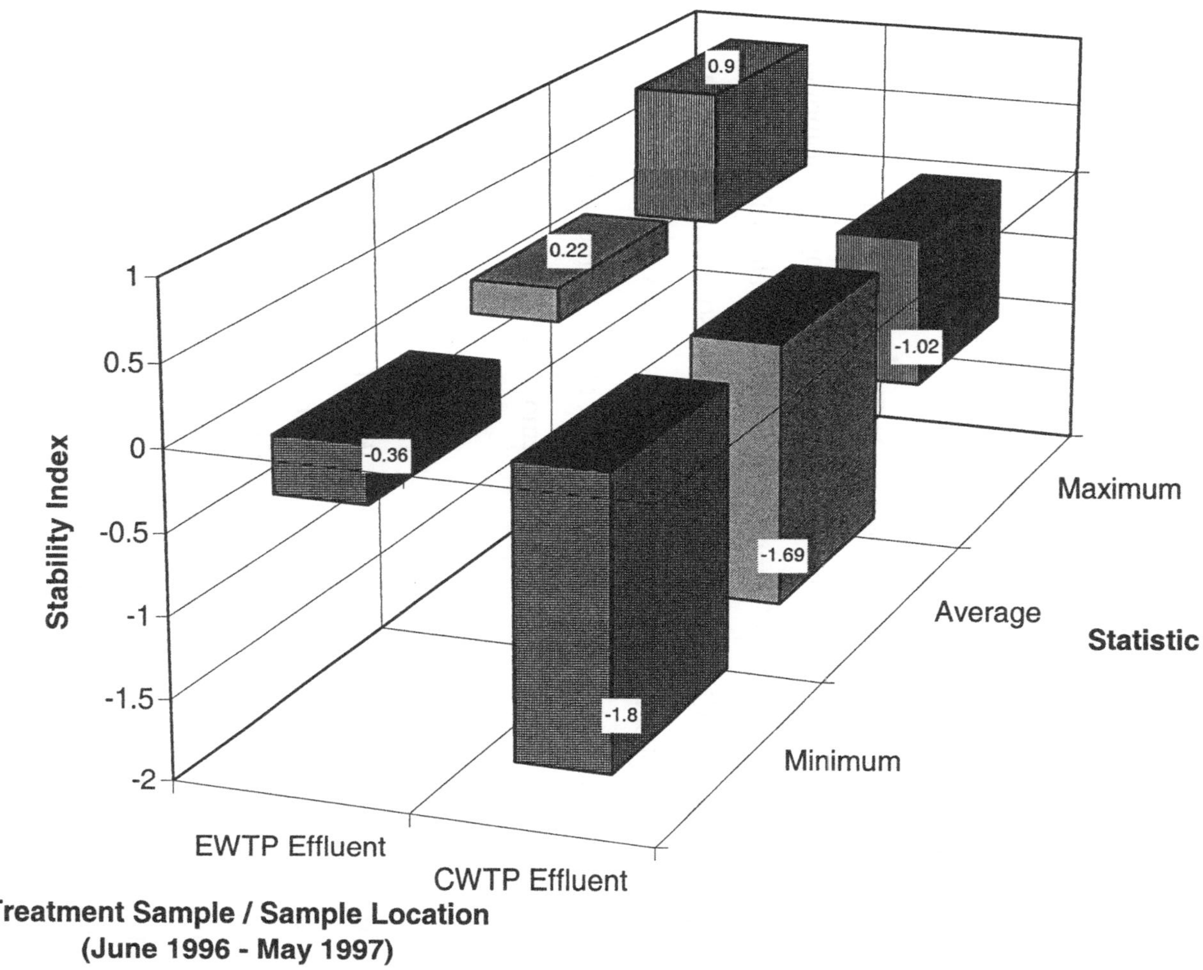

Figure 4.22 Effect of membrane treatment at CWTP and lime softening at EWTP on the stability index of the groundwater treated by the City of Plantation

Iron. As shown in Table 4.11, both lime softening and the nanofilter/RO membrane in combination with an iron sequestration agent were very successful in removing iron (approximately 97 percent on average). The final effluent levels of 0.02 mg/L (EWTP) and 0.05 mg/L (CWTP) are well below the secondary MCL of 0.3 mg/L.

Color. As can be seen from Table 4.11, the influent water quality for the EWTP and CWTP are very similar except in the instance of color, where the CWTP influent was more highly colored (100-150 pcu) compared to the EWTP influent (45-85 pcu). However, the membrane process did remove color (effluent = 3-4 pcu) to a larger degree (Figure 4.21) as compared to the lime softening/chlorination process (effluent = 4-13 pcu). Both plant values are well below the secondary drinking water standard of 15 pcu. Moreover, the removal of color at the membrane plant also results in the removal of DBP precursors.

TOC. Very limited data are available with regard to TOC measurements. From the data available, the influent to the CWTP had TOC values of 18-30 mg/L. During one historical sampling of the CWTP, the influent had a TOC value of 22 mg/L and an effluent value of 1.1 mg/L, indicating a removal efficiency of 95 percent. TOC data sampled on September 30, 1997 (Table 4.11) indicated that the TOC concentration was higher at the CWTP influent (17.8 mg/L) than at the EWTP influent (11.3 mg/L). The membrane process was much more efficient at TOC removal (98.7 percent) than the lime softening process (61 percent). However, TOC removal at the EWTP was quite high for a lime softening facility (Thompson et al. 1997).

UV. The UV data collected on September 30, 1997 (Table 4.11) mirrored the TOC data in that the level of UV absorbance was higher at the CWTP (0.635 cm^{-1}) than at the EWTP (0.328 cm^{-1}). The membrane plant removed UV-absorbing substances to a greater degree (98 percent) than the lime softening plant (60 percent), which was practicing prechlorination on the day of sampling. SUVA—which was calculated using the formula UV x 100/DOC, and it was assumed that the influent TOC was equivalent to the DOC in this low-turbidity groundwater—was 2.9 L/mg-m at the EWTP and 3.6 L/mg-m at the CWTP. These values are indicative of intermediate levels of humic substances in the influent waters (Edzwald and Van Benschoten 1990).

THMs. Samples from eight distribution-system locations from the old lime softening/ chloramination CWTP and the EWTP (February 1989) had a range of TTHM values from 28 to 58 μg/L, with an average of 36 μg/L. This indicates that they would be in compliance with the

Stage 1 TTHM MCL of 80 μg/L with just the lime softening process. The quarterly TTHM results with the membrane plant (March 1997) had a range of values from 0.5 to 29 μg/L for the eight distribution-system samples, with an average value of 5.5 μg/L. Six of the samples had <2.0 μg/L TTHMs. Because of the relative flows of water from the two plants, the majority of water in the distribution system was from the membrane softening plant (i.e., very little of the water was from the lime softening plant, which produced on the order of 29 μg/L TTHMs). The TTHM results for the September 30, 1997 sampling had a range of values from 1.0 to 3.5 μg/L for the eight distribution-system samples. Again, because of the relative flows of water from the two plants, the overwhelming majority of water in the distribution system was from the membrane softening plant (i.e., the TTHMs from the lime softening plant were diluted out). These data indicate that the change to a membrane softening process will meet the goal of complying with future, more stringent regulations (i.e., previously proposed placeholder Stage 2 TTHM MCL of 40 μg/L).

HAAs. On September 30, 1997, eight distribution-system locations were sampled. However, because of the relative flows of water from the two plants, the majority of water in the distribution system was from the membrane softening plant. HAAs were only detected at one sample site in the distribution-system fed (in part) by the lime-softening plant. The sum of the nine HAAs (HAA9) was 4.7 μg/L, which is well below the previously proposed placeholder Stage 2 MCL (for HAA5) of 30 μg/L. The only two HAA species detected were dichloro- and bromochloroacetic acid (at 3.6 and 1.1 μg/L, respectively). These data are consistent with previous research that has shown that chloramination reduces the formation of THMs and trichloroacetic acid, but still allows for the formation of dihalogenated HAAs (Symons et al. 1998).

CNCl. During the September 30, 1997 sampling, CNCl was not detected in the distribution system. However, analyses were not performed for cyanogen bromide, the brominated analogue of CNCl.

Total chlorine. The level of total chlorine (Table 4.11) in the effluents of the EWTP and the CWTP varied from 2.1 to 4.1 mg/L and 3.1 to 4.1 mg/L, respectively. The average effluent values at the EWTP and the CWTP are 3.2 mg/L and 3.5 mg/L respectively. The MRDL for chloramines is 4.0 mg/L (USEPA 1998a). Compliance with the MRDL is based on an annual average for the distribution-system samples. Total chlorine in the distribution system during the March 1997 THM monitoring ranged from 0.6 to 2.6 mg/L, with an average of 2.0 mg/L, well below the MRDL. Total chlorine in the distribution system during the September 1997 monitoring ranged from 1.9 to 2.6 mg/L, which was below the MRDL.

Ammonia: As shown in Table 4.11, ammonia-nitrogen in the groundwater source was high (0.86 mg/L in the EWTP influent and 1.57 mg/L in the CWTP influent on September 30, 1997). The membrane/degassification/chlorination process partially reduced the ammonia concentration (to 0.87 mg/L). At both plants, sufficient ammonia was present for the formation of chloramines.

pH. The influent pH for both plants are similar, but the effluent pH levels are vastly different. At the EWTP, the addition of lime raises the pH (average = 9.2) to precipitate Ca^{2+} ions, whereas during membrane softening the pH is lowered (average = 6.2) to prevent carbonate scaling and aid in the retention of Ca^{2+} and Mg^{2+} ions. Subsequent to the membrane process, the CWTP brings the pH of the effluent water up to 7.4 (on average) by stripping out carbon dioxide and by the addition of NaOH, whereas the EWTP brings the pH of the effluent down to 8.6 (on average). Many water utilities distribute their water at a pH around 8 (or higher) as a part of their corrosion-control program.

Alkalinity. The influent alkalinity at the EWTP ranged between 224 and 260 mg/L, with an average of 244 mg/L (Table 4.11). Lime softening resulted in an effluent alkalinity that ranged between 29 and 82 mg/L, with an average of 38 mg/L. The influent alkalinity at the CWTP ranged between 242 and 316 mg/L, with an average of 272 mg/L. Membrane softening with acid addition pretreatment resulted in an effluent alkalinity that ranged from 27 to 31 mg/L. Most lime softening plants produce water with an alkalinity between 30 and 60 mg/L. During the development of the D/DBP Rule, it was recognized that utilities should not be required to lower their alkalinity below 40-60 mg/L, as too low of an alkalinity would tend to make the water potentially corrosive (USEPA 1997).

Stability index. The recommended value for Langelier's index is that it be anywhere between zero to slightly positive (0 to 0.2). The effluent from the lime softening plant adheres to this guideline (Figure 4.22) as the average value is 0.22. On the other hand, the effluent from the membrane softening plant always has a negative Langelier's index, with an average value of -1.69. These negative values indicate a potential for corrosion. However, this can be offset by the addition of corrosion inhibitors.

Turbidity. The filtered-water turbidity goal is <0.25 NTU. The turbidity of the CWTP finished water has never exceeded 0.25 NTU to date (Table 4.11). Within the analytical variability of turbidity measurements, the two plants produced comparable effluent turbidities. Because the raw-water turbidity is so low, turbidity removal is not a significant challenge at either plant.

Summary of System Performance

It can be summarized that the membrane process at the CWTP is much better than the lime softening process at the EWTP in removing hardness and DBP precursors, as well as color. The membrane process—and removal of hydrogen sulfide in the degassing towers—frees the system to use chlorine (in actuality chloramines) as the primary disinfectant in a product water low in chlorine demand, which should allow the City to meet future groundwater microbial regulations while complying with the Stage 1 and previously proposed placeholder Stage 2 D/DBP regulations.

The lime softening process meets the Stage 1 TTHM MCL of the D/DBP Rule, but it would not reliably enable compliance with the previously proposed placeholder Stage 2 TTHM MCL. However, the overall distribution system served by the two plants does have a TTHM level below the previously proposed placeholder Stage 2 MCL. The effluent water produced by the lime softening plant has a more desirable stability index, whereas the addition of a corrosion inhibitor is required for the effluent produced by the membrane plant.

CHAPTER 5
FACTORS AFFECTING DECISIONS TO IMPLEMENT TECHNOLOGIES

INTRODUCTION

It is important to note that the decision-making process for each of the case studies included in this report was conducted during a period in which the specific requirements and associated schedules for the Microbial/DBP Rule cluster were uncertain. Some were conducted before the first regulatory negotiation in 1992 and 1993, some after the initial rules were proposed in 1994, and some during the Federal Advisory Committee meetings in 1997 through 2000. All were completed before the Agreement in Principle for the Stage 2 D/DBP Rule and LT2ESWTR was signed in September 2000. As such, assumptions regarding regulatory outcomes and schedules were made during the decision-making process for each of the case studies.

In addition, new information became available over the course of the rule-making process. For example, bromate was not being considered as a by-product of concern prior to the early 1990s. This DBP now plays a crucial role in the feasibility of implementing ozone disinfection, especially in waters containing elevated levels of bromide. From the standpoint of health effects, potential adverse reproductive/developmental outcomes associated with elevated DBP levels became available in the mid to late 1990s (Waller et al. 1998), adding short-term/acute issues to the already present long-term/chronic concerns regarding DBP exposure. Other contaminants and associated health effects are being identified on an ongoing basis, such as *N*-nitrosodimethylamine (NDMA), a potential by-product of chloramination (Najm and Trussell 2000). Finally, the water profession is also continually learning about the ability of old and new treatment processes to meet water quality objectives, as evidenced by the recognition in the late 1990s that UV light could inactivate *Cryptosporidium* at low dosages (Clancy et al. 1998).

The purpose of these case studies is to unfold the story of how various drinking-water systems arrived at decisions on technology implementation and what they learned through the process. It is incumbent upon any utility considering changes in their treatment process to have a full understanding of the regulatory framework and to be aware of the latest information regarding the advantages and disadvantages of various technologies.

Some of the factors that affect the decision-making process include meeting current and anticipated regulations, meeting multiple water quality objectives, making changes on an incremental basis or all at once, and site-specific considerations. This chapter addresses the role of these different factors and provides a summary of the decision-making process of the ten case studies in this project.

MULTIPLE OBJECTIVES

Meeting multiple water-quality objectives plays an important role in the decision-making process when a utility selects a treatment modification for compliance with the D/DBP Rule. The objective of this section of the report is to identify the water quality factors (in addition to DBPs) and other factors behind a utility's decision to implement these treatment modifications.

Besides compliance with the DBP Rule, utilities must meet other regulatory requirements (e.g., SWTR, IESWTR, LT2ESWTR, TCR, LCR) and secondary drinking-water standards (e.g., color, taste and odor, TDS). In addition, there are operational, financial, and engineering issues that affect the selection of treatment technologies.

The case studies conducted for this project demonstrate the role played by multiple water-quality objectives and other factors in treatment-modification selection (Table 5.1). A major water-quality objective in selecting treatment modifications was achieving a balance between microbial and DBP control. Another major objective was selecting a cost-effective DBP control strategy. The control of tastes and odors was also an important consideration in the selection of a strong oxidant and alternative disinfectant (i.e., ozone, chlorine dioxide) and/or the installation of GAC in the filters.

Compliance forecasts in the RIA for Stage 1 of the D/DBP Rule and for the previously proposed Stage 2 (Krasner et al. 1996) predicted that a significant number of utilities would require advanced treatment modifications such as ozonation, membranes, and GAC to comply with the previously proposed placeholder Stage 2 MCLs, whereas the majority of modifications for Stage 1 would involve enhanced coagulation and/or chloramination.

Table 5.1

Multiple water-quality objectives and other factors in treatment-modification selection

Case study	Water-quality factors and other objectives	Solution to each objective
NNWW	Balance microbial and DBP control	Pre-chlorination, post-chloramination
	Cryptosporidium inactivation	Ozone
	Modification that could be fast-tracked in time for the Stage 1 D/DBP Rule	Chloramines
	Cost-effective DBP control	Chloramines for Stage 1 D/DBP Rule, ozone for Stage 2 D/DBP Rule
	Need to replace a plant	Chloramine conversion project and installation of ozone were part of larger decision process that included a new replacement plant
Poway	Modification compatible with other systems in the same county	Chloramines
	Learn from others' experiences in treating similar source waters	Chloramines
	Cost-effective DBP control	Chloramines
	Escalating cost of imported water	Maximize use of local source
GCWA	Cost-effective DBP control	Primary disinfection with chlorine dioxide, post-chloramination
	Effective primary disinfectant	Chlorine dioxide
	Tastes and odors	Chlorine dioxide

(continued)

Table 5.1 (Continued)

Case study	Water-quality factors and other objectives	Solution to each objective
Skagit Co. PUD	Optimal disinfection while minimizing DBP formation	Primary disinfection with chlorine dioxide, intermediate chlorination, post-chloramination
	Cost-effective DBP control	Chlorine dioxide/chloramines
	Corrosion control without significantly increasing DBP formation	Raise treated-water pH and use chloramines for distribution system residual maintenance
Ann Arbor	SWTR virus requirements	Ozone
	DBPs	Ozone/chloramines
	Tastes and odors	Ozone/GAC on top of the filters
	Cost and space issues associated with meeting *CT* requirements	Ozone
	AOC	GAC on top of the filters
ACWD	THMs	Ozone/chloramines
	Turbidity and TOC	Ozone, coagulation with ferric chloride
	Tastes and odors	Ozone (plus hydrogen peroxide)
	Microbial pathogens	Ozone/chlorine
	Aldehydes	Biofiltration

(continued)

Table 5.1 (Continued)

Case study	Water-quality factors and other objectives	Solution to each objective
Phoenix	THMs	Enhanced coagulation, move the point of chlorination
	Cost-effective DBP precursor control	Enhanced coagulation
	Blending well water throughout the distribution system	Use chlorine for distribution system residual maintenance
Tampa	Residuals disposal problems	Switch coagulant to ferric sulfate, develop beneficial reuse of ferric humate residual
	DBPs	Enhanced coagulation
	Color	Enhanced coagulation
	Turbidity	Enhanced coagulation
Hopewell	Tastes and odors	GAC in filters
	Bacterial regrowth	Use of post-chloramination
	HAAs	Biodegradation on GAC
	THMs	Seasonal pre-chloramination
Plantation	Infrastructure replacement	New plant
	Increased capacity	New plant
	DBPs	Membranes/chloramines
	Hardness	Membranes
	Color	Membranes
	Iron	Membranes/iron sequestration agent

In the case studies, utilities were concerned with choosing advanced treatment processes (e.g., ozonation, membranes) that would enable them to satisfy other current and anticipated future regulations (e.g., microbial inactivation) or other water quality objectives (e.g., color and hardness removal) in addition to the D/DBP Rule.

Balancing Microbial Control

Although many of the utilities in this project use chloramines for secondary disinfection and to minimize DBP formation, a stronger disinfectant (chlorine, chlorine dioxide, ozone) was required for primary disinfection. When chloramines are used for primary disinfection, compliance with the SWTR may be compromised. Moreover, some utilities may need—as part of a LT2ESWTR—to inactivate *Cryptosporidium*, so ozone is being installed. More recently, some utilities have begun to consider UV disinfection for *Cryptosporidium* control.

Taste-and-Odor Control

Although chlorine dioxide and ozone are both alternative disinfectants that can achieve good primary disinfection while typically not producing THMs, some systems will choose ozone—which is more expensive—because of the ability of ozone to destroy certain taste- and odor-causing contaminants in water. (Note, the UV dosages that are used for disinfection will not destroy taste- and odor-causing compounds.) Likewise, GAC is being added to filters for taste-and-control, but it is typically not being regenerated with a frequency required for DBP precursor control. However, the use of GAC in filters can promote biological activity that can degrade AOC and has also been found to degrade HAAs.

Color and Hardness Removal

Other aesthetic concerns include color and hardness. Enhanced coagulation has been used to remove color, which is also a major source of DBP precursors. For highly colored and hard groundwaters, membranes can be a cost-effective means of treating such waters. In fact, the removal of color and hardness may be needed to justify the costs of membrane technologies.

Other Considerations

A common element in corrosion-control programs is raising the pH of the treated water. However, in the presence of free chlorine, THM formation increases at an elevated pH. Alternatively, in the presence of chloramines, monochloramine formation is optimized at an alkaline pH and THM formation is minimized.

Very importantly, many utilities may have site-specific issues that significantly affect the selection of treatment modifications. In a region in which neighboring utilities use chloramines, there are advantages in switching to chloramines, especially if any of the systems have any interconnections. Alternatively, if there are blended distribution systems with chlorinated groundwaters, the use of precursor removal technologies instead of using post-chloramination may be preferable.

Space and retrofit considerations will affect technology choices. In some cases, a relatively expensive technology such as ozonation may be deemed cost-effective if space is limited and other options are not feasible. Moreover, in the case of infrastructure replacement, the choice of membrane technology may be cost-effective when compared to retrofitting an existing conventional plant. Finally, residual disposal issues may affect treatment selection options.

STAGED IMPLEMENTATION

In addition to Stages 1 and 2 of the D/DBP Rule, the SWTR has been staged (i.e., the original rule, the IESWTR, and the LT2ESWTR). Because of the uncertainty of how stringent certain regulations will be and the high costs of advanced treatment technologies, many utilities have implemented treatment modifications in stages. Typically, most utilities have made treatment modifications to meet Stage 1 of the D/DBP Rule that have been cost-effective to meet their site-specific needs and objectives, while continuing to study or implement long-term treatment changes to meet future more stringent regulations.

Typically, enhanced coagulation and/or a switch to chloramines for secondary disinfection are being used as part of a utility's first stage of treatment changes. The next treatment modification that is typically being incorporated is a switch to chlorine dioxide or ozone for primary disinfection. Utilities that are going straight to the more expensive advanced treatment modifications (e.g., ozone, membranes) to meet the Stage 1 D/DBP Rule are typically those that have relatively high levels of DBP precursors (TOC and/or bromide) and/or other contaminants (e.g., color, hardness). These utilities recognize that they ultimately will need such technologies to handle future anticipated more stringent regulations.

Other advantages of staged implementation are (1) allowing for advances in a particular area of technology and (2) learning from others that have implemented these treatment modifications sooner. For example, our understanding of the formation and control of ozone by-products (e.g., bromate and low-pH ozonation, AOC and biofiltration) has advanced since utilities in the U.S. first implemented ozonation. In addition, advances in the development of membranes are benefiting utilities that are now implementing this precursor-removal technology. Furthermore, a new alternative to *Cryptosporidium* control, UV disinfection, will be used by some systems that held off on ozone installation.

As part of the RIA for the D/DBP Rule, a "regrets analysis" was conducted (Cromwell 2001). Regrets analysis is a technique that can shed some light on the relative merits of alternative strategies even in the face of stark uncertainties. One of the conclusions from this analysis was the following: When it is not possible to confidently assess the central tendency of the risk, regrets analysis returns a result indicating that it is better to do something than it is to do nothing, and that it is better to just do something than it is to try to do everything. This is why a staged approach to DBP regulation was a rational choice.

The major disadvantage to staged implementation is that the treatment process must be re-optimized each time treatment modifications are made. However, this problem can be minimized if the full range of changes in treatment is envisioned in advance and if incremental modifications are made that will be part of and consistent with long-term modifications. Cost is typically an important driver that results in advancing incrementally. Also, avoidance of unknown problems and tradeoffs is another reason for incremental changes.

Another consideration is where an existing facility is in its "life cycle." If the need for a replacement plant is imminent, it may be more cost-effective to build a new plant with advanced treatment technologies than it is to incrementally retrofit the existing plant.

FINANCING OPTIONS

As discussed above, cost is a major consideration in selecting and implementing treatment modifications. Site-specific and retrofit issues affect these costs. Other water-quality objectives that can be met by implementing the new technology are important in selling the treatment change to the utility management, the City Council, and the consumers.

Demographics of Consumers

In each of the case studies in this project, demographic information was obtained for consumers in each region served. This information was typically obtained from the 1990 U.S. Census, as many of the decisions to implement treatment changes by the participating utilities in the study were made in the later 1980s or early 1990s. Table 5.2 gives a summary of the demographic information.

Although this set of utilities represents a small subsection of the systems at large, some preliminary information can be obtained from these case studies. Consumers that have a higher income and/or educational attainment may be more willing to support the implementation of expensive advanced technologies, whereas in other communities there may be more of a tendency to implement modifications in a staged manner so that lower cost modifications can be utilized initially.

Other Financial Considerations

In addition to a community's "willingness to pay," these case studies have demonstrated that there may be unique site-specific considerations that affect the financial feasibility of implementing advanced technologies.

For example, access to capital was not a problem for Plantation. The membrane softening plant was paid for from a $20 million cash reserve. (A bond was floated to replenish the reserves, but they were able to pay cash.) The service area benefits from a strong economy and vigorous population growth, as reflected in a strong financial position. As noted in the City

Table 5.2

Demographic information on consumers in regions served by participating utilities*

		Median values				
		Household income ($)		Cost of housing	Age	For the population ≥25 yr, % that
Utility	City or county	1990	1997	($)	(years)	had bachelor's degree or higher
NNWW	Newport News	27,469	34,306	84,400	29	18
Poway	Poway	53,252		219,400	~30	34
GCWA	Texas City;†	26,144		49,700	~30	10
	Bacliff†	26,892		43,500	~30	9
Skagit Co. PUD	Skagit Co.	28,389	38,148	80,800	~34	16
Ann Arbor	Ann Arbor	34,594		111,996	28.5	59
ACWD	Fremont;	51,231;		263,400;‡	32;	30,
	Newark;	50,471;		224,900;‡	30;	21,
	Union City	46,988		228,900‡	31	21
Phoenix	Phoenix	29,291		76,600	~30	20
Tampa	Tampa	22,772		58,600	~33	19
Hopewell	Hopewell	26,934	32,781	54,500	~30	10
Plantation	Plantation	41,832		130,060	37	32

*Most of the data are from the 1990 U.S. Census

†GCWA sells water to seven municipalities in Texas between Houston and Galveston, which include Texas City and Bacliff

‡Housing costs in table are from 1990 census; costs in case study for this community (Chapter 3) were for 1998

of Plantation's web site, "Plantation is an upscale community that features a unique blend of residential living and business enterprise."

Alternatively, one of the utilities that chose to incrementally make modifications to comply with various SDWA regulations—in particular the DBP rules—did so because of a unique history. At the outset of this compliance process, Skagit Co. PUD was faced with the $10 million cost of a filtration plant as the first step. That represented a large adjustment in rates for a city of this size moving from a very low-cost unfiltered status. Ozone would have added perhaps another $3 million to that total, which made it much more expensive than the alternative of chlorine dioxide.

In addition to concerns over capital costs, there are issues with the increase in costs of operations. In some cases, utilities in this study have been able to take advantage of unique opportunities to minimize these additional costs. For example, the City of Tampa realized an economic advantage with their implementation of enhanced coagulation. They switched from alum to ferric sulfate, in part, because it produced a ferric humate residual with a beneficial reuse. Other utilities have also explored the implementation of new water treatment processes in which a residual with a beneficial use could be obtained. For example, a pellet softening process in the Netherlands reduced the calcium hardness of the water while producing a residual that could be used in cattle feed and in other industrial and agricultural operations (van Eekeren 1992).

For some utilities, power costs may be a major obstacle to the implementation of ozone. However, at ACWD, the unique topography of the plant site resulted in the incorporation of hydroelectric power generation facilities into the design, which are capable of providing all of the power needed to run the plant, including ozone generation.

However, it must be recognized that the implementation of expensive advanced technologies can have significant affects on rates. The executive management of Ann Arbor Water Utilities was attempting to implement a gradual rate increase program to fund their $15 million ozone project rather than having to increase rates in a single sudden change. However, action on implementation of this project was triggered by SWTR compliance issues, which resulted in a formal consent order.

DECISION-MAKING STRATEGIES

The decision-making process to implement treatment changes in order to comply with the D/DBP Rule typically began with an examination of anticipated rule requirements, as the regulation has been developed in stages. Information on the advantages and disadvantages of different treatment modifications were then gathered from the literature and from the experiences of other utilities and/or consulting engineering firms.

Most utilities than embarked on a series of bench-, pilot-, and/or full-scale tests to evaluate the cost and feasibility of one or more treatment processes that could be used to comply with the regulations. Based on this information, an implementation strategy was designed to either incrementally meet changes in regulations one at a time or to make a "quantum leap," thus requiring no additional efforts to meet any future anticipated regulations. This strategy was based on the following elements:

- Utilities selecting treatment processes for compliance with the D/DBP Rule were taking into account other water quality objectives/constraints of the plants.
- Some of the significant regulations affecting treatment-process selection besides the D/DBP Rule were the SWTR (and ESWTRs), the TCR, and the LCR.
- Aesthetic issues such as taste and odor, color, and TDS played a major role in influencing treatment-modification selection at some utilities.
- Utilities had to also take into consideration financial, space-limitation, and site-specific objectives while evaluating treatment modifications.
- Although financial considerations were a major part of the treatment-process selection, the utilities studied did not necessarily choose the least-cost alternative when selection was based only on DBP control.

SUMMARY

The main issues that affected the decision-making process involved (1) compliance, (2) utility business, and (3) customer acceptance:

- Utilities need to comply with multiple (current) regulatory requirements.
  - In addition, systems must consider future changes in the D/DBP Rule.
    - The Rule has and may continue to include requirements for new DBPs (e.g., bromate).
    - Changes in compliance requirements are being developed (e.g., change from an RAA to an LRAA basis).
  - Other new regulations are being proposed (e.g., LT2ESWTR).
  - New technologies (e.g., UV) are being proposed to enable compliance with these new regulations.
- Various factors relate to utility business issues:
  - Where is the utility in the life cycle of the existing facility (e.g., incremental change or new infrastructure needed)?
  - Cost (capital, operations and maintenance) of the treatment modifications need to be evaluated, especially in the context of complying with a suite of regulations.
  - There are site-specific issues that must be considered.
  - New treatment processes require training for the operating staff.
  - The relationship of the utility to their governing body is an important issue; systems need to work well with that group.
- Customer acceptance is important.
  - Multiple benefits with the technology modification can help.
  - Aesthetic issues (e.g., taste and odor) can be critical factors.
  - Distribution system water quality issues need to be considered.
  - Rate payer buy-in is needed.
  - Public involvement and outreach must be pursued.

CHAPTER 6
CONCLUSIONS

GETTING TO A DECISION - A PRIMER

Technology Selection

During the development of the Regulatory Impact Analysis (RIA) for the D/DBP Rule and the ESWTR, technology selection was based on a least-cost sorting routine. Although the use of chloramines as an alternative disinfectant is an inexpensive technology choice for DBP control, there are various reasons utilities might choose to continue to use free chlorine in the distribution system and, thus, need to explore other DBP-control measures. For example, there can be concerns over (1) taste-and-odor issues associated with chloramines; (2) operational problems in maintaining the correct ratio of chlorine to ammonia throughout a distribution system that is blended with chlorinated groundwater; and (3) the likelihood of encountering nitrification problems. Thus, the RIA considered the premise that some systems would look for the least-cost options while continuing to use free chlorine for secondary disinfection.

Table 6.1 shows the least-cost sorting routine for technology selection for large surface-water systems that filter but do not soften, which was used during the development of the Stage 1 D/DBP Rule (USEPA 1994a). Various combinations of DBP precursor removal (e.g., enhanced coagulation, GAC) with or without alternative disinfectants (e.g., ozone, chloramines) are available. Based on least-cost options, some combination of enhanced coagulation, ozone, and/or chloramines would be more cost effective than a GAC or membrane option (Krasner et al. 1996).

Figure 6.1 shows the decision tree for large systems that filter, which was used during the development of the Stage 2 D/DBP Rule and the LT2ESWTR (Chowdhury and Haymore 2000). This decision tree was based on a low-cost sorting routine for simultaneous compliance with various Stage 2 D/DBP Rule and LT2ESWTR options under consideration. This decision tree expanded on the Stage 1 D/DBP Rule effort (Table 6.1) in several ways:

Table 6.1

Least-cost sorting routine for technology selection for large surface-water systems that filter but do not soften, which was used during the development of the Stage 1 D/DBP Rule

Treatment technology*
Cl_2/conventional
Cl_2/NH_2Cl
Enhanced coagulation
Enhanced coagulation + Cl_2/NH_2Cl†
O_3/NH_2Cl‡
Enhanced coagulation + O_3/NH_2Cl§
GAC10
Enhanced coagulation + GAC10
Enhanced coagulation + GAC10 + Cl_2/NH_2Cl
GAC10 + O_3/NH_2Cl
Enhanced coagulation + GAC10 + O_3/NH_2Cl
GAC20
Enhanced coagulation + GAC20
GAC20 + O_3/NH_2Cl
Enhanced coagulation + GAC20 + O_3/NH_2Cl
Enhanced coagulation + membranes

*Primary and secondary disinfectants were chlorine unless noted otherwise.

†Chlorine applied as primary disinfectant to settled water.

‡Although chlorine dioxide was not specifically examined in this list of technology choices, it was expected that some of the systems utilizing alternative disinfectants would choose chlorine dioxide.

§Ozone applied as primary disinfectant to either raw or settled water.

Source: Adapted from USEPA 1994a.

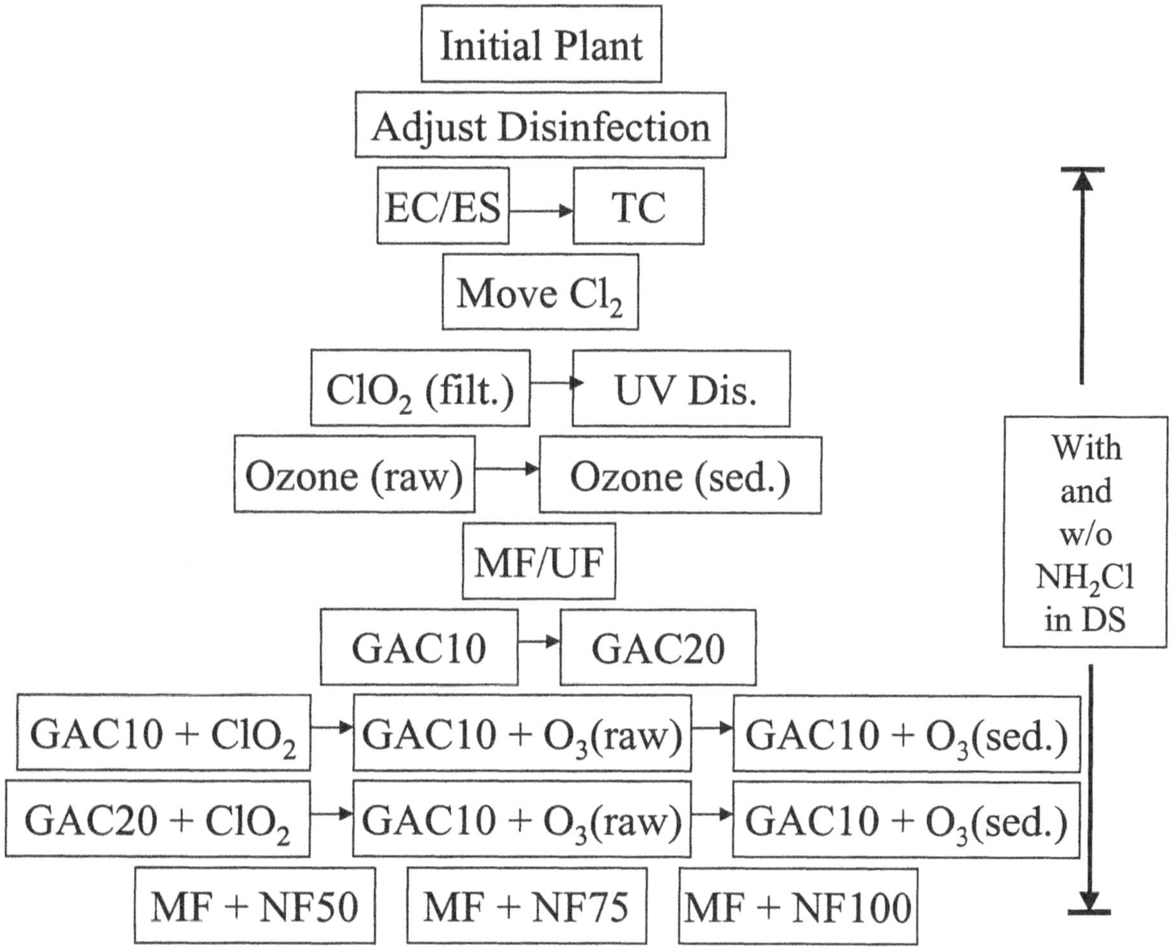

Figure 6.1 Large-system decision tree for Stage 2 D/DBP Rule and LT2ESWTR used in RIA (*Source*: Chowdhury and Haymore 2000)

- It examined enhanced softening (ES) in addition to enhanced coagulation (EC).
- It included a "turbo" coagulation (TC) option, in which a utility might achieve more TOC removal than was required in the TOC-removal requirements (Table 3.20) to better minimize THM or HAA formation.
- It examined chlorine dioxide as well as ozone.
- It examined different locations for the application of alternative primary disinfectants: chlorine dioxide at the filters (filt.); ozone in the raw water or in the sedimentation (sed.) basins.
- It included UV disinfection, as well as MF or ultrafiltration (UF) for pathogen removal.

- It examined nanofiltration (NF) with different percentages (i.e., 50, 75, 100 percent) of the plant flow being treated by NF membranes.
- It considered these treatment choices with or without (w/o) chloramines as the disinfectant in the distribution system (DS).
- Although not shown in this figure, some utilities will implement other treatment modifications—as part of the microbial toolbox—in order to comply with the LT2ESWTR.

However, the case studies in this project have demonstrated that there are multiple water-quality objectives and issues that affect technology selection. The decision-making process needs to consider different driving forces, which affect the choice of technology, in order to meet multiple water-quality objectives. In addition, there are other operational, engineering, and financial issues that need to be considered. Use of such a decision-making process presents a starting framework for consideration.

The first question that must be addressed is whether the DBP levels are less than the Stage 1 (or 2) MCLs or not. Even if the current DBP levels are less than regulatory limits, the consequences of complying with the ESWTR must be considered. For example, Ann Arbor was in compliance with the THM Rule, but was sometimes out of compliance with the SWTR. They had to select a means of improving disinfection that still allowed DBPs levels to be within regulatory limits. Ann Arbor selected ozonation, which introduced bromate formation as a new DBP issue. If the level of bromate had been greater than the Stage 1 MCL, Ann Arbor would have had to adjust the disinfection/treatment processes to simultaneously comply with the D/DBP Rule and the SWTR.

If a utility needs to make changes to meet the D/DBP Rule, simultaneous compliance with the ESWTR must be examined. Moreover, for many utilities, there are other water quality objectives that need to be met, which affect the choice of technology. For example, in the Stage 2 D/DBP Rule/LT2ESWTR decision tree (Figure 6.1), chlorine dioxide was selected before ozone as the lower-cost alternative disinfectant. However, as demonstrated in the case studies in this project, some utilities will choose ozone because of multiple water-quality objectives. Likewise, expensive technologies such as GAC or membranes may be selected if there are multiple objectives that can be met with these treatment processes.

However, meeting multiple water-quality objectives does not necessarily require more expensive technologies. A utility may be able to better optimize an existing process, especially if this action allows for a cost-effective solution to many issues. For example, Tampa changed their coagulation process and achieved greater TOC removal than was required in the Stage 1 D/DBP Rule (Table 4.7) (comparable to TC in Figure 6.1). Other driving forces in this decision-making process were improved color removal and a need to better address residuals disposal (Figure 4.8).

As part of the decision-making process, utility staff and consultants need to consider the control of a wide range of contaminants, as well as trade-off issues associated with each technology (Table 6.2). A table that summarizes the ability of different technologies to control contaminants of particular concern to a utility can be an important tool in technology selection. A technology that is the lowest cost but controls the least number of contaminants may not result in the lowest overall cost of treatment modifications when a utility is attempting to address many regulations and water-quality objectives simultaneously.

However, various regulations are being promulgated at different points in time, and some regulations are being implemented in stages (e.g., the D/DBP Rule). Thus, a utility needs to consider either putting in an advanced technology (e.g., ozone, GAC, membranes) that looks as if it will handle current and future anticipated regulations (as Plantation did with membranes) or they need to modify the treatment processes one at a time (as Skagit Co. PUD did). Site-specific issues and access to capital will dictate, in part, which route a utility will take.

Table 6.3 lists some of the constraints and issues that a utility must consider during the decision-making process. In addition to meeting primary and secondary drinking-water standards, there are operational, engineering, and financial issues that must be addressed. In most cases, utilities must consider treatment modifications that pose retrofit issues. Space availability and hydraulics of the plant may limit cost-effective retrofit choices (as was the case with Ann Arbor). In other cases, a utility may be either considering an infrastructure replacement (as was the situation with Plantation) or adding additional parallel treatment trains in order to meet new capacity demands. In those cases, the installation of advanced treatment processes may be more cost-effective to consider than in a retrofit situation. Because of the uncertainty of future regulations, a utility must consider future anticipated requirements in addition to promulgated standards during the decision-making process.

Table 6.2

Comparison of different technologies for the control of multiple water-quality objectives[*]

Issue	EC	ES	NH_2Cl	ClO_2	O_3	UV	GAC	NF/RO
Regulatory use								
Control THM/HAA formation	Yes	Yes	Yes	Yes	Yes	Yes	Yes	Yes
Remove/inactivate *Giardia*	Yes	---	---[†]	Yes	Yes	Yes	---	Yes
Remove/inactivate viruses	Yes	---	---[†]	Yes	Yes	Yes	---	Yes
Remove/inactivate *Cryptosporidium*	Yes	---	No	Yes	Yes	Yes	---	Yes
Remove/oxidize SOCs	No	No	No	---	Yes	No	Yes	Yes
Remove/biodegrade AOC	Yes	---	No	No	No	No	Yes	---
Biodegrade HAAs	No	No	No	No	No	No	Yes	No
Other water-quality objectives								
Remove/oxidize tastes and odors	No	No	No	Yes	Yes	No	Yes	Yes
Remove/oxidize color	Yes	---	No	---	Yes	No	Yes	Yes
Remove hardness	No	Yes	No	No	No	No	No	Yes
Possible tradeoffs								
Form other regulated DBPs	No	No	No	Yes	Yes	No	No	No
Form other DBPs (non-regulated)	No	No	Yes	Yes	Yes	No	No	No
Form AOC	No	No	No	No	Yes	No	No	No
Kidney dialysis issues	No	No	Yes	Yes	No	No	No	No
Nitrification issues	No	No	Yes	No	No	No	No	No
Residuals issues	Yes	Yes	No	No	No	No	---[‡]	Yes
Corrosion issues	Yes	Yes	Yes	No	No	No	No	Yes

[*]This is a general, overall comparison. For specific contaminants, each technology has a different level of efficacy.

[†]Chloramines are not a strong disinfectant for the inactivation of *Giardia* or viruses.

[‡]Depending on the GAC usage, there is the need for periodic replacement or reactivation of the GAC.

Table 6.3

Other issues that affect the decision-making process*

Issue	Sub issues
Operational and engineering constraints/issues	
	Retrofit issues
	Plant hydraulics
	Plant layout
	Space availability
	Electrical requirements
	Infrastructure replacement
	Capacity of facility
	Compatible with rest of treatment process
	Compatible with other plants in system/area
Financial constraints/issues	
	Access to capital
	Impact on rates
	Staged implementation
	Uncertainty of future regulations

*List of some of the issues that are considered in addition to meeting primary and secondary drinking-water standards.

Implementation

Once a technology has been selected, there are a number of steps required for the successful implementation of that technology:

- The information used in the decision-making process will need to be summarized and presented to the utility's governing body (e.g., Board of Directors, City Council) and to the consumers.
  - The decision-making process will need to be described.
  - The regulatory and other water-quality issues that affected the decision-making process will need to documented.

- Other important driving factors/variables that shaped the decision will need to be discussed.
- The information and analysis that was used in getting to a certain decision will need to be presented.
- Because there are uncertainties associated with every decision, how these uncertainties were coped with must be addressed.
- The decision will need to be supported by research (bench-, pilot-, and/or full-scale; computer modeled or simulated) that was conducted by the utility, its consultants, and/or the experiences of others.
- The financial analysis should include the major break points in terms of capital costs and how these affected the decision-making process.
- It will also be important to discuss the major trip wires in terms of operations and system reliability associated with the new technology.
- Moreover, the kind of buffer (margin of safety) that was allowed to ensure compliance and where this is expected to be most critical should be presented.
- Furthermore, the extent to which other water-quality issues were addressed, in addition to compliance with the D/DBP Rule, will need to be raised.
- Because the implementation strategy may have been designed to incrementally meet changes in regulations one at a time rather than making a "quantum leap" or a "one-shot" strategy (i.e., installing a technology that will require no additional efforts to meet any future regulations), the strategy will need to be highlighted.

- In order to persuade the governing body to agree to implement the modifications, a series of presentations will need to be made to provide them with the background and the underlying logic for the project.
- A public information program will need to be used to ensure consumer buy-in. The customers need to be supportive of the changes, including the increase in water rates. An extensive public relations campaign may be need to be undertaken to educate and inform the customers as to the water-quality benefits from the treatment change.
- After the treatment change has been approved, additional training for the operating staff will be needed. Often the manufacturers of the new treatment technology can provide technical support in this effort.

- An implementation plan will need to be developed and supported by a multi-disciplinary, interdepartmental implementation team.
- Contingency planning will need to address how to handle expected and unexpected problems during the implementation of the treatment modifications.
- There may need to be shifts in the overall operational strategy or operating paradigms to make the modifications successful (e.g., changing the frequency of standard rounds).
- If there are implementation problems or tradeoffs, some will need to be resolved immediately, whereas for others there may be a period of adjustment.
- It will probably take a period of time for the operational staff to be "comfortable" with the new or modified technology and for the treatment plant to operate like an "integral unit," rather than as a treatment plant with a "new process."

To facilitate implementation of the Stage 1 D/DBP Rule and the IESWTR, the USEPA developed a number of guidance manuals:

- Technical guidance manuals:
  - Alternative Disinfectants and Oxidants Guidance Manual (USEPA 1999a)
  - Disinfection Profiling and Benchmarking Guidance Manual (USEPA 1999b)
  - Enhanced Coagulation and Enhanced Precipitative Softening Guidance Manual (USEPA 1999c)
  - Guidance Manual for Compliance with the Interim Enhanced Surface Water Treatment Rule: Turbidity Provisions (USEPA 1999d)
  - Guidance Manual for Conducting Sanitary Surveys of Public Water Systems; Surface Water and Ground Water Under the Direct Influence (GWUDI) (USEPA 1999e)
  - Microbial and Disinfection Byproduct Rules Simultaneous Compliance Manual (USEPA 1999f)
  - Uncovered Finished Water Reservoirs Guidance Manual (USEPA 1999g)
- Implementation guidance manuals:
  - Implementation Guidance for the Interim Enhanced Surface Water Treatment Rule (USEPA 2001a)
  - Implementation Guidance for the Stage 1 Disinfectants/Disinfection Byproducts Rule (USEPA 2001b)

- Plain English guidance manuals:
    - The Interim Enhanced Surface Water Treatment Rule - What Does It Mean to You? (USEPA 2001c)
    - The Stage 1 Disinfectants and Disinfection Byproducts Rule - What Does It Mean to You? (USEPA 2001d)

SUMMARY

In this project, case studies were conducted at 10 representative U.S. utilities that have made treatment modifications to comply with the Stage 1 D/DBP Rule (and the previously proposed Stage 2 D/DBP Rule) and the ESWTR. In these studies, there was a documentation of (1) problems and tradeoffs, (2) operational experience (which gave a long-term picture of the strengths and weaknesses, as well as solutions to problems, for each treatment modification), (3) risk/risk balancing, and (4) the decision-making process.

To collect the information needed, a series of information collection instruments (questionnaires) were designed that asked for information on water quality, operational experiences, implementation issues, costs, and the decision-making process. Copies of relevant reports and memorandum, which were used to support information provided on the questionnaires, were obtained. These included water quality master plans, conceptual design reports, and decision analysis alternatives reports.

Information was obtained from a diverse group of water industry professionals at each utility in order to meet the project's data objectives. Separate questionnaires were prepared and filled out by different departments at many of the participating utilities. Furthermore, follow-up questions were necessary in assessing the operational experience of participating utilities.

Finally, an important aspect of the case-study project was to document the decision-making process. Elements that were examined included processes or steps taken and approaches followed, the roles and responsibilities of different members of the organization in the decision making, and varying perspectives and weighting of criteria that were considered. In addition, the project documented each utility's understandings of "known" and "unknown" elements. This included the following elements:

(1) information and/or beliefs used in the decision making (e.g., sources of information, beliefs and level of confidence),
(2) major uncertainties/risks identified (e.g., Stage 1 versus Stage 2 compliance requirements in the D/DBP Rule, overall system reliability and other tradeoff relationships),
(3) efforts to cope with uncertainties/risks (e.g., deferral of action, hedging and contingency approaches; capital costs/operations-and-maintenance costs/control systems tradeoffs; the margin of safety used in design to assure compliance; development of better information such as research and pilot-plant testing)

Interviewees on the decision-making process included various utility staff, such as the general manager, engineering staff, plant superintendent, operations staff, water quality laboratory staff, and water production and treatment staff.

The main objective of this project was to document "real-world" information on the implementation of D/DBP control strategies. In this study, information was gathered and interpreted in a manner that should aid other utilities that are planning on implementing such DBP control strategies. The main issues that affected the decision-making process involved (1) compliance, (2) utility business, and (3) customer acceptance (Chapter 5).

FUTURE RESEARCH

Many participating utilities in this case-study project began their decision-making process in the later 1980s or early 1990s. In most cases in this study, decisions were being made and implemented prior to the promulgation of the Stage 1 D/DBP Rule. Moreover, most utilities utilized "placeholder" MCLs in the proposed Stage 2 D/DBP Rule for long-term planning. In the near future, the USEPA will propose a newly negotiated Stage 2 D/DBP Rule.

As other utilities select and implement treatment modifications in order to comply with the promulgated Stage 1 D/DBP Rule and/or the soon-to-be-proposed Stage 2 D/DBP Rule, a research opportunity exists to document the decision-making process in the new millennium. Moreover, with an even greater number of utilities exploring alternative technologies, a great opportunity exists for the sharing and exchange of implementation experiences. For example, with the greater use of the Internet for information exchange, a potential effort could be to establish electronic bulletin boards to facilitate the sharing of the knowledge gained from these experiences.

However, even with the anticipated promulgation of a Stage 2 D/DBP Rule, the uncertainty over the health effects of DBPs and the likelihood of even future DBP regulations remains. Thus, there is still a research need to document how utilities can make decisions in the face of uncertainty.

Utilities are selecting treatment modifications that will enable compliance with the D/DBP Rule, as well as meet multiple water quality objectives. However, in assessing the costs and benefits of the D/DBP Rule, compliance with only the D/DBP Rule and the ESWTR have been considered. A research effort should be considered to develop tools to do a cost/benefit analysis that considers the multiple benefits of installing these treatment modifications. Moreover, such an analysis will also need to more fully consider the direct and indirect costs of tradeoffs associated with the implementation of these treatment changes. Such an analysis that considers the various benefits and costs of these treatment changes will enable utilities, consultants, and the regulators a better means of judging the cost-effectiveness of the implementation of new technology beyond compliance with one (or two) regulations at a time.

APPENDIX A

SYNOPSIS OF CASE-STUDY QUESTIONNAIRE

BACKGROUND UTILITY INFORMATION:
TO BE PROVIDED BY PRINCIPAL UTILITY CONTACT

Utility Information

Name of utility; address.

1. Name of the treatment plant that has been modified to meet the D/DBP Rule.

2a. Plant flow at design and at typical flow.

2b. The population served by this treatment plant.

Source Water/Plant Influent Information

3a. Specific source(s) of raw water at this plant. If more than one source water, typical blends (minimum, average, maximum) over the course of the year.

3b. If groundwater is treated: Is it classified as a groundwater under the direct influence of a surface water?

WATER QUALITY INFORMATION:
TO BE PROVIDED BY WATER QUALITY LABORATORY

Raw Water Quality Information

1a. Typical levels of the following measurements of DBP precursors in raw water over the course of the year (minimum, average, maximum, frequency of monitoring): TOC, DOC, particulate organic carbon, UV, SUVA, bromide, THMFP.

1b. Seasonal variations in DBP precursor concentrations.

1c. Dates for period of sampling.

2a. Typical levels of the following water quality parameters and microbial contaminants in raw water over the course of the year (minimum, average, maximum, frequency of monitoring): water temperature, pH, total alkalinity, total hardness, calcium hardness, ammonia, turbidity, HPC bacteria, total coliforms, fecal coliforms, *Giardia*, *Cryptosporidium*, viruses. List of other parameters (e.g., algae, tastes and odors, synthetic organic contaminants, etc., that may affect the overall process modifications, especially if the modifications were implemented to meet multiple treatment objectives).

2b. Seasonal variations in levels of water quality parameters or microbial contaminants.

2c. Dates for period of sampling.

Baseline Finished Water Quality Information

3a. Typical finished (distribution-system) water quality for the following parameters before modifications were made to meet the D/DBP Rule (minimum, average, maximum, frequency of monitoring): water temperature, pH, total alkalinity, total hardness, calcium hardness, turbidity, TOC, TOC removal, TTHMs, HAA5, total HAAs, TOX.

3b. Seasonal variations in levels of finished water quality parameters.

3c. Dates for period of sampling.

Modified-Treatment Finished Water Quality Information

4a. Typical finished (distribution-system) water quality for the following parameters after modifications were implemented to meet the D/DBP Rule (minimum, average, maximum, frequency of monitoring): water temperature, pH, total alkalinity, total hardness, calcium hardness, turbidity, TOC, TOC removal, TTHMs, HAA5, total HAAs, TOX, cyanogen chloride, chlorite, chlorate, bromate, formaldehyde, sum of aldehydes, AOC, AOC removal, biodegradable organic carbon, removal of biodegradable organic carbon.

4b. Seasonal variations in levels of finished water quality parameters.

4c. Dates for period of sampling.

5a. For systems that use ozone: range and average level of HPC bacteria in the filter effluent before postfilter disinfection.

5b. Prior to implementation of ozone: range and average level of HPC bacteria in the filter effluent before postfilter disinfection.

6. Range and average level of particle concentration in finished water.
7. Changes in microbial quality of finished water after implementation of treatment changes.

Additional Information

8. Are you doing DBP or TOC monitoring for the ICR?
9. If you have any reports, etc., that document more details on any of the above questions, please include two copies of each report with your completed questionnaire.

INFORMATION ON TREATMENT-MODIFICATION SELECTION PROCESS: TO BE PROVIDED BY PROCESS ENGINEERING DEPARTMENT

Information on Treatment-Modification Selection Process

1a. How did you decide on selecting process improvements at your treatment plant? Did you conduct any bench-, pilot-, or full-scale evaluation of new or modified treatment processes?

1b. Who conducted the study? Level of participation by the plant staff? Year in which the evaluation was conducted? Duration of the evaluation? Cost of conducting the evaluation?

2. A brief summary of the results and implications of these evaluations:
    - If the evaluation included DBP precursor (TOC) removal experiments by coagulation: Which coagulant(s)? The pH of coagulation evaluated? What pH-adjustment chemicals were evaluated for pH depression and subsequent base addition?
    - If the evaluation included DBP precursor (TOC) removal experiments by lime softening: Softening pH evaluated? What coagulants were evaluated, and what were their dosages? Removal of magnesium hardness in these experiments?
    - If alternative disinfectants (ozone, chlorine dioxide, and/or chloramines) were evaluated: Which one(s)? Details on the disinfection process(es) studied (e.g., disinfectant dose and contact time, staging of primary and secondary disinfectants)?

- If GAC was evaluated: EBCT of the GAC? Regeneration frequency of the GAC? Other details on the GAC study?
- If membranes were evaluated: Membrane type/composition? Average MWCO of membrane(s) evaluated? Operating pressure of membrane(s) evaluated? Were the membranes evaluated for removal of microbial pathogens, DBP precursors, color, hardness or some other contaminant?

Additional Information

3. If you have any reports, etc., that document more details on any of the above questions, please include two copies of each report with your completed questionnaire.

TREATMENT/OPERATIONS INFORMATION:
TO BE PROVIDED BY OPERATIONS DIVISION

Baseline Treatment Information

1. A schematic of the plant before modifications were made to meet the D/DBP Rule, with the locations of chemical addition and the names of chemicals used—in particular, those chemicals used for disinfection, oxidation, coagulation, softening, and pH adjustment.

2a. Typical coagulation or softening conditions before modifications (minimum, average, maximum): Name of coagulant or softening chemicals. Typical dose (and formula) previously used. The pH in the coagulation or softening basin.

2b. Corrosion-control practices before modifications (use of phosphate-based corrosion inhibitors, target pH for finished water, chemical dosages of corrosion inhibitors and pH-adjustment chemicals, etc.).

3. Details on the disinfection process—typical values and range—before modifications were made (e.g., disinfectant dose and contact time, *CT* achieved, log *Giardia* inactivation, information on how the log inactivation was calculated, etc.).

Modified Treatment Information

4. A schematic of the plant after modifications were made to meet the D/DBP Rule, with the locations of chemical addition and the names of chemicals used—in particular, those chemicals used for disinfection, oxidation, coagulation, softening, and pH adjustment.

5a. Typical coagulation or softening conditions after modifications (minimum, average, maximum): Name (and formula) of coagulant or softening chemicals. Typical dose now used. The pH in the coagulation or softening basin.

5b. Corrosion-control practices after modifications (use of phosphate-based corrosion inhibitors, target pH for finished water, chemical dosages of corrosion inhibitors and pH-adjustment chemicals, etc.).

6. First point of disinfectant addition in your plant? If no prechlorination is practiced (in particular, if no chlorine is applied before the coagulation or softening process), what other oxidant are used at the plant to control growths in the basins and the filters?

7a. Details on the disinfection process—typical values and range—after modifications were made (e.g., disinfectant dose and contact time for each point of disinfection, *CT* achieved, log *Giardia* inactivation, information on how the log inactivation was calculated, etc.).

7b. If ozonation is used: Ozone contact time (contact time for the overall contactor as well as the contact time in each chamber). Ozone contactor hydraulic efficiency. Ozone dose and residual (minimum, average, maximum) (if the ozone dose is split between more than one chamber, the dose in each chamber; if the residual is measured in the effluent of more than one chamber, the residual in each chamber effluent). The pH of the ozonated water.

8. If ozonation is used: Do you allow/encourage biological activity in your filters? Are the filters receiving water with a chlorine residual or chloramine residual? Do you use chlorinated or chloraminated water for filter backwashing?

9. If ozone is applied before the filters, do you carry an ozone residual on top of the filters? If so, how much of a residual?

10. If ozone is used, do you disinfect the filter effluent to inactivate HPC bacteria shed from the filters? What disinfectant do you use for this purpose and what is the dose and contact time?

11. What filter media do you have?

12. If GAC is one of the filter media: Was GAC as a filter media put in to prevent taste-and-odor problems in the finished water, for DBP precursor (TOC) removal, for removal of some other micropollutant, or for providing a better medium for biological activity? GAC used as a filter media replacement or in post-filter contactor? Type, size, and uniformity coefficient of GAC used. Depth of GAC in filter. Loading rate of GAC filter. Backwash frequency of GAC filter. Criterion for backwashing GAC filter. EBCT of GAC at design flow and at typical flow. Frequency needed to regenerate or replace GAC. Criterion for regenerating GAC.
13. If membranes are used: Application(s) (e.g., organics/DBP precursors, color, microbials). Membrane type/composition. Average MWCO. Operating pressure. Plant recovery at the membrane plant (i.e., what percentage of the feedwater is recovered as product water?). Chemicals (with formulas) used to pretreat the feedwater and their dosages (minimum, average, maximum). How are the membranes cleaned, and at what frequency? Waste disposal methods for the concentrate.
14. Filtered-water turbidity goal. How often is the goal exceeded?

Additional Information

15. If you have any reports, etc., that document more details on any of the above questions, please include two copies of each report with your completed questionnaire.

TREATMENT MODIFICATIONS—IMPLEMENTATION INFORMATION: TO BE PROVIDED BY ENGINEERING OR OPERATIONS DIVISION

1. When were the modifications brought on-line at full scale?

Cost Information

2. The capital cost of these modifications (a detailed breakdown of these costs, including costs for land acquisition, extensive site preparation, etc.).

3a. Additional annual cost of operations due to these modifications (a detailed breakdown of these costs, including labor, electrical, chemicals, etc.).

3b. Additional annual cost of capital recovery due to these modifications (a detailed breakdown of these costs, including amortization period, interest rate, etc.).

Implementation Issues

4. Did the implementation of the technology have the expected impact on DBP formation? If not, what additional actions were taken to provide the expected result?
5. Were there other, unanticipated consequences as a result of implementing the treatment changes (e.g., did implementation of enhanced coagulation or ozonation affect the control of iron and manganese in your plant)?
6. Were there any significant problems in implementing these modifications? E.g., if ozonation was implemented, did it cause a filter air binding problem? Was ozone degassing in the filters addressed during design? E.g., if membranes were installed, what are your experiences with membrane fouling, how often does it occur, and what methods are employed to deter fouling?
7. Were there any significant "tradeoffs" in implementing these modifications? E.g., if ozonation and biological filtration were implemented, have there been any effects on the filtered water quality and filtration operations?
8. If there were implementation problems or tradeoffs, how did you resolve them? E.g., did implementation of the modifications require a change in your corrosion-control practices, and, if so, what was the cost of the modified corrosion-control program?

Operational Issues

9. What period of time was required for the operational staff to be "comfortable" with the new or modified technology and for the treatment plant to operate like an "integral unit," rather than as a treatment plant with a "new process"?
10. Did the treatment strategy implemented significantly change operational needs at the plant? Was additional staff added? If so, how many persons and in what job classifications?

11. Did the treatment strategy implemented require additional training for the operating staff? How did you plan for these needs? Was the initial training conducted adequate or was additional training required after start-up?
12. Was there resistance by the operating staff to the treatment modifications?
13. Were any shifts in overall operational strategy or operating paradigms necessary to make the modifications successful (e.g., changing the frequency of standard rounds)?

Additional Information

14. If you have any reports, etc., that document more details on any of the above questions, please include two copies of each report with your completed questionnaire.

INFORMATION ON THE DECISION-MAKING PROCESS:
PERSPECTIVES TO BE PROVIDED BY
SELECTED MEMBERS OF THE ORGANIZATION

Decision-Making Process

1. Describe the decision-making process and your role in it.
2. What were the most important driving factors/variables shaping this decision?
3. What decision have you reached?
4. What is the path that got you there? What information did you use? What analysis?
5. What were your greatest areas of uncertainty along the way? What have you done to cope with these uncertainties?
6. What research have you performed? What AWWARF research have you used? What additional research do you believe would be worthwhile to improve decision making for your utility or for the industry in general?
7. Where are the major breakpoints in terms of capital costs? How did these affect the decision-making process?
8. Where are the major "tripwires" in terms of operations and system reliability? How did these affect the decision-making process?

9. What kind of buffer (margin of safety) did you allow to ensure compliance? Where is this most critical?
10. To what extent did compliance with the D/DBP Rule raise other water quality issues that must be addressed: source-water protection, source-water selection and availability, plant replacement, treatment optimization (including for the control of other contaminants), sludge disposal, distribution-system operations and maintenance, distribution-system replacement, wholesale relationships?
11. Were the changes implemented to address issues other than DBPs? If so, what other water quality issues were addressed by the modifications?
12. Was the implementation strategy designed to incrementally meet changes in regulations one at a time or was the strategy intended to make a "quantum leap," thus requiring no additional efforts to meet any future regulations (e.g., putting in ozone, GAC and/or membranes as a "one shot" strategy)?
13. What persuaded the governing body (e.g., Board of Directors, City Council) to agree to implement the modifications?
14. Were the customers fully supportive of the changes, including the increase in water rates (assuming the rates increased)? What type of public information program was used to ensure buy-in?
15. In selling the strategy to the customers/Board/City Council, was it acceptable that the modification was required "to meet regulatory requirements," or was there substantial resistance to "unfunded mandates"? If the latter, how were the stakeholders ultimately persuaded to support the modifications?
16. If you had the opportunity to make your decision again, would you implement the same strategy and modifications? If not, what would you do differently?

Additional Information

17. If you have any reports, etc., that document more details on any of the above questions, please include two copies of each report with your completed questionnaire.

REFERENCES

Aieta, E.M., and J.D. Berg. 1986. A Review of Chlorine Dioxide in Drinking Water Treatment. *Jour. AWWA*, 78(6):62-72.

Amirtharajah, A., and K.M. Mills. 1982. Rapid-Mix Design for Mechanisms of Alum Coagulation. *Jour. AWWA*, 74(4):210-216.

Amy, G.; M. Siddiqui, W. Zhai, J. Debroux, and W. Odem. 1994. *Survey of Bromide in Drinking Water and Impacts on DBP Formation.* Denver, Colo.: AWWARF and AWWA.

AWWA. 2001. Partnership for Safe Water: Annual Data Summary Report. Denver, Colo.: AWWA.

Baribeau, H., S.W. Krasner, R. Chinn, and P.C. Singer. Impact of Biomass on the Stability of Haloacetic Acids and Trihalomethanes in a Simulated Distribution System. In *Proc. of the 2000 AWWA Water Quality Technology Conference*. Denver, Colo.: AWWA.

Becker, W.C., and C.R. O'Melia. 1996. *Optimizing Ozonation for Turbidity and Organics (TOC) Removal by Coagulation and Filtration*. Denver, Colo: AWWARF and AWWA.

Bellamy, W.D. 1995. *Full-Scale Ozone Contactor Study*. Denver, Colo: AWWARF and AWWA.

Bergman, R.A. 1996. Cost of Membrane Softening in Florida. *Jour. AWWA*, 88(5):32-43.

Bolyard, M., P.S. Fair, and D.P. Hautman. 1992. Occurrence of Chlorate in Hypochlorite Solutions Used for Drinking Water Disinfection. *Environ. Sci. Technol.*, 26(8):1663-1665

Bontrager, S.L., Z. Chowdhury, and J. Hayes. 1992. Optimization of Conventional Treatment for Disinfection By-Product Control. In *Proc. of the 1992 AWWA Annual Conference*. Denver, Colo.: AWWA.

Bull, R.J. 1995. Carcinogenic Properties of Brominated Haloacetates. In *Workshop Report—Disinfection By-Products in Drinking Water. Critical Issues in Health Effects Research.* Washington, D.C.: International Life Science Institute (ILSI) Health and Environmental Sciences Institute.

Cheng, R.C., S.W. Krasner, J.F. Green, and K.L. Wattier. 1995. Enhanced Coagulation: A Preliminary Evaluation. *Jour. AWWA*, 87(2):91-103.

Chowdhury, Z, and C. Haymore. 2000. Compliance Forecasts to Costs. Presented at Microbial/DBP Technology Workgroup meeting, Washington, D.C., March 28, 2000.

Clancy, J.L., T.M. Hargy, M.M. Marshall, and J.E. Dyksen. 1998. UV Light Inactivation of Cryptosporidium Oocysts. *Jour. AWWA*, 90(9):92-102.

Clark, S.W. 1989. Discussion of Strawman Rule for Disinfectants and Disinfection By-Products. Presented at Science Advisory Board Drinking Water Committee meeting, Washington, D.C., October 11, 1989.

Coffey, B.M., S.W. Krasner, C.-y. Kuo, and J.F. Green. 1998. Bromate Formation and Control During Ozonation to Achieve *Giardia* and *Cryptosporidium* Inactivation. In *Proc. of the 1998 International Ozone Assocation (IOA) Pan American Group Conference*. Stamford, Conn.: IOA.

Coffey, B.M., S.W. Krasner, M.J. Sclimenti, P.A. Hacker, and J.T. Gramith. 1996. A Comparison of Biologically Active Filters for the Removal of Ozone By-Products, Turbidity, and Particles. In *Proc. of the 1995 AWWA Water Quality Technology Conference*. Denver, Colo.: AWWA.

Conlon, W.J., and S.A. McClellan. 1989. Membrane Softening: A Treatment Process Comes of Age. *Jour. AWWA*, 81(11):47-51

Cowman, G.A., and P.C. Singer. 1996. Effect of Bromide Ion on Haloacetic Acid Speciation Resulting from Chlorination and Chloramination of Aquatic Humic Substances. *Environ. Sci. Technol.*, 30(1):16-24.

Cromwell, J.E., III. 2001. Regrets Analysis of Disinfection By-Product Control. In *Microbial Pathogens and Disinfection By-products in Drinking Water: Health Effects and Management of Risks*. Washington, D.C.: ILSI.

Daniel, P.A. 1998. *Balancing Multiple Water Quality Objectives*. Denver, Colo.: AWWARF and AWWA.

Dietrich, A.M, and R.C. Hoehn. 1991. *Taste-and-Odor Problems Associated with Chlorine Dioxide*. Denver, Colo.: AWWARF and AWWA.

Dykes, G.M., and W.J. Conlon. 1989. Use of Membrane Technology in Florida. *Jour. AWWA*, 81(11):43-46

Dyksen, J.E., R. Ford, K. Raman, J.K. Schaefer, L. Fung, B. Schwartz, and M. Barnes. 1992. *In-line Ozone and Hydrogen Peroxide Treatment for Removal of Organic Chemicals*. Denver, Colo: AWWARF and AWWA.

Edwards, G.A., and A. Amirtharajah. 1985. Removing Color Caused by Humic Acids. *Jour. AWWA*, 77(3):50-57.

Edzwald, J.K., and J.E. Van Benschoten. 1990. Aluminum Coagulation of Natural Organic Matter. In *Chemical Water and Wastewater Treatment.* Edited by H.H. Hahn and R. Klute. Berlin: Springer-Verlag.

Ferguson, D.W., J.T. Gramith, and M.J. McGuire. 1991. Applying Ozone for Organics Control and Disinfection: A Utility Perspective. *Jour. AWWA*, 83(5):32-39.

Finch, G.R., C.W. Labatiuk, R.D. Helmer, and M. Belosevic. 1992. *Ozone and Ozone-Peroxide Disinfection of Giardia and Viruses.* Denver, Colo: AWWARF and AWWA.

Finch, G.R., E.K. Black, L. Gÿurk, and M. Belosevic. 1994. *Ozone Disinfection of Giardia and Cryptosporidium.* Denver, Colo: AWWARF and AWWA.

Fu, P., H. Ruiz, K. Thompson, and C. Spangenberg. 1994. Selecting Membranes for Removing NOM and DBP Precursors. *Jour. AWWA*, 86(12):55-72.

George, D.B., V.D. Adams, S.A. Huddleston, K.L. Roberts, and M.B. Borup. 1990. *Case Studies of Modified Disinfection Practices for Trihalomethane Control.* Denver, Colo.: AWWARF and AWWA.

Gianatasio, J.M., M.A. Hook, and L.J. Hjersted. 1996. Enhanced Coagulation: An Operational Perspective; Enhanced Ferric Sulfate Treatment for Color and TOC Removal at the City of Tampa Hillsborough River Water Treatment Plant. In *Proc. of the 1995 AWWA Water Quality Technology Conference.* Denver, Colo.: AWWA.

Glaze, W.H., and H.S. Weinberg. 1993. *Identification and Occurrence of Ozonation By-Products in Drinking Water.* Denver, Colo: AWWARF and AWWA.

Griese, M.H., J.J. Kaczur, and G. Gordon. 1992. Combining Methods for the Reduction of Oxychlorine Residuals in Drinking Water. *Jour. AWWA*, 84(11):69-77.

Hall, E.S., and R.F. Packham. 1965. Coagulation of Organic Color With Hydrolyzing Coagulants. *Jour. AWWA*, 57(9):1149-1166.

Hess, A.F., G.L. Myers, D.M. Owen, and J. Baker. 1985. Control of Chlorine Dioxide Application at Galveston County Water Authority. In *Proc. of the 1985 Instrument Society of America International Conference.*

Iatrou, A., and W.R. Knocke. 1992. Removing Chlorite by the Addition of Ferrous Iron. *Jour. AWWA*, 84(11):63-68.

Jacangelo, J.G., N.L. Patania, K.M. Reagan, E.M. Aieta, S.W. Krasner, and M.J. McGuire. 1989. Ozonation—Assessing Its Role in the Formation and Control of Disinfection By-products. *Jour. AWWA*, 81(8):74-84.

Glaze, W.H., and H.S. Weinberg. 1993. *Identification and Occurrence of Ozonation By-products in Drinking Water.* Denver, Colo.: AWWARF and AWWA.

Gordon, G., L. Adam, and B. Bubnis. 1995. *Minimizing Chlorate Ion Formation in Drinking Water When Hypochlorite Ion is the Chlorinating Agent.* Denver, Colo.: AWWARF and AWWA.

Krasner, S.W., and S.E. Barrett. 1985. Aroma and Flavor Characteristics of Free Chlorine and Chloramines. In *Proc. of the 1984 AWWA Water Quality Technology Conference.* Denver, Colo.: AWWA.

Krasner, S.W., M.J. McGuire, J.G. Jacangelo, N.L. Patania, K.M. Reagan, and E.M. Aieta. 1989. The Occurrence of Disinfection By-products in US Drinking Water. *Jour. AWWA*, 81(8):41-53.

Krasner, S.W., W.H. Glaze, H.S. Weinberg, P.A. Daniel, and I.N. Najm. 1993. Formation and Control of Bromate During Ozonation of Waters Containing Bromide. *Jour. AWWA*, 85(1):73-81.

Krasner, S.W., M.J. Sclimenti, and E.G. Means. 1994. Quality Degradation: Implications for DBP Formation. *Jour. AWWA*, 86(6):34-47.

Krasner, S.W., and G.L. Amy. 1995. Jar-Test Evaluations of Enhanced Coagulation. *Jour. AWWA,* 87(10):93-107.

Krasner, S.W., D.M. Owen, and J.E. Cromwell. 1996. Regulatory Impact Analysis of the D/DBP Rule. In *Water Disinfection and Natural Organic Matter: Characterization and Control.* Edited by R.A. Minear and G.L. Amy. Washington, D.C.: American Chemical Society.

Krasner, S.W., Z.K. Chowdhury, M.A. Edwards, and K.A. Bell. 1997. Use of SUVA in Developing Revised TOC Removal Requirements. In *Proc. of the 1997 AWWA Water Quality Technology Conference.* Denver, Colo.: AWWA.

Krasner, S.W., S. Pastor, R. Chinn, M.J. Sclimenti, H.S. Weinberg, S.D. Richardson, and A.D. Thruston Jr. 2001. The Occurrence of a New Generation of DBPs (Beyond the ICR). In *Proc. of the 2001 AWWA Water Quality Technology Conference.* Denver, Colo.: AWWA.

Lykins, B.W. Jr., R.M. Clark, and J.Q. Adams. 1988. Granular Activated Carbon for Controlling THMs. *Jour. AWWA*, 80(5):85-92.

Malcolm Pirnie, Inc. 1988. Basis of Design Memorandum; Membrane Softening Facility; Central Water Treatment Plant; Plantation, Florida. Engineering report for City of Plantation, Fla., Project 1279-01-1. Orland, Fla.: Malcolm Pirnie, Inc.

Malcolm Pirnie, Inc. 1993. Bay-Delta Water Quality Monitoring. Report for Metropolitan Water District of Southern California, No. 1500-041. Carlsbad, Calif.: Malcolm Pirnie, Inc.

McGuire, M.J., and R.G. Meadow. 1988. AWWARF Trihalomethane Survey. *Jour. AWWA*, 80(1):61-68.

McGuire, M.J., M.K. Davis, S. Liang, C.H. Tate, E.M. Aieta, I.E. Wallace, D.R. Wilkes, J.C. Crittenden, and K. Vaith. 1989. *Optimization and Economic Evaluation of Granular Activated Carbon for Organic Removal*. Denver, Colo.: AWWARF and AWWA.

McGuire, M.J., M.K. Davis, C.H. Tate, E.M. Aieta, E.W. Howe, and J.C. Crittenden. 1991. Evaluating GAC for Trihalomethane Control. *Jour. AWWA*, 83(1):38-48.

Means, E.G., III, and S.W. Krasner. 1993. D-DBP Regulation: Issues and Ramifications. *Jour. AWWA*, 85(2):68-73.

Najm, I.N., and R.R. Trussell. 2000. NDMA Formation in Water & Wastewater. In *Proc. of the 2000 AWWA Water Quality Technology Conference*. Denver, Colo.: AWWA.

Norman, T.S., L.L. Harms, and R.W. Looyenga. 1980. The Use of Chloramines To Prevent Trihalomethane Formation. *Jour. AWWA*, 72(3):176-180.

Norton, C.D., and M.W. LeChevallier. 1997. Chloramination: Its Effect on Distribution System Water Quality. *Jour. AWWA*, 89(7):66-77.

Owen, D.M., Z.K. Chowdhury, S.W. Krasner, and E.G. Means III. 1993a. Predicting THM Formation in Bromide-Rich Waters. In *Proc. of the 1992 AWWA Water Quality Technology Conference*. Denver, Colo.: AWWA.

Owen, D.M., G.L. Amy, and Z.K. Chowdhury. 1993b. *Characterization of Natural Organic Matter and Its Relationship to Treatability*. Denver, Colo.: AWWARF and AWWA.

Pedersen, E.J., III, E.T. Urbansky, B.J. Mariñas, and D.W. Margerum. 1999. Formation of Cyanogen Chloride from the Reaction of Monochloramine and Formaldehyde. *Environ. Sci. Technol.*, 33(23):4239-4249.

Pegram, R. A. 1995. Use of Mechanistic and Pharmacokinetic Data: Bromodichloromethane as a Case Study. In *Workshop Report—Disinfection By-Products in Drinking Water. Critical Issues in Health Effects Research*. Washington, D.C.: ILSI Health and Environmental Sciences Institute.

Randtke, S.J., R.C. Hoehn, W.R. Knocke, A.M. Dietrich, B.W. Long, and N. Wang,. 1994. A Comprehensive Assessment of DBP Precursor Removal by Enhanced Coagulation and Softening. In *Proc. of the 1994 AWWA Annual Conference (Water Quality)*. Denver, Colo.: AWWA.

Reckhow, D.A., and P.C. Singer. 1990. Chlorination By-products in Drinking Waters: From Formation Potentials to Finished Water Concentrations. *Jour. AWWA*, 82(4):173-180.

Singer, P.C., H. Arora, E. Dundore, K. Brophy, and H.S. Weinberg. 1999. Control of Haloacetic Acid Concentrations by Biofiltration: A Case Study. In *Proc. of the 1999 AWWA Water Quality Technology Conference*. Denver, Colo.: AWWA.

Skadsen, J. 1993. Nitrification in a Distribution System. *Jour. AWWA*, 85(7):95-103

Skadsen, J., T.G. Myers, and B. Bellamy. 1997. Optimization of Ozonation for DDBP and Microbial Control. In *Proc. of the 1997 AWWA Water Quality Technology Conference*. Denver, Colo., AWWA.

Snoeyink, V., and D. Jenkins. 1980. *Water Chemistry*. New York, N.Y.: John Wiley & Sons.

Stevens, A.A., L.A. Moore, and R.J. Miltner. 1989. Formation and Control of Non-Trihalomethane Disinfection By-products. *Jour. AWWA*, 81:8:54.

Symons, J.M., R. Xia, G.E. Speitel Jr., A.C. Diehl, C.J. Hwang, S.W. Krasner, and S.E. Barrett. 1998. *Factors Affecting Disinfection By-Product Formation During Chloramination*. Denver, Colo.: AWWARF and AWWA.

Taylor, J.S., D.M. Thompson, and J.K. Carswell. 1987. Applying Membrane Processes to Groundwater Sources for Trihalomethane Precursor Control. *Jour. AWWA*, 79(8):72-82.

Thompson, A., N.O. Matthews, G. Myers, and D.M. Owen. 1988. Selection of an Analytical Technique to Measure Chlorine Dioxide in the Field and Determination of Residual Effectiveness. In *Proc. of the 1987 AWWA Water Quality Technology Conference*. Denver, Colo.: AWWA.

Thompson, A., N.O. Matthews, A.J. Mittl, and D.M. Owen. 1989. Chlorite and Chlorate Concentration in the Distribution System. In *Proc. of the 1988 AWWA Water Quality Technology Conference*. Denver, Colo.: AWWA.

Thompson, J.D., M.C. White, G.W. Harrington, and P.C. Singer. 1997. Enhanced Softening: Factors Influencing DBP Precursor Removal. *Jour. AWWA*, 89(6):94-105

USEPA. 1979. National Interim Primary Drinking Water Regulations: Control of Trihalomethanes in Drinking Water; Final Rule. *Fed. Reg*., 44(231):68624-68707.

USEPA. 1989a. National Primary Drinking Water Regulations: Disinfection; Turbidity, *Giardia lamblia*, Viruses, Legionella, and Heterotrophic Bacteria; Final Rule. *Fed. Reg.*, 54:27486.

USEPA. 1989b. National Primary Drinking Water Regulations: Total Coliforms (Including Fecal Coliform and *E. coli*); Final Rule. *Fed. Reg.*, 54(124):27544.

USEPA. 1991. Maximum Contaminant Level Goals and National Primary Drinking Water Regulations for Lead and Copper; Final Rule. *Fed. Reg.*, 59:26460.

USEPA. 1994a. National Primary Drinking Water Regulations: Disinfectants and Disinfection Byproducts Rule; Proposed Rule. *Fed. Reg.*, 59(145):38668-38829.

USEPA. 1994b. National Primary Drinking Water Regulations: Enhanced Surface Water Treatment Requirements; Proposed Rule. *Fed. Reg.*, 59(145):38832.

USEPA. 1996a. National Primary Drinking Water Regulations: Monitoring Requirements for Public Drinking Water Supplies; Final Rule. *Fed. Reg.*, 61(94):24354-24388.

USEPA. 1996b. *DBP/ICR Analytical Methods Manual*. EPA 814-B-96-002. Cincinnati, Ohio: USEPA.

USEPA. 1997. National Primary Drinking Water Regulations: Disinfectants and Disinfection Byproducts; Notice of Data Availability, Proposed Rule. *Fed. Reg.*, 62 (212), 59387-59484.

USEPA. 1998a. National Primary Drinking Water Regulations: Disinfectants and Disinfection Byproducts; Final Rule. *Fed. Reg.*, 63(241):69390-69476.

USEPA. 1998b. National Primary Drinking Water Regulations: Interim Enhanced Surface Water Treatment; Final Rule. *Fed. Reg.*, 63(241):69477-69521.

USEPA. 1999a. Alternative Disinfectants and Oxidants Guidance Manual. EPA 815-R-99-014. Washington, D.C.: USEPA.

USEPA. 1999b. Disinfection Profiling and Benchmarking Guidance Manual. EPA-815-R-99-013. Washington, D.C.: USEPA.

USEPA. 1999c. Enhanced Coagulation and Enhanced Precipitative Softening Guidance Manual. EPA-815-R-99-012. Washington, D.C.: USEPA.

USEPA. 1999d. Guidance Manual for Compliance with the Interim Enhanced Surface Water Treatment Rule: Turbidity Provisions. EPA 815-R-99-010. Washington, D.C.: USEPA.

USEPA. 1999e. Guidance Manual for Conducting Sanitary Surveys of Public Water Systems; Surface Water and Ground Water Under the Direct Influence (GWUDI). EPA 815-R-99-016. Washington, D.C.: USEPA.

USEPA. 1999f. Microbial and Disinfection Byproduct Rules Simultaneous Compliance Manual. EPA-815-R-99-015. Washington, D.C.: USEPA.

USEPA. 1999g. Uncovered Finished Water Reservoirs Guidance Manual. EPA 815-R-99-011. Washington, D.C.: USEPA.

USEPA. 2000. Stage 2 Microbial and Disinfection Byproducts Federal Advisory Committee Agreement in Principle. *Fed. Reg.*, 65(251):83015-83024.

USEPA. 2001a. Implementation Guidance for the Interim Enhanced Surface Water Treatment Rule. EPA 816-R-01-011. Washington, D.C.: USEPA.

USEPA. 2001b. Implementation Guidance for the Stage 1 Disinfectants/Disinfection Byproducts Rule. EPA 816-R-01-012. Washington, D.C.: USEPA.

USEPA. 2001c. The Interim Enhanced Surface Water Treatment Rule - What Does It Mean to You? EPA 816-R-01-014. Washington, D.C.: USEPA.

USEPA. 2001d. The Stage 1 Disinfectants and Disinfection Byproducts Rule - What Does It Mean to You? EPA 816-R-01-014. Washington, D.C.: USEPA.

van der Kooij, D., A. Visser, and W.A.M. Hijnen. 1982. Determining the Concentration of Easily Assimilable Organic Carbon in Drinking Water. *Jour. AWWA*, 74(10):540-545.

Van Eekeren, M.W.M. 1992. Pellet Softening in the Netherlands. SWE 92.021. Nieuwegein, Netherlands: The Netherlands Waterworks' Testing and Research Institute KIWA N.V.

Waller, K., S.H. Swan, G. DeLorenze, and B. Hopkins. 1998. Trihalomethanes in Drinking Water and Spontaneous Abortion. *Epidemiology*, 9:2:134-140.

White, G.C. 1999. *Handbook of Chlorination and Alternative Disinfectants*. 4th ed. New York: John Wiley and Sons.

Wolfe, R.L., N.I. Lieu, G. Izaguirre, and E.G Means. 1990. Ammonia-Oxidizing Bacteria in a Chloraminated Distribution System: Seasonal Occurrence, Distribution and Disinfection Resistance. *Appl. Environ. Microbiol.*, 56(2):451-462.

ABBREVIATIONS

ACWD	Alameda County Water District
&	and
AOC	assimilable organic carbon
@	at
AWWA	American Water Works Association
AwwaRF	Awwa Research Foundation
BAF	biologically active filtration
BAT	best available technology
^{o}C	degrees Celsius
Ca	calcium
Ca^{2+}	calcium ion
CA	cellulose acetate
$CaCO_3$	calcium carbonate
CaO	calcium oxide
CAP	Central Arizona Project
C/C_o	concentration divided by initial concentration
CDM	Camp Dresser & McKee
cfu/mL	colony forming units per milliliter
Cl_2	chlorine
ClO_2	chlorine dioxide
cm^{-1}	inverse centimeters
CNCl	cyanogen chloride
Co.	county
CO_2	carbon dioxide
col/100 mL	colonies per 100 milliliters
CRW	Colorado River Water

CT	concentration (of disinfectant residual) multiplied by (disinfectant contact) time
cu ft	cubic feet
CWA	County Water Authority
CWTP	Central Water Treatment Plant
DBP	disinfection by-product
D/DBP	disinfectants/disinfection by-products
DHS	Department of Health Services
Dis.	disinfection
DL	detection limit
DOC	dissolved organic carbon
$	dollars
$/MG	dollars per million gallons
$/yr	dollars per year
DS	distribution system
E.	East
EBCT	empty bed contact time
EC	enhanced coagulation
e.g.	for example
ES	enhanced softening
ESWTR	Enhanced Surface Water Treatment Rule
EWTP	East Water Treatment Plant
°F	degrees Fahrenheit
FACA	Federal Advisory Committee Act
Fe	iron
Fed. Reg.	*Federal Register*
filt.	filter
ft	foot

FY	fiscal year
GAC	granular activated carbon
GAC10	granular activated carbon treatment with a 10-min empty bed contact time and a regeneration frequency of 180 days
GAC20	granular activated carbon treatment with a 20-min empty bed contact time and a regeneration frequency of 60 days
gal	gallons
gal/month	gallons per month
GCWA	Gulf Coast Water Authority
g/L	grams per liter
g/mL	grams per milliliter
gpcd	gallons per capita per day
gpd/ft	gallons per day per foot
gpd/sf	gallons per day per square foot
gpm	gallons per minute
gpm/sf	gallons per minute per square foot
GWUDI	ground water under the direct influence
h	hour
HAA	haloacetic acid
HAA5	sum of five haloacetic acids (i.e., monochloro-, dichloro-, trichloro-, monobromo-, and dibromoacetic acid)
HAA6	sum of six haloacetic acids (i.e., HAA5 + bromochloroacetic acid)
HAA9	sum of all nine HAAs
HCuFT	hundred cubic feet
HMWTP	Harwood's Mill Water Treatment Plant
Hp	horsepower
HPC	heterotrophic plate count
HRWTP	Hillsborough River Water Treatment Plant
H_2SO_4	sulfuric acid

ICR	Information Collection Rule
i.e.	that is
IESWTR	Interim Enhanced Surface Water Treatment Rule
ILSI	International Life Science Institute
in.	inch
in./min	inches per minute
”	inches
IOA	International Ozone Assocation
Jour.	Journal
$KMnO_4$	potassium permanganate
kW•h	kilowatt-hour
lb	pounds
lb/day	pounds per day
lb/MG	pounds per million gallons
LCR	Lead and Copper Rule
L/mg-m	liters per milligram-meters
LOX	liquid oxygen
LRAA	locational running annual average
LT2ESWTR	Long-Term 2 Enhanced Surface Water Treatment Rule
MCL	maximum contaminant level
MF	membrame filtration
Mg	magnesium
Mg^{2+}	magnesium ion
MG	million gallons
mgd	million gallons per day
mg/L	milligrams per liter

mg/L as N	milligrams per liter as nitrogen
mg-min/L	milligram-minutes per liter
min	minutes
mL/min	milliliters per minute
mm	millimeter
MSJWTP	Mission San Jose Water Treatment Plant
MOU	Memorandum of Understanding
MRDL	maximum residual disinfectant limit
MWCO	molecular-weight cutoff
MWDSC	Metropolitan Water District of Southern California
μg/L	micrograms per liter
μg Cl^-/L	micrograms (total organic halogen as) chloride per liter
N.	North
NA	not available
NaOCl	sodium hypochlorite
NaOH	sodium hydroxide
ND	not detected or nondetect
NF	nanofiltration
NF50	nanofiltration of 50 percent of the plant flow
NF75	nanofiltration of 75 percent of the plant flow
NF100	nanofiltration of 100 percent of the plant flow
NH_2Cl	monochloramine, chloramines
NH_3	ammonia
NH_3-N	ammonia-nitrogen
NM	not measured
NNWW	Newport News Waterworks
No.	number
NOM	natural organic matter
NS	not sampled
NSF	National Sanitation Foundation

NTU	nephelometric turbidity unit
#/100 mL	number per 100 milliliters
# cysts/100 gal	number of cysts per 100 gallons
# cysts/100 L	number of cysts per 100 liters
# oocysts/100 gal	number of oocysts per 100 gallons
# oocysts/100 L	number of oocysts per 100 liters
O_3	ozone
PAC	powdered activated carbon
PC	personal computer
PC/PLC	personal computer/programmable logic controller
pcu	platinum-cobalt color unit
P.E.	professional engineer
pfu/100 L	plaque-forming units per 100 liter
PLC	programmable logic controller
ppm	parts per million
Proc.	Proceedings
psi	pounds per square inch
PUD	Public Utility District
PVC	polyvinyl chloride
RAA	running annual average
Rd.	Road
RIA	regulatory impact analysis
RO	reverse osmosis
SCADA	supervisory control and data acquisition
SDS	simulated distribution system
SDWA	Safe Drinking Water Act
sec	seconds

sed.	sedimentation basin
sf	square foot
SOC	synthetic organic contaminant
sp.	species
SPW	State Project Water
SRP	Salt River Project
St.	Street
SUVA	specific ultraviolet absorbance
SWTR	Surface Water Treatment Rule
T_{10}	the point at which 90 percent of the water is in contact with the disinfectant residual for time T
T_{50}	average detention time
TC	turbo coagulation
TCR	Total Coliform Rule
TDS	total dissolved solids
TFC	thin film composite
THM	trihalomethane
THMFP	trihalomethane formation potential
TNRCC	Texas Natural Resources Conservation Commission
TOC	total organic carbon
TOX	total organic halogen
TP1	treatment plant 1
TP2	treatment plant 2
TTHMs	total trihalomethanes
UF	ultrafiltration
USEPA	United States Environmental Protection Agency
UV	ultraviolet (absorbance at 254 nanometers) or ultraviolet disinfection
WD	water district

w/o or W/O	without
WQ2000	Water Quality 2000 program
WQMP	Water Quality Master Plan
WTP	water treatment plant
WTP2	Water Treatment Plant Number 2
yr	year

IWA Publishing's authorised EU representative for General Product Safety Regulations is Diane D'Arras, 15 rue Duret, 75116 Paris, France, e-mail: safety@iwap.co.uk.

Printed and bound by CPI Group (UK) Ltd, Croydon, CR0 4YY
12/05/2026
02108892-0005